de Gruyter Expositions in Mathematics 3

Editors

The Stefan Problem

by

Anvarbek M. Meirmanov

Translated from the Russian

by

Marek Niezgódka and Anna Crowley

Walter de Gruyter · Berlin · New York 1992

Author

Anvarbek M. Meirmanov, Institut für Angewandte Mathematik, Universität Bonn,
Wegeler Str. 6, D-5300 Bonn 1, Germany

Translators

Marek Niezgódka, Institute of Applied Mathematics and Mechanics, Warsaw University,
Banacha 2, 00-913 Warsaw, Poland
Anna Crowley, Department of Mathematics, Royal Military College of Sciences,
Shrivenham, Swindon Wilts SN6 8LA, England

Title of the Russian original edition: Zadacha Stefana. Publisher: Nauka, Novosibirsk 1986

1991 Mathematics Subject Classification: 35–02; 35R35, 35K20. 80–02; 80A22.

☺ Printed on acid free paper which falls within the guidelines of the ANSI to ensure permanence and durability.

Library of Congress Cataloging-in-Publication Data

Meirmanov, A. M. (Anvarbek Mukatovich)
  [Zadacha Stefana, English]
  The Stefan problem / by Anvarbek M. Meirmanov ; translated
from the Russian by Marek Niezgódka and Anna Crowley.
    p.   cm.   (De Gruyter expositions in mathematics,
ISSN 0938-6572 ; 3)
  Translation of: Zadacha Stefana.
  Includes bibliographical references and index.
  ISBN 3-11-011479-8 (alk. paper)
  1. Heat--Transmission.   2. Boundary value problems.   I.Title.
II. Series.
QC321.M4513   1992   91-39310
536'.2--dc20   CIP

Die Deutsche Bibliothek – Cataloging-in-Publication Data

Meirmanov, Anvarbek M.:
The Stefan problem / by Anvarbek M. Meirmanov. Transl. from
the Russian by Marek Niezgódka and Anna Crowley. –
Berlin ; New York : de Gruyter, 1992
  (De Gruyter expositions in mathematics ; 3)
  Einheitssacht.: Zadača Stefana <engl.>
  ISBN 3-11-011479-8
NE: GT

Preface to the English edition

This English edition is the exact translation of the Russian original publication except for two points: Theorem 11 on pages 32–36 of Chapter 1 and the Appendix.

Very few new results on the Stefan problem have appeared since the publication of this book in 1986. The most important results in my opinion concern the behaviour of the mushy region and are due to Berger & Rogers and Götz & Zaltsman. I have included the results of Götz & Zaltsman for the reason that their proof is simpler and clearer for the reader.

Another point concerning the Appendix. The Russian original publication of the book presents a very short version of the model which describes phase transitions in a binary alloy. Since this book appeared we have published a full version of the Appendix in "Zhurnal Prikladnoi Mekhaniki i Tekhnicheskoi Fiziki", *No. 4* (1989), pp. 39–45 (English translation: I.G. Götz, A.M. Meirmanov, Modelling crystallization of a binary alloy, J. Appl. Mech. Tech. Phys., *No. 4* (1989), pp. 545–550). The Appendix in the English edition is based on this paper. In my opinion it is more understandable for the reader to have this full version of the mathematical model of phase transitions in a binary alloy.

Bonn, November 1991   *A.M. Meirmanov*

Preface

More than twenty years have passed since the appearance of the first monograph on the Stefan problem. In the meantime, many new ideas have been discovered and research techniques set up that have contributed to solving numerous complex problems. To some extent, the Stefan problem in its primary setting has settled itself, and time has come to consolidate the material collected.

In this book, the primary concerns are existence and uniqueness questions for the Stefan problem and a study of its structure. The choice of material presented was at first dictated by the author's interests, for a couple of years focussed on the Stefan problem. A large part of the book presents results obtained by the author and his colleagues. Many important aspects remain beyond the scope of the book, including numerical methods for solving the Stefan problem, questions of optimal control for phase change problems and the multidimensional quasi-steady Stefan problem.

A historical overview and bibliography of the Stefan problem up to 1967 can be found in [198]. At the end of this book, a current bibliography on the presented material is given. The bibliography does not pretend to any completeness, for this we refer the reader to the comprehensive review papers by Magenes [144] and Tarzia [216].

The author wishes to take advantage of this opportunity to express his gratitude to B.M. Anisyutin who has offered the material for Section III.1, to I.A. Kaliev who has written Sections V.5, VI.4 and VII.4, as well as to N.A. Kulagina and A.G. Petrova for assistance in collecting the list of references.

Table of Contents

Chapter I
Preliminaries

Chapter II
Classical solution of the multidimensional Stefan problem

Chapter III
Existence of the classical solution to the multidimensional
Stefan problem on an arbitrary time interval

Chapter IV
Lagrange variables in the multidimensional one-phase Stefan problem

Chapter V
Classical solution of the one-dimensional Stefan
problem for the homogeneous heat equation

Chapter VI
Structure of the generalized solution to the one-phase
Stefan problem. Existence of a mushy region

Chapter VII
Time-periodic solutions of the one-dimensional Stefan problem

Chapter VIII
Approximate approaches to the two-phase Stefan problem

Appendix
I.G. Götz, A.M. Meirmanov: Modelling of binary alloy crystallization . . . 222

Introduction

The Stefan problem consists of determining a temperature field and phase change boundaries in a pure material. The aggregate state of a medium is assumed to vary on account of heat conduction subject to the action of external and internal heat sources. Energy transfer in each phase of the material involved is described by the heat equation and the behaviour of the phase change boundary, referred to as a *free boundary*, by the Stefan condition. The latter expresses the energy balance at transition of the medium from one aggregate state to another.

We assume throughout this book that a medium may be in only two aggregate states, liquid and solid, for instance. A key condition on the free boundary, complementary to the Stefan condition, prescribes the medium temperature as equal to the melting point of the appropriate material, assumed to be a known constant. This condition has an axiomatic character, in as much as it does not follow from any fundamental law but rather reflects numerous real situations sufficiently exactly. In terms of the thermodynamic parameters of the state of medium, the hypothesis of fixed phase change temperature is equivalent to assuming the specific internal energy U as a single-valued function of temperature, with discontinuity of the first kind at the melting point, sufficiently smooth elsewhere. A schematic form of the above dependence is depicted in Figure 1. The correspondence below is of such a nature.

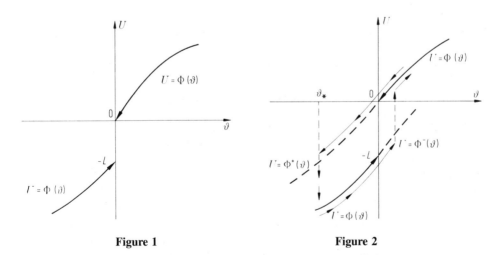

Figure 1	Figure 2

In many processes the melting temperature of pure material at a given external action depends on the process evolution. In particular, gallium may remain in the liquid phase at temperatures well below its mean melting temperature. Turning again to the thermodynamic state parameters, we can see that the specific internal energy of the material in a neighbourhood of the mean melting temperature is then a double-valued function of temperature. A representative form of this correspondence is shown in Figure 2.

At a given point of the material in the liquid phase, the dependence of the specific internal energy on temperature is described by the upper branch of the function. After decreasing the temperature of the medium to a critical value ϑ_*, to its left in the graph relating ϑ and U to each other, the dependence of the specific internal energy on temperature is described by the lower branch, corresponding to the solid phase. It is by no means obvious that, upon increasing the temperature, the phase change will occur exactly at the same point. In other words, in some situations one can expect the occurence of a hysteresis effect.

The value ϑ_* represents the genuine melting temperature at which the phase transition proceeds. In the corresponding model, this value is unknown. To complete the mathematical setting, a reasonable axiom which would supply an additional equation for determining the melting temperature is necessary.

One such model has been proposed by A. Visintin during the Maubuisson symposium on free boundary problems. His approach, more accurately describing the behaviour of a continuum, seems to outline future trends. These models themselves require a refined construction, detailed mathematical foundations and experimental confirmation.

In the sequel, as already mentioned, the melting temperature of a pure material will be assumed to be a given constant. This axiom completes the mathematical model referred to as a *Stefan problem*.

A study of the Stefan problem can be developed in several directions: existence and uniqueness of a solution in one- and multidimensional cases, an analysis of the structure and qualitative properties of the solution, including its asymptotic behaviour as $t \to \infty$, a quasi-steady formulation of the multidimensional Stefan problem, numerical methods for its solving and optimal control.

The one-dimensional Stefan problem has been almost completely explored. The earliest existence results for the classical solution were local in time. L.I. Rubinstein [197] has proved the global existence in time of the classical solution under an analyticity hypothesis for the free boundary. Without such an assumption, under rather natural hypotheses, the same result has been obtained by J. Li-Shang in 1965 [141]. Using different arguments it was proved later by J.R. Cannon, J. Douglas Jr., C.D. Hill, D.B. Henry, D.B. Kotlow, M. Primicerio, and the author; cf., [43], [45] – [47], [50] – [52], [148]. All these results refer to the case of temperature prescribed at the fixed boundaries of the domain. Conditions that ensure the existence of the classical solution for problems with heat flux prescribed at the fixed boundaries were obtained by J.R. Cannon and M. Primicerio [53].

In the papers by J.R. Cannon et al., qualitative properties of the solution and the asymptotic behaviour of the free boundary as $t \to \infty$ were discussed in detail. The first complete results on large time behaviour of the solutions to the Stefan problem for nonlinear parabolic equations were obtained by N.V. Khusnutdinova [122].

As regards the multidimensional Stefan problem, for a long time the only results available on existence and uniqueness of the generalized solution were due to O.A. Oleinik [184] and S.L. Kamenomotskaya [112] (obtained in 1960). Let us remark that the class of functions where the solution of the Stefan problem is unique turns out to be larger than the class containing its generalized solutions. As a rule, the situation is the opposite: uniqueness is ensured for solutions that fulfil additional regularity hypotheses, stronger than those provided by the existence proof. Each classical solution of the Stefan problem is at the same time its generalized solution, therefore the uniqueness question has been completely answered. The question remains, under what conditions the generalized solution of the multidimensional Stefan problem will become classical.

Here, the first advances date back to the publication of the paper [76] by G. Duvaut (in 1973), where a one-phase Stefan problem was transformed to a variational inequality with the existence of a weak solution provided. In 1975, A. Friedman and D. Kinderlehrer [97], using Duvaut's transformation, proved the Lipschitz continuity of the free boundary in the one-phase Stefan problem. It was still not enough for concluding the existence of the classical solution. This has been shown by D. Kinderlehrer and L. Nirenberg (cf., [127]) in 1978, exploiting results by L.A. Caffarelli [35] – [37] on the smoothness of the free boundary in elliptic variational inequalities and introducing Legendre variables in which an a priori fixed plane corresponds to the free boundary.

An equivalent variational inequality formulation for the two-phase Stefan problem was given by M. Frémond [89] in 1974, but, unlike the one-phase situation, in this case only continuity of the temperature was proved (cf., [39]).

An alternative approach to the construction of a classical solution to the multidimensional two-phase Stefan problem has been proposed by the author in 1979 (cf., [150]). By constructing a special regularization of the Stefan condition and introducing von Mises variables, where a fixed surface corresponds to the free boundary, the existence of a classical solution has been proved on a small time interval (cf., [150]). Well-known in hydrodynamics (cf., [169]), von Mises variables were for the first time applied to the Stefan problem by I.I. Danilyuk. In 1981, E.I. Hanzawa [104] obtained an existence result analogous to that of [150] by applying the Nash-Moser theory.

A large number of papers on the multidimensional Stefan problem were devoted to generalized solutions in the case of nonlinear parabolic equations and various nonlinear boundary conditions imposed on the fixed boundaries. We refer in particular to results by A. Damlamian [61], G. Duvaut [75] – [78], E. Magenes, C. Verdi and A. Visintin [145], M. Niezgódka and I. Pawlow [174], M. Niezgódka, I. Pawlow and A. Visintin [175], I. Pawlow [186], A. Visintin [220], [221].

One of the most significant numerical approaches to solving the Stefan problem is due to B.D. Moiseenko and A.A. Samarskii (cf., [168]), who applied the generalized problem setting developed by O.A. Oleinik. During the last decade, many authors applied the finite element method to solving the Stefan problem (cf., [22], [23], [170] – [172], [178] – [180] to list a few).

The quasi-steady Stefan problem differs from the non-stationary settings we have mentioned in as much as the heat transfer in each phase is described by an elliptic equation. In consequence, a study of such a problem requires tools different from those used in the present book. For the quasi-steady Stefan problem, E.V. Radkevich and A.S. Melikulov [194] proved the existence of a classical solution in the two-dimensional case by adapting the method of boundary variation, set up by Radkevich for filtration problems with free boundaries. An extensive study of the quasi-steady Stefan problem has been developed by I.I. Danilyuk [67] – [70] and his co-workers (cf., [12], [26] – [28]).

Recently, much attention has been paid to the so-called *mushy phase* (mushy region, according to Atthey [3]), arising in the Stefan problem under some conditions. What is the mushy phase? Let us recall the definition of a generalized solution. By its construction, every classical solution with the liquid and solid phases separated by a smooth hypersurface is also a generalized solution of the Stefan problem. For a long time, the reverse question was approached by an analysis of the differential properties of the phase change boundary. In the one-dimensional case, such a study was performed by A. Friedman [93], and J.R. Cannon, D. Henry and D.B. Kotlow [45] for the homogeneous heat equation. The apparently natural supposition that if the initial state contains two phases, liquid and solid, separated by a smooth surface, then the generalized solution of the Stefan problem will in fact be classical, turned out to be misleading. The notion of the generalized solution to the Stefan problem proved to be essentially more comprehensive than its classical solution.

In 1981, the author [154] constructed the generalized solution with a mushy phase, absent at the initial time instant. In the mushy phase, the temperature ϑ is identically at the melting point ϑ_* and the specific internal energy U assumes values in (U_-, U_+), where

$$ U_\pm = \lim_{\vartheta \to \vartheta_* \pm 0} \Phi(\vartheta), \quad U = \Phi(\vartheta), \quad \vartheta \neq \vartheta_*. $$

In that example, there was no phase change boundary at all, instead one could observe a domain occupied by the mushy phase. What was the origin of the latter? In the primary formulation of the Stefan problem, the classical solution was to be determined. Thus, the non-uniqueness of the dependence of U on ϑ at the melting point remained outside the scope. On the contrary, an implicit axiom introduced in defining the generalized solution allowed the specific internal energy U to assume values in the interval (U_-, U_+) at the melting temperature, hence as a matter of fact accepting the possibility of creating there states of the system different from the solid and liquid phases.

The possibility of forming a new phase, admitted in the definition of the generalized solution to the Stefan problem, has been confirmed by an example constructed by the author. The driving factors in that example were internal heat sources in the heat equations and the vanishing of the temperature gradient on the free boundary at the initial time instant. This contradicts the condition underlying the classical solvability of the multi-dimensional two-phase Stefan problem (cf., [150]) which requires that the temperature gradient on the free boundary is everywhere different from zero.

In 1982, M. Primicerio [189] constructed an analogous example of the generalized solution to the Stefan problem that was non-classical. The example resulted from a discussion during the Montecatini symposium (cf., [85]). In particular, it was noticed there that a mushy phase had already been observed by D. Atthey [3] in a numerical solution of the Stefan problem.

In subsequent papers on the Stefan problem, the axiom permitting the existence of an intermediate state of the system, different from both liquid and solid, did not cause any objections. Of course, one can apply an opposite approach which excludes values of the specific internal energy in (U_-, U_+) at the melting temperature and, by this, eliminates generalized solutions with a mushy phase. The former treatment is preferable, because it emphasises the necessity of more accurate modelling in the mushy phase, observed in real processes.

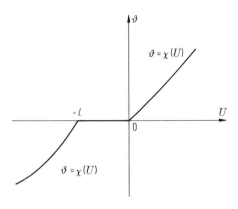

Figure 3

Summarizing, we can claim that the Stefan problem is equivalent to an initial-boundary value problem for the degenerate equation

$$\frac{\partial U}{\partial t} = \Delta \chi(U),$$

with the single-valued function $\chi(U)$ inverse to $U = \Phi(\vartheta)$, schematically depicted in Figure 3.

We now briefly outline the contents of the book.

In Section I.1 (Section 1 of Chapter I), the problem is formulated. In Section I.2, necessary results from functional analysis and the theory of differential equations are recalled. Section I.3 includes proofs of the existence and uniqueness, and comparison results for the generalized solutions of the Stefan problem.

Chapter II reports results of the author (cf., [150], [151]) on the local in time classical solution of the multidimensional Stefan problem.

In Chapter III, the existence of the classical solution to the multidimensional Stefan problem on an arbitrary time interval is proved under some monotonicity conditions for the initial data in a one-phase problem, and provided that the initial data are close to the stationary solution in the two-phase problem. In Section III.1, results by B.M. Anisyutin for the one-phase Stefan problem are presented. He used Caffarelli's theorem [36] on the smoothness of the free boundary in the filtration problem, as developed by A. Friedman [96], and applied the construction of Legendre variables introduced by D. Kinderlehrer and L. Nirenberg [128]. Although the method is presented for the heat equation, the results remain true for the more general nonlinear heat equation

$$a(\vartheta) \, \frac{\partial \vartheta}{\partial t} = \Delta \vartheta,$$

where $a'(\vartheta) > 0$. The results of Section III.2 were published in [152].

In Chapter IV, the one-phase Stefan problem is studied in Lagrange coordinates, where the solution is defined on an a priori given domain and satisfies a strongly nonlinear system of non-standard differential equations.

The introduction of Lagrange variables permitted a complete analysis of the one-dimensional Stefan problem with a nonlinear condition of a special type imposed on the known boundary (cf., [149]), while in the multidimensional case the global existence in time of the solution was proved only for the linearized problem. The results of Chapter IV were obtained by the author and V.V. Pukhnachev [160]. In Section V.1, exact non-selfsimilar solutions of the Stefan problem are constructed for a special class of nonlinear heat equations (cf., [160]).

Chapter V is devoted to the one-dimensional Stefan problem. For the classical solution, existence theorems are proved for one- and two-phase problems with sufficiently weak hypotheses on the input data. Moreover, qualitative properties of the solutions, including their asymptotic behaviour as $t \to \infty$, are considered.

In Section V.6, the filtration problem for a viscid compressible liquid in a vertical porous layer, equivalent to the one-phase Stefan problem, is studied. Filtration proceeds in the domain $\Omega(t) = \{x \; : \; 0 < x < R(t)\}$, whose right boundary $x = R(t)$ is unknown and the liquid is supplied through the left boundary $x = 0$. In case of the linearized equations, the problem formulated there was studied by A. Friedman and R. Jensen [97] in 1975. The author studied that problem in its exact formulation in 1977 (cf., [149]). Already the proof of an estimate of the solution in the maximum norm was more difficult than for the linearized equations. Estimates for the derivatives then follow from [137] after introducing Lagrange variables.

We mention an interesting feature of the problem in its exact formulation. Let the mass of the liquid supplied through the left boundary grow infinitely with time. It is natural to expect then that the corresponding free boundary $x = R(t)$ will grow to infinity; indeed, such a result was obtained in [97]. However, in the exact setting, there are initial states such that the domain $\Omega(t)$ remains uniformly bounded as $t \to \infty$ despite an unbounded growth of the total mass of the liquid in $\Omega(t)$.

Chapter VI summarizes results obtained by the author [154], partially jointly with I.A. Kaliev [111], as well as original results by I.A. Kaliev on the structure of the generalized solution of the one-dimensional Stefan problem with arbitrary initial distribution of the specific internal energy. We mention Kaliev's theorem for the homogeneous heat equation: suppose that the constant temperature prescribed at one of the end-points of the domain Ω has a value above the melting point, while the corresponding temperature at the other end-point is below the melting point; then, regardless of the initial distribution of the specific internal energy (the number of connected components of the liquid, solid and mushy phase can even be infinite), there exists a finite time instant such that, starting from it, Ω will split into just one connected component of each of both phases, liquid and solid, i.e., each generalized solution of the Stefan problem will become classical.

The results of Chapter VI are used in Chapter VII to study time-periodic solutions of the one-dimensional two-phase Stefan problem in a bounded domain Ω. In the case of the temperature prescribed on the boundary of Ω close to a constant, existence of the classical solution was proved by M. Štědrý and O. Vejvoda [212] in 1981. The author (cf., [158]) achieved a stronger result in the case of arbitrary boundary temperature, provided it was everywhere different from the melting point. A general situation was considered by I.A. Kaliev and the author.

A construction of approximate models of the two-phase Stefan problem and an analysis of their correctness are given in Chapter VIII. A crystallization process is modelled for a cylindrical ingot whose heat conductivity along directions orthogonal to its axis significantly exceeds the heat conductivity along the axis itself. As a rule, the approximate models depend on some (say, small) parameter so that the approximate solution in a certain norm approaches the exact one as the parameter tends to zero. The exact solutions depend then on the small parameter and, as the parameter goes to zero, they approach the solution of the approximate model in a certain norm. According to the terminology of [197], the medium occupied by the ingot is called a "concentrated capacity", and the approximate model is referred to as the Stefan problem for a medium with "concentrated capacity". The results of this chapter were published in [147], [156], [159].

Theorems and lemmas are separately indexed in each chapter, formulae are numbered within each section. So, (2.3) corresponds to the third indexed formula of Section 2. All notation remains valid within one section, but the same symbols can have a different meaning in different sections.

Chapter I

Preliminaries

1. Problem statement

In this book, a class of problems is studied, traditionally assumed to be mathematical models of crystallization (freezing) or melting in pure (one-component) materials. These models admit the following formulation, usually referred to as the Stefan problem.

Let Ω be a domain occupied by the material under consideration. At a time $t > 0$, Ω is composed of two subdomains $\Omega^+(t)$ and $\Omega^-(t)$, appropriately occupied by the liquid and solid phases of the material. The subdomains are separated by an unknown regular surface $\Gamma(t)$ to be determined.

No rigorous notions of the solid and liquid phases are introduced. It is assumed that if a constant ϑ_* represents the appropriate melting temperature then the temperature is not lower than ϑ_* in liquid phase and not higher than ϑ_* in the solid.

In each of the subdomains $\Omega^\pm(t)$, the temperature $\vartheta(x,t)$ of the medium satisfies the heat equation

$$\varrho c \frac{\partial \vartheta}{\partial t} = \text{div}\,(\kappa \nabla \vartheta) + f, \tag{1.1}$$

where $f(x,t)$ is a given function describing heat sources or heat flow, ϱ denotes the mass density, κ the heat conductivity, and c the heat capacity at constant volume. As a rule, all these coefficients are assumed constant, different in each of the phases. For instance, let $c = c_L$ in the liquid phase and $c = c_S$ in the solid. Everywhere below $c_L, c_S, \kappa_L, \kappa_S$ are assumed to be given strictly positive functions of temperature.

On the unknown surface $\Gamma(t)$ between phases, two conditions are imposed: that the temperature is at the melting point

$$\vartheta(x,t) = \vartheta_*, \tag{1.2}$$

and the relation, traditionally referred to as the Stefan condition,

$$+\varrho L V_\nu = \left[\kappa \frac{\partial \vartheta}{\partial \nu}\right], \tag{1.3}$$

holds with V_ν the normal velocity of the surface $\Gamma(t)$.

Here, if the surface $\Gamma(t)$ is described by $h(x,t) = 0$, then

$$V_\nu = -\frac{\partial h}{\partial t}(\nabla h \cdot \nu)^{-1}$$

and the symbol $[\varphi]$ on the right-hand side of (1.3) represents the jump of the function φ across the phase change surface:

$$[\varphi] \;=\; \lim_{\substack{(x,t)\to\Gamma(t)\\ \vartheta<\vartheta_*}} \varphi(x,t) \;-\; \lim_{\substack{(x,t)\to\Gamma(t)\\ \vartheta>\vartheta_*}} \varphi(x,t) \;.$$

L is a positive constant representing the *specific latent heat of melting* (of freezing or crystallization), equal to the minimal portion of energy necessary for transforming the bulk of unitary mass of the substance from the solid state to liquid at the constant melting temperature ϑ_*.

The Stefan condition follows from the energy conservation law by its application to elementary volumes that contain both phases at the same time.

To complete the formulation of the problem, external actions at the fixed boundary of domain Ω (boundary conditions) and an initial state of the medium (initial condition) are to be prescribed. As a rule, either the temperature or the heat flux are assumed to be given at the boundary of Ω and the temperature is assumed to be known at the initial time instant.

The above formulation is referred to as the *two-phase one-front Stefan problem* and represents one of the simplest problem settings. Modifications are possible in both directions: either to the one-phase Stefan problem on the simplest side, or to multi-phase multi-front Stefan problems in the most complex situation.

Multi-phase Stefan problems refer to materials capable of assuming any of three or more different states (solid, liquid and gaseous, in particular). Since all the principal difficulties characteristic of phase transition problems occur in the two-phase situation, we shall not consider multi-phase formulations separately.

In turn, the one-phase Stefan problem represents a special case of the two-phase formulation, with the temperature constant in one of the phases, assuming the melting value.

The corresponding mathematical model follows in an obvious way, including the heat equation to be fulfilled in the phase with variable temperature and the Stefan condition on the unknown phase boundary, with the heat flux

$$q = -\kappa\nabla\vartheta$$

vanishing in the phase with constant temperature.

A sufficiently regular solution $\Gamma(t), \vartheta$ (i.e., a solution continuous together with all derivatives of ϑ that enter equation (1.1) and conditions (1.2),(1.3)) is called the *classical solution of the two-phase* (one-phase) *Stefan problem*.

In turn, the Stefan problem is the simplest model of the dynamics of first-order phase transitions. It does not take account of variations of medium characteristics such as mass density, velocity, stresses, etc.

Whilst the Stefan problem has been extensively studied, both by physicists and mathematicians, problems that take account of the medium motion still await the construction of any mathematically closed model. This situation requires a separate study, therefore we shall everywhere neglect variations of the mass density and instead consider it normalized to unity.

In 1960, S.L. Kamenomotskaya and O.A. Oleinik proposed a new formulation of the Stefan problem, comprehending its classical form as a special case. The Stefan problem was treated as an initial-boundary value problem, with a continuous function $\vartheta(x, t)$ to be determined so that the equation

$$\frac{\partial U}{\partial t} = \operatorname{div}(\kappa \nabla \vartheta) + f \tag{1.4}$$

is satisfied everywhere in $\Omega_T = \Omega \times (0, T)$, where $U = \Phi(\vartheta)$ is a function having a discontinuity of the first type at $\vartheta = \vartheta_*$. In addition, the multi-valued relations

$$U = \begin{cases} \int_0^\vartheta c_S \, d\tau - L & \text{if} \quad \vartheta < \vartheta_* \\ \in [-L, 0] & \text{if} \quad \vartheta = \vartheta_* \\ \int_0^\vartheta c_L \, d\tau & \text{if} \quad \vartheta > \vartheta_* \end{cases}$$

are to be satisfied everywhere, including the discontinuity points. Equation (1.4) is then to be interpreted in the sense of distributions.

Let $\varphi(x, t)$ be a sufficiently regular function, vanishing on the boundary of Ω and at $t = T$. Then equation (1.4) is equivalent to the integral identity

$$\int_{\Omega_T} \{\kappa \nabla \vartheta \nabla \varphi - U \frac{\partial \varphi}{\partial t} - f\varphi\} dx dt = \int_\Omega U\varphi \mid_{t=0} dx, \tag{1.5}$$

being fulfilled for any function φ with the above properties.

As will be shown in Section 3 of this chapter, in order to assure the uniqueness of a solution to equation (1.4) (or to identity (1.5)) a distribution of the specific internal energy $U(x, 0) = U_0(x)$ has to be prescribed at the initial time instant rather than one of the temperature $\vartheta_0(x)$. If the critical set $\{\vartheta(x, 0) = \vartheta_*\}$ has measure zero, then the function $U_0(x)$ is almost everywhere uniquely specified by $\vartheta_0(x)$ and it matters not which one of them is prescribed at $t = 0$. The situation changes though as soon as the above measure becomes positive. We shall resume this topic after introducing a rigorous notion of the solid and liquid phases.

A solution of the integral identity (1.5) (or, equation (1.4) subject to appropriate boundary and initial conditions) will be called a *generalized solution* of the Stefan problem.

Clearly, any classical solution of the Stefan problem is its generalized solution, too. In general, the reverse is false. But, as long as the level set $\{\vartheta(x, t) = \vartheta_*\}$ is a regular hypersurface separating the subdomains $\{\vartheta(x, t) > \vartheta_*\}$ and $\{\vartheta(x, t) < \vartheta_*\}$, identity (1.5) implies both equation (1.1) and boundary condition (1.3).

The above construction of generalized solutions seemed new to mathematicians studying the Stefan problem, although it had already been used for a quite long time in gas dynamics [185].

It is well-known that the fundamental conservation laws of continuum mechanics can be formulated as a system of equations in divergence form

$$\frac{\partial F}{\partial t} + \operatorname{div}(Fv - G) = X. \tag{1.6}$$

In particular, in the energy conservation law $F = \varrho U$ where ϱ is the mass density of the medium, U is the specific internal energy; $G = -q$ where $q = -\kappa \nabla \vartheta$ is the heat flow vector; $X = P : D$ the double convolution of tensors P and D, where P is the stress tensor and D the deformation rate tensor.

If all functions that enter equations (1.6) are continuously differentiable, then the resulting motion of the medium is called *continuous*. Since the continuous motions do not encompass all possible physical phenomena, a notion of the *generalized motion* is introduced, when all the functions in equation (1.6) are only bounded. In particular, in phase change problems the specific internal energy has a finite jump across the boundary separating various phases. Or, if a medium consists of various components with different thermophysical properties, then also heat conductivities, specific heat coefficients and, sometimes, temperature gradients have jump discontinuities across the interface between different components. In the case of the generalized motion, equations (1.6) are to be understood in the sense of distributions.

Among the generalized motions, the subclass of strongly discontinuous motions has been studied most extensively. Defined over the domain Ω_T, this class comprises motions that are continuous except on some regular hypersurface Γ_T where the functions F, v and G have finite one-sided limits, in general different. These limits on the surface Γ_T (which is called a *strong discontinuity surface*) have to satisfy the following system of strong discontinuity conditions

$$[F(v \cdot \nu - V_\nu) - G\nu] = 0, \tag{1.7}$$

where V_ν represents the normal velocity of the intersection $\Gamma(t)$ of the hypersurface Γ_T with the plane $\{t = \text{const}\}$. Equations (1.7) follow from (1.6) due to the assumed type of motion. Indeed, each equation of the form (1.6) is equivalent to the integral identity

$$\int_{\Omega_T} \left\{ (G - Fv)\nabla\varphi - F\frac{\partial\varphi}{\partial t} - X\varphi \right\} dx dt = 0, \tag{1.8}$$

for any arbitrary function φ, regular and finite on Ω_T. Let Q be an arbitrary domain containing Γ_T and

$$Q = Q^+ \cup Q^- \cup \sigma,$$

where $\sigma \subset \Gamma_T$, functions F, G, v are continuously differentiable in Q^+ and in Q^-, and there fulfil equation (1.6) in the classical sense.

Let us take a function φ with support in Q. Multiply equation (1.6) by φ, then integrate over each of the subdomains Q^+ and Q^-. Let us transfer the derivatives from functions F, G, v in these integrals onto φ, then take the sum of the terms integrated over Q^+ and Q^-. The resulting sum equals zero and can be split into an integral over the domain Q and another integral over the surface σ. The first term coincides with the left-hand side of (1.8) and thus is equal to zero. The integrand in the second term coincides with the left-hand side of (1.7) up to an arbitrary multiplier. As φ was taken as arbitrary and the integral over σ vanished, the whole integrand in (1.8) vanishes, too.

Further we shall confine ourselves to the energy conservation law, with the medium fixed (its velocity equal zero), mass density equal to one and with the dissipative component $P : D$ substituted by a given function f. In such a case, the corresponding law reduces to equation (1.4). On the strong discontinuity surface, the usual Stefan condition

$$[U]V_\nu + [q\nu] = 0 \; ,$$

with $[U] = -L$, $q = -\kappa\nabla\vartheta$, follows then as a consequence.

Hence, the function U in the problem formulation due to Kamenomotskaya and Oleinik represents just the specific internal energy, with a discontinuity of the first type accompanying the transition from the liquid phase to the solid and conversely.

Precisely this observation underlies the definition of the solid and liquid phases. More precisely, we shall consider a medium to be in the liquid phase at a given time moment and at a given point, if its specific internal energy is non-negative, and to be in the solid phase, if the corresponding value is less than $-L$. We have assumed there that the medium has vanishing internal energy at the melting point (melting temperature) in the liquid phase. This hypothesis is non-essential because the internal energy is specified up to an additive constant in the energy conservation law.

The definition of the solid and liquid phases we have just introduced is more precise than that given earlier, since it distinguishes between different phases at melting temperature.

In general, there are no reasons which prevent the specific internal energy from assuming values in $(-L, 0)$ over a set of positive measure. In such a case we shall say that the medium is in the *mushy phase (mushy state)*.

In some cases, it is convenient to interpret the mushy phase as the set of all points with temperature at melting point, i.e., with the specific internal energy in the interval $[-L, 0]$.

So far we have treated the temperature ϑ as a solution of the Stefan problem, with the specific internal energy U specified by the thermodynamic state equation $U = \Phi(\vartheta)$. But, this relation does not specify U uniquely at discontinuity points $\vartheta = \vartheta_*$ of function Φ (in mushy phase). At the same time, the inverse dependence $\vartheta = \chi(U)$ is single-valued at all values of U and (1.4) can be interpreted as a nonlinear equation with respect to U (degenerate in the mushy phase). Therefore, it seems more reasonable to consider the specific internal energy U rather than the temperature ϑ as the solution of the Stefan problem.

We are now ready to refine the definition of a classical solution of the Stefan problem once more – as a strongly discontinuous motion such that the corresponding mushy phase occupies a set of measure zero.

Are there any generalized motions possible with the mushy phase occupying a set of positive measure? The answer is yes. In this section, we shall construct an appropriate mathematical model which follows from (1.5). The correctness of this model will be shown in Chapter VI.

For simplicity, we shall confine ourselves to a one-dimensional situation. Let us take $\Omega = (-1, 1)$ and consider a process such that the domain $\Omega_T^- = \{(x, t) \; : \; -1 < x <$

$R^-(t)$, $0 < t < T$} is filled with the solid phase, the domain $\Omega_T^+ = \{(x,t) \; : \; R^+(t) < x < 1,\ 0 < t < T\}$ by the liquid phase and $\Omega_T^* = \{(x,t) \; : \; R^-(t) < x < R^+(t),\ 0 < t < T\}$ by the mushy phase. The functions $R^-(t), R^+(t)$ are assumed to be continuously differentiable.

By the definition of motions with strong discontinuity, the appropriate motion is continuous everywhere except on the lines $x = R^-(t)$ and $x = R^+(t)$. Hence, in the domains Ω_T^- and Ω_T^+ equation (1.4) is equivalent to (1.1), and in Ω_T^* reduces to the equation

$$\frac{\partial U}{\partial t} = f \,. \tag{1.9}$$

All derivatives of the functions ϑ and U which enter equations (1.1) and (1.9) are continuous. The a priori hypothesis on the regularity of the functions ϑ and U in these domains is non-essential. Such regularity follows from the assumptions on the structure of the solution to equation (1.4) and is due to the differentiability of solutions to parabolic equations with sufficiently regular κ, c and f (cf., [137]). If the velocity of the surface $x = R(t)$ is defined as the time derivative of the function $R(t)$ and the limits of the specific internal energy are taken as 0 on the curve $x = R^+(t)$ from the side of liquid phase, and equal to $-L$ on $x = R^-(t)$ from the side of the solid, then it follows from (1.7) (where $v = 0, F = U, G = \kappa \dfrac{\partial \vartheta}{\partial x}$) that

$$\left(U(R^-(t) + 0, t) + L\right)\frac{dR^-}{dt}(t) = \kappa_S \frac{\partial \vartheta}{\partial x}(R^-(t) - 0, t) \,, \tag{1.10}$$

$$U(R^+(t) - 0, t)\frac{dR^+}{dt}(t) = \kappa_L \frac{\partial \vartheta}{\partial x}(R^+(t) + 0, t) \,. \tag{1.11}$$

The above relations, together with initial conditions at $t = 0$ and boundary conditions at $x = \pm 1$ are not sufficient to provide the correctness of the model. Additional hypotheses are necessary here.

Let us consider the formation of a mushy phase, i.e., the case

$$R^-(0) = R^+(0) = x_0 \quad \text{and} \quad R^-(t) < R^+(t) \qquad \text{for} \quad t > 0 \,. \tag{1.12}$$

It turns out that, under some special hypotheses on data, boundary condition (1.10) or (1.11) breaks up into two independent conditions and the resulting boundary value problem splits into three autonomous problems to be solved successively.

Let $f(x,t)$ be continuous and $f(x_0, 0) > 0$. Without loss of generality we can assume $\vartheta_* = 0$.

By assumption, the temperature ϑ is negative in the subdomain Ω_T^- and vanishes on the right-hand part of its boundary $\{x = R^-(t)\}$. Thus,

$$\frac{\partial \vartheta}{\partial x}(R^-(t) - 0, t) \geq 0. \tag{1.13}$$

Analogously,

$$\frac{\partial \vartheta}{\partial x}(R^+(t) + 0, t) \geq 0. \tag{1.14}$$

By definition, in the mushy phase the specific internal energy assumes values from the interval $(-L, 0)$. Therefore

$$U(R^-(t) + 0, t) + L \geq 0 , \quad U(R^+(t) - 0, t) \leq 0. \tag{1.15}$$

By comparing (1.11) with (1.14) and (1.15) we see that $\dfrac{dR^+}{dt}(t) \leq 0$. Hence, according to (1.12), $\dfrac{dR^-}{dt}(t) < 0$ at least locally in time. By (1.10), (1.13) and (1.15), this can happen only if

$$U(R^-(t) + 0, t) + L = 0 , \quad \frac{\partial \vartheta}{\partial x}(R^-(t) - 0, t) = 0 . \tag{1.16}$$

In an alternative case, (1.10), (1.13) and (1.15) imply that $R^-(t)$ is non-decreasing. Therefore, due to (1.12), $R^+(t)$ is then strictly increasing, but this is possible only provided that

$$U(R^+(t) - 0, t) = 0 , \quad \frac{\partial \vartheta}{\partial x}(R^+(t) + 0, t) = 0 . \tag{1.17}$$

By differentiating the identity $\vartheta(R^-(t) - 0, t) = 0$ with respect to t, we can discover which case occurs. We then obtain

$$\frac{\partial \vartheta}{\partial x}(R^-(t) - 0, t)\frac{dR^-}{dt}(t) + \frac{\partial \vartheta}{\partial t}(R^-(t) - 0, t) = 0.$$

If (1.16) are satisfied, then the last equality, together with the second of the relations (1.16) and equation (1.1), yields

$$\frac{\partial^2 \vartheta}{\partial x^2}(R^-(t) - 0, t) = -\frac{1}{\kappa}f(R^-(t), t).$$

For any fixed t, the function $\vartheta(x, t)$ admits the representation

$$\vartheta(x, t) = \frac{\partial^2 \vartheta}{\partial x^2}(R^-(t) - 0, t)\frac{|x - R^-(t)|^2}{2} + o(|x - R^-(t)|^2)$$

for $x < R^-(t)$.

Since the temperature is negative for $x < R^-(t)$, due to the last two relations, the first case (equations (1.16)) occurs if $f(x_0, 0) > 0$. Analogously, if $f(x_0, 0) < 0$ the second case (relations (1.17)) holds.

In the situation we consider conditions (1.16) are satisfied. The function $R^-(t)$ which determines the domain Ω_T^-, and the temperature ϑ in this domain are specified by equation (1.1), condition (1.2), the second of the relations (1.16), the boundary condition at $x = -1$ and the initial condition at $t = 0$. This model is closed.

Suppose the function $R^-(t)$ has been found. Then, by integrating equation (1.9) and applying the first of the relations (1.16) we determine the function $U(x, t)$ everywhere above the curve $x = R^-(t)$ as

$$U = -L + \int_{r^-(x)}^t f(x, \tau)d\tau , \tag{1.18}$$

with the equation $t = r^-(x)$ describing the same curve $x = R^-(t)$.

The function $R^+(t)$ which specifies the domain Ω_T^+ as well as the temperature ϑ in this domain are prescribed by equation (1.1), conditions (1.2), (1.11), the boundary condition at $x = 1$ and the initial condition at $t = 0$. In (1.11), the function U is given by (1.18).

Now consider the homogeneous heat equation (i.e., $f = 0$) and let the initial temperature ϑ_0 be zero on an interval (x_0^-, x_0^+) in Ω. As already mentioned, in order to specify a unique solution of the Stefan problem, one has to prescribe an initial distribution $U_0(x)$ of the specific internal energy. But what are the values of $U_0(x)$ on (x_0^-, x_0^+) where $\vartheta_0 = 0$?

In this situation a common solution was to introduce a sharp interface x_0 separating the liquid phase from the solid. This implicitly prescribed the function $U_0(x)$ such that $U_0 = -L$ on the interval (x_0^-, x_0) (solid phase at vanishing temperature) and $U_0 = 0$ on (x_0, x_0^+) (liquid phase at vanishing temperature). With such a definition of the function $U_0(x)$, the classical solution is the unique solution of the Stefan problem.

But there are alternative possibilities for specifying the initial distribution of the specific internal energy. For instance, let the mushy phase occupy the whole interval (x_0^-, x_0^+) at $t = 0$:

$$-L < U_0(x) < 0 \qquad \text{for} \qquad x \in (x_0^-, x_0^+) ,$$

with the solid to the left and the liquid to the right. What can one then say about the solution? It is natural to assume that the mushy phase will not disappear instantaneously and, at least on a small time interval, two curves $\{x = R^-(t)\}$ and $\{x = R^+(t)\}$ will exist which split the domain Ω_T into three subdomains Ω_T^-, Ω_T^* and Ω_T^+, occupied by the solid, mushy and liquid phases, respectively.

Let the process we consider be a strongly discontinuous motion, i.e., let the functions $R^-(t), R^+(t)$ be continuously differentiable and ϑ, U be sufficiently regular in the closure of each of the domains Ω_T^-, Ω_T^* and Ω_T^+.

In Ω_T^- and Ω_T^+, the temperature ϑ satisfies equation (1.1) with $f = 0$, the specific internal energy U satisfies equation (1.9) with $f = 0$ in Ω_T^*, and conditions (1.2), (1.10) and (1.11) are satisfied on the boundaries $\{x = R^-(t)\}$ and $\{x = R^+(t)\}$.

By arguments similar to those we have used above, due to conditions (1.10), (1.11) and the hypotheses on the structure of the solution it follows that $R^-(t)$ is non-decreasing and $R^+(t)$ non-increasing, i.e., for each point in Ω_T^* there exists a linear segment, parallel to the t-axis, connecting this point with the line $\{t = 0\}$. Thus, given an initial distribution $U_0(x)$, equation (1.9) can be integrated over Ω_T^* to determine the value of the specific internal energy U at any $(x,t) \in \Omega_T^*$ by

$$U(x,t) = U_0(x) \qquad \text{for} \quad (x,t) \in \Omega_T^*.$$

Hence (1.10) and (1.11) can be given the form

$$(U_0(R^-(t)) + L)\frac{dR^-}{dt}(t) = \frac{\partial \vartheta}{\partial x}(R^-(t) - 0, t), \qquad (1.19)$$

$$U_0(R^+(t))\frac{dR^+}{dt}(t) = \frac{\partial \vartheta}{\partial x}(R^+(t) + 0, t), \qquad (1.20)$$

respectively, and the original problem admits a decomposition into two autonomous problems:

(i) determination of the function $R^-(t)$ and the temperature ϑ in the domain Ω_T^- from the system (1.1) (with $f = 0$), (1.2), (1.19), the boundary condition at $x = -1$ and the initial condition at $t = 0$,

(ii) determination of $R^+(t)$ and ϑ in the domain Ω_T^+ from the system (1.1) (with $f = 0$), (1.2), (1.20), the boundary condition at $x = 1$ and the initial condition at $t = 0$.

The problems are to be solved until the curves $x = R^-(t)$ and $x = R^+(t)$ cross. Then, the mushy phase will disappear (or, more precisely, will degenerate to a point) and thereafter the standard two-phase Stefan problem is to be solved.

In the sequel we shall everywhere take $L = 1$. Except in the last chapter, we shall everywhere assume the thermal conductivity is equal to 1 . This will not affect the generality of the considerations, because it is sufficient to introduce the *reduced temperature*

$$\tilde{\vartheta} = \int_0^\vartheta \kappa(s)\,ds,$$

which satisfies the heat conduction equation with the appropriate coefficient $\tilde{\kappa} = 1$.

In contrast, in the last chapter we shall consider the more general equation of heat conduction in anisotropic media

$$\frac{\partial U}{\partial t} = \sum_{i=1}^3 \frac{\partial}{\partial x_i}\left(\kappa_i \frac{\partial \vartheta}{\partial x_i}\right). \tag{1.21}$$

Let us now discuss the latter case in more detail. To be more specific, we shall consider a model of the crystallization of a cylindrical slab occupying a given domain $\Omega \subset \mathbb{R}^3$, activated by heat conduction in a surrounding medium Q. This means that equation (1.21) is fulfilled in sense of distributions in the domain $\overline{\Omega}_T \cup \overline{Q}_T$, provided the thermal conductivities κ_i and the specific internal energy U are defined in an appropriate way.

Since the phase change occurs only in the domain Ω (in the slab), the internal energy U explicitly depends both on temperature and the point in $\mathbb{R}^3$. U is twice continuously differentiable as a function of temperature everywhere in Q_T (for simplicity we shall assume that U coincides there with ϑ) and in Ω_T it is equal to the discontinuous function Φ introduced above.

Let $\eta(x)$ be the characteristic function of the domain Ω. Then

$$U = \eta(x)\Phi(\vartheta) + \big(1 - \eta(x)\big)\,\vartheta\ .$$

We shall consider the special situation where the heat conductivity of the slab in directions orthogonal to its axis (let it be the axis x_3) dominates both over the values corresponding to the direction of the axis and over the values for the surrounding medium:

$$\kappa_1 = \kappa_2 = \frac{1}{\varepsilon}\eta(x) + 1 - \eta(x), \quad \kappa_3 = 1, \quad \varepsilon \ll 1\ .$$

We shall assume that the generalized solution corresponding to equation (1.21) is a strongly discontinuous motion. One of the strong discontinuity surfaces is a priori given as the common part of the boundary of Ω and Q. Such strong discontinuities are referred to as *contact discontinuities*. If $n = (n_1, n_2, 0)$ is the normal to cylindrical surface Π, then the condition on the contact discontinuity Π that follows from (1.21) takes the form

$$\left[\frac{\partial \vartheta}{\partial N}\right] = 0 , \tag{1.22}$$

where $\dfrac{\partial \vartheta}{\partial N} = \nabla \vartheta \cdot n$, if the limit is taken from the side of Q and $\dfrac{\partial \vartheta}{\partial N} = \dfrac{1}{\varepsilon} \nabla \vartheta \cdot n$, if the limit is taken from the side of Ω. The condition (1.22) prescribed on the surface Π is not sufficient for completing the problem, an additional relation is necessary. In order to establish this, let us return to equation (1.21). Actually, it consists of two equations: the heat balance equation

$$\frac{\partial U}{\partial t} = \text{div } q \tag{1.23}$$

and Fourier's law

$$q = -\Lambda \langle \nabla \vartheta \rangle , \tag{1.24}$$

where $\Lambda = \text{diag } (\kappa_1, \kappa_2 \kappa_3)$ is the thermal conductivity matrix. The condition (1.22) on Π is a consequence of the balance equation (1.23), whereas Fourier's law lacks its counterpart. It can be interpreted as a system of three equations of the form (1.6), with $F = 0$, G is a vector whose i-th component is equal to ϑ and all other components vanish, $X = (1/\kappa_i) \, q_i$ for $i = 1, 2, 3$. The corresponding conditions on the strong discontinuity surface Π are

$$[\vartheta] \cdot n_i = 0 , \quad i = 1, 2, 3$$

and therefore

$$[\vartheta] = 0 . \tag{1.25}$$

In many cases, the continuity condition (1.25) for the temperature on the strong discontinuity surface is evident and it does not require any justification. A different situation arises if the medium that fills Q does not conduct heat, i.e., $q = 0$ in Q. Then the continuity of the temperature across Π (condition (1.25)) does not follow from any equation.

If there is a phase change surface $\Gamma(t)$ within the slab, then the Stefan condition

$$\left[\frac{\partial \vartheta}{\partial x_3} \nu_3 + \frac{1}{\varepsilon}\left(\frac{\partial \vartheta}{\partial x_1} \nu_1 + \frac{\partial \vartheta}{\partial x_2} \nu_2\right)\right] = -L V_\nu \tag{1.26}$$

holds, with V_ν representing the normal velocity of the motion of the surface $\Gamma(t)$, along $\nu = (\nu_1, \nu_2, \nu_3)$.

By the definition of a strongly discontinuous motion away from the surfaces Π and $\Gamma(t)$, equation (1.21) is satisfied in the classical sense, i.e., the temperature ϑ in the domain Q_T fulfils the heat equation

$$\frac{\partial \vartheta}{\partial t} = \Delta \vartheta , \tag{1.27}$$

and it satisfies the equation

$$\frac{\partial \Phi(\vartheta)}{\partial t} = \frac{\partial^2 \vartheta}{\partial x_3^2} + \frac{1}{\varepsilon}\left(\frac{\partial^2 \vartheta}{\partial x_1^2} + \frac{\partial^2 \vartheta}{\partial x_2^2}\right) \tag{1.28}$$

in the outer domain Ω_T which surrounds the surface $\Gamma(t)$.

To complete the problem, we still have to impose a boundary condition on the boundary of the closed domain $\overline{Q} \cup \overline{\Omega}$ and an initial condition at $t = 0$.

2. Assumed notation. Auxiliary notation

2.1. Notation

IR^n	is the n–dimensional Euclidean space of points $x = (x_1, x_2, \ldots, x_n)$;
Ω, Q, G	are domains in IR^n;
IR^{n+1}	is the $(n+1)$–dimensional Euclidean space of points (x, t), where $x \in \mathrm{IR}^n$ and $t \in (-\infty, \infty)$ represents time;
$\Omega_T \subset \mathrm{IR}^{n+1}$	is the set of points (x, t) with $t \in (0, T)$, in particular,
IR_T^{n+1}	$:= \mathrm{IR}^n \times (0, T)$; $\Omega(t) \subset \mathrm{IR}^{n+1}$ is the intersection of the domain Ω_T with the plane $\{t = \mathrm{const.}\}$;
$\overline{\Omega}\,(\overline{\Omega}_T)$	denotes the closure of the set $\Omega\,(\Omega_T)$. If $u = u(x, t)$, then
$D_i u$	$:= \dfrac{\partial u}{\partial x_i}, \; i = 1, \ldots, n; \; \dfrac{\partial u}{\partial t} = D_t u;$
$D_x u$	$:= \nabla u = (D_1 u, \ldots, D_n u);$
$D_x^{\bar{m}} u$	$:= D_1^{m_1} \ldots D_n^{m_n} u, \; \bar{m} = (m_1, \ldots, m_n), \; |\bar{m}| = m_1 + \ldots + m_n;$
$D_0^{\bar{m}} u$	$:= D_t^{m_0} D_1^{m_1} \ldots D_{n-1}^{m_{n-1}} u, \; |\bar{m}| = m_0 + m_1 + \ldots + m_{n-1};$
$\hat{D}^{\bar{m}}$	$:= D_t^{m_0} D_1^{m_1} \ldots D_n^{m_n} u, \; |\bar{m}| = m_0 + m_1 + \ldots + m_n\,.$

2.2. Basic function spaces

$L_q(\Omega)\,(L_q(\Omega_T))$ denotes the Banach space of all measurable functions on Ω (on Ω_T), q-integrable on Ω (on Ω_T), $q \geq 1$, with the norm

$$\|u\|_{q,\Omega} = \left(\int_\Omega |u|^q \, dx\right)^{1/q}, \qquad \left(\|u\|_{q,\Omega_T} = \left(\int_{\Omega_T} |u|^q \, dx \, dt\right)^{1/q}\right).$$

$W_q^l(\Omega)$, l integer, $(1 \leq q \leq \infty)$ denotes the Banach space of all functions in $L_q(\Omega)$ which have the generalized derivatives up to order l inclusive, with the norm

$$\|u\|_{q,\Omega}^{(l)} = \sum_{|\bar{m}|=0}^{l} \|D_x^{\bar{m}} u\|_{q,\Omega}.$$

$\mathring{W}_q^1(\Omega)$ denotes the closure in $W_q^1(\Omega)$-norm of the set of all finite functions defined in Ω.

$W_{q,\text{loc}}^l(\Omega)$ denotes the space of functions which belong to $W_q^l(Q)$ where $Q \subset \Omega$ is an arbitrary bounded domain.

$W_q^{2l,l}(\Omega_T)$, l integer, $q \geq 1$ is the closed subspace of $L_q(\Omega_T)$, including functions whose generalized derivatives $D_t^r D_x^s$ with any r and s, $2r + |s| \leq 2l$ have finite norm

$$\|u\|_{q,\Omega_T}^{(2l)} = \sum_{2r+|s|=0}^{2l} \|D_t^r D_x^s u\|_{q,\Omega_T} .$$

$W_2^{1,0}(\Omega_T)$ is the Hilbert space with the scalar product

$$(u,v)_{W_2^{1,0}(\Omega_T)} = \int_{\Omega_T} (u\,v + \nabla u \cdot \nabla v)\, dx\, dt .$$

$W_2^{1,1}(\Omega_T)$ is the Hilbert space with the scalar product

$$(u,v)_{W_2^{1,1}(\Omega_T)} = \int_{\Omega_T} (uv + D_t u D_t v + \nabla u \cdot \nabla v)\, dx\, dt .$$

$C(\overline{\Omega})$ ($C(\overline{\Omega}_T)$) denotes the space of functions continuous on $\overline{\Omega}$ (on $\overline{\Omega}_T$), equipped with the norm

$$|u|_\Omega^{(0)} = \max_{x \in \Omega} |u(x)| \quad \left(|u|_{\Omega_T}^{(0)} = \max_{(x,t) \in \Omega_T} |u(x,t)|\right) .$$

$H^\alpha(\overline{\Omega})$, $0 < \alpha < 1$, is the space of Hölder continuous functions, with the norm

$$|u|_\Omega^{(\alpha)} = |u|_\Omega^{(0)} + \langle u \rangle_\Omega^{(\alpha)},$$

where $\langle u \rangle_\Omega^{(\alpha)}$ is the *Hölder constant*, defined by

$$\langle u \rangle_\Omega^{(\alpha)} = \sup_{x_1,x_2 \in \Omega} \left\{ |u(x_1) - u(x_2)|\, |x_1 - x_2|^{-\alpha} \right\}.$$

$H^l(\overline{\Omega})$, $l > 0$ non-integer, is a Banach space, equipped with the norm

$$|u|_\Omega^{(l)} = \sum_{|\overline{m}|=0}^{[l]} |D_x^{\overline{m}} u|_\Omega^{(0)} + \sum_{|\overline{m}|=[l]} \langle D_x^{\overline{m}} u \rangle_\Omega^{(l-[l])}.$$

$H^{l,l/2}(\overline{\Omega}_T)$, $l > 0$ non-integer, is the space of functions $u(x,t)$ continuous on $\overline{\Omega}_T$ together with the derivatives $D_t^r D_x^s u$ at $2r + |s| \leq l$, equipped with the norm

$$|u|_{\Omega_T}^{(l)} = \sum_{2r+|s|=0}^{[l]} |D_t^r D_x^s u|_{\Omega_T}^{(0)} + \sum_{2r+|s|=[l]} \langle D_t^r D_x^s u \rangle_{x,\Omega_T}^{(l-[l])}$$

$$+ \sum_{0<l-2r-|s|<2} \langle D_t^r D_x^s u \rangle_{t,\Omega_T}^{((l-2r-|s|)/2)},$$

where, for $0 < \alpha < 1$,

$$\langle v \rangle^{(\alpha)}_{x,\Omega_T} = \sup_{(x,t),(x',t)\in\Omega_T} \left\{ |v(x,t) - (v(x',t)| \, |x - x'|^{-\alpha} \right\},$$

$$\langle v \rangle^{(\alpha)}_{t,\Omega_T} = \sup_{(x,t),(x,t')\in\Omega_T} \left\{ |v(x,t) - v(x,t')| \, |t - t'|^{-\alpha} \right\}.$$

$C^{2l,l}(\overline{\Omega}_T)$, l integer, is the set of all functions continuous on $\overline{\Omega}_T$ together with the derivatives $D_t^r D_x^s$ for $2r + |s| \le 2l$.

Let $\mathbf{X}(\overline{\Omega})$ be any of the above spaces of differentiable functions. Then $\mathbf{X}(\Omega)$ will denote the set of all elements of $\mathbf{X}(\overline{\Omega'})$ for any closed subdomain $\overline{\Omega'} \subset \Omega$.

Functions dependent on $x \in \Omega$ and $t \in (0, T)$ will often be treated as functions of t with values in the Banach space $\mathbf{X}(\Omega)$. For instance, $L_q(0, T; W_p^l(\Omega))$ is the space of functions $u(t)$ defined on $(0, T)$ with values in $W_p^l(\Omega)$, equipped with the norm

$$\|u\| = \left(\int_0^T \left(\|u(t)\|^{(l)}_{p,\Omega} \right)^q dt \right)^{1/q},$$

where $\|u\|_{\mathbf{X}(\Omega)}$ represents the norm of $u(t)$ in $\mathbf{X}(\Omega)$.

2.3. Auxiliary inequalities and embedding theorems

We now recall a few inequalities that we shall often use in the sequel. The *Cauchy inequality*

$$\left| \sum_{i,j=1}^n a_{ij}\xi_i\eta_j \right| \le \left(\sum_{i,j=1}^n a_{ij}\xi_i\xi_j \right)^{1/2} \left(\sum_{i,j=1}^n a_{ij}\eta_i\eta_j \right)^{1/2}$$

holds for any non-negative matrix $\mathbf{a} = \{a_{ij}\}$ and arbitrary $\xi, \eta \in \mathbb{R}^n$.

Young's inequality

$$ab \le \frac{1}{p}\varepsilon^p a^p + \frac{1}{q}\varepsilon^{-q} b^q, \quad \varepsilon > 0, \quad \frac{1}{p} + \frac{1}{q} = 1, \, p > 1,$$

is true for all non-negative a, b.

Hölder's inequality has the form

$$\|uv\|_{1,\Omega} \le \|u\|_{p,\Omega}\|v\|_{q,\Omega}, \quad \frac{1}{p} + \frac{1}{q} = 1, \, p \ge 1.$$

In deriving a priori estimates, one often makes use of *Gronwall's inequality*:

Lemma 1. *Let $y(t)$ be an absolutely continuous function, non-negative on $[0, T]$, with $y(0) = 0$ and such that*

$$\frac{dy}{dt}(t) \le c(t)\, y(t) + \mathcal{F}(t),$$

with functions $c(t)$ and $\mathcal{F}(t)$ non-negative and integrable on $[0, T]$. Then

$$y(t) \leq \int_0^t \mathcal{F}(\tau)\, d\tau \, \exp\left(\int_0^t c(\tau)\, d\tau\right).$$

By an *embedding operator* from a certain space $\mathbf{X}$ into a larger space $\mathbf{Y}$ we shall mean the operator that maps each element of $\mathbf{X}$ into itself, but treated as element of $\mathbf{Y}$.

Theorem 1. *Let $\Omega \subset \mathbb{R}^n$ be a bounded domain with piecewise regular boundary. Then the embedding operator from $W_p^l(\Omega)$ into $H^\alpha(\overline{\Omega})$ is bounded for $\alpha p \leq lp - n$ and completely continuous for $\alpha p < lp - n$ $(\alpha < 1)$.*
If $l = 1$ and $n > p$, then the embedding operator is completely continuous from $W_p^1(\Omega)$ into $L_q(\Omega)$, as long as $q < pn\,(n-p)^{-1}$.

More refined relations between different function spaces are expressed by the following *interpolation inequalities.*

Lemma 2. *Let $\Omega \subset \mathbb{R}^n$ and $u \in W_p^l(\Omega) \cap L_q(\Omega)$, $1 \leq p, q \leq \infty$. Then $u \in W_r^k(\Omega)$ and*

$$\|u\|_{r,\Omega}^{(k)} \leq c\big(\|u\|_{p,\Omega}^{(l)}\big)^\alpha \big(\|u\|_{q,\Omega}\big)^{1-\alpha}, \tag{2.1}$$

where

$$\frac{1}{r} = \frac{k}{n} + \alpha\Big(\frac{1}{p} - \frac{l}{n}\Big) + (1-\alpha)\frac{1}{q}, \quad \frac{k}{l} \leq \alpha \leq 1.$$

The above does not hold if $l - k - n/p$ is a non-negative integer and $1 < p < \infty$. In this case, $\alpha < 1$.

Upon applying Young's inequality to the right-hand side, inequality (2.1) can be given the form

$$\|u\|_{r,\Omega}^{(k)} \leq c\alpha\varepsilon^{1/\alpha}\|u\|_{p,\Omega}^{(l)} + c(1-\alpha)\varepsilon^{-1/(1-\alpha)}\|u\|_{q,\Omega}, \quad \varepsilon > 0.$$

Lemma 3. *Let either $u \in \mathring{W}_m^1(\Omega)$ or $u \in W_m^1(\Omega)$ with $\displaystyle\int_\Omega u(x)\, dx = 0$ in the second case. Then*

$$\|u\|_{q,\Omega} \leq c\|\nabla u\|_{m,\Omega}^\alpha \|u\|_{r,\Omega}^{1-\alpha}, \tag{2.2}$$

where

$$\alpha = \Big(\frac{1}{r} - \frac{1}{q}\Big)\Big(\frac{1}{r} - \frac{n-m}{nm}\Big)^{-1}.$$

If $m < n$, inequality (2.2) holds for $q \in [r, mn/(n-m)]$ when $r \leq mn/(n-m)$, and for $q \in [mn/(n-m), r]$ when $r > mn/(n-m)$. If $m \geq n$, then (2.2) is true for all $q \geq r$, including $q = \infty$ when $m > n$.

Lemma 4. *Let $\Omega \subset \mathbb{R}^n$ be a domain satisfying the cone condition, i.e., for each point of $\partial\Omega$ there exists a non-trivial radial cone K of height d, completely contained in $\overline{\Omega}$. Then for each function $u \in H^{l,l/2}(\overline{\Omega}_T)$, where $\Omega_T = \Omega \times (0,T)$,*

$$|u|_{\Omega_T}^{(r)} \le c_1 \delta^{l-r} |u|_{\Omega_T}^{(l)} + c_2 \delta^{-r} |u|_{\Omega_T}^{(0)}, \tag{2.3}$$

for arbitrary $r \in (0,l]$ and $\delta \in (0, \min\{d, T^{1/2}\})$ where c_1, c_2 are constants depending on r, l and the angle at the vertex of K.

Lemma 5. *Let Ω be the same as in Lemma 4. Then, for each function $u \in W_p^{2,1}(\Omega_T)$,*

$$|D_x^s u|_{\Omega_T}^{(\lambda)} \le c_3 \delta^{2-s-(n+2)/p-\lambda} \|u\|_{p,\Omega_T}^{(2)} + c_4 \delta^{-(s+(n+2)/p+\lambda)} \|u\|_{p,\Omega_T}, \tag{2.4}$$

provided that the constant δ satisfies the condition

$$0 < \delta < \min\{d, T^{1/2}\} \quad and \quad 2 - s - (n+2)/p > \lambda.$$

Lemma 6. *Let $\Omega \subset \mathbb{R}^n$ satisfy the cone condition, and let the function $u(x,t)$ defined on $\Omega_T - \Omega \times (0,T)$ fulfil the Hölder condition in t with index α and Hölder constant μ_1. Moreover, let $D_x u$ exist and satisfy the Hölder condition in x with index β and Hölder constant μ_2, for every $t \in [0,T]$. Then $D_x u$ satisfies the Hölder condition in t with index $\delta = \alpha\beta/(1 + \beta)$ and Hölder constant μ which depends only on $\alpha, \beta, \mu_1, \mu_2, d$ and the conical angle at the vertex of K.*

Lemmas 1–6 and Theorem 1 have been formulated and proved in [2], [137].

2.4. Auxiliary facts from analysis

The results of this section can be found in [130] and [223].

Let $\mathbf{X}$ be a real normed space, $\mathbf{X}^*$ its *dual space* and let (v, u) denote the value of a linear bounded functional $v \in \mathbf{X}^*$ for $u \in \mathbf{X}$. The dual space to $L_p(\Omega)$, $1 < p < \infty$, is isomorphic to $L_q(\Omega)$ where $\dfrac{1}{p} + \dfrac{1}{q} = 1$.

Let $\mathbf{X}^{**}$ be the *second dual* to $\mathbf{X}$ and let $\mathcal{J}$ denote the *canonical isomorphic embedding* of $\mathbf{X}$ into $\mathbf{X}^{**}$, $\mathcal{J}\mathbf{X} \subset \mathbf{X}^{**}$. If $\mathcal{J}\mathbf{X} = \mathbf{X}^{**}$, the space $\mathbf{X}$ is *reflexive*. All Hilbert spaces and $L_p(\Omega)$ with $1 < p < \infty$ are reflexive.

A sequence $\{u_n\}$ of elements of $\mathbf{X}$ is *weakly convergent* to $u \in \mathbf{X}$, if for every $v \in \mathbf{X}^*$,

$$\lim_{n \to \infty} (v, u_n) = (v, u).$$

A subset K of Banach space $\mathbf{X}$ is *compact* if every sequence of its elements contains a subsequence convergent to an element of K.

In particular, every closed bounded subset of $H^l(\overline{\Omega})$ ($H^{l,l/2}(\overline{\Omega}_T)$) is compact in the spaces $H^r(\overline{\Omega})$ (in $H^{r,r/2}(\overline{\Omega}_T)$) for $r < l$.

Analogously, $K \subset \mathbf{X}$ is *weakly compact* if every sequence of its elements contains a subsequence which converges weakly to an element of K.

Lemma 7. *All bounded closed subsets of reflexive Banach spaces are weakly compact. If u is the weak limit of the sequence $\{u_n\}$, then*

$$\|u\| \leq \varliminf_{n \to \infty} \|u_n\|.$$

For an operator Ψ defined on the Banach space $\mathbf{X}$, any solution of the equation $u = \Psi u$ is called a *fixed point* of this operator. An operator from the Banach space $\mathbf{X}$ into the Banach space $\mathbf{Y}$ is *completely continuous* if it maps bounded closed subsets of $\mathbf{X}$ onto sets which are compact in $\mathbf{Y}$. In particular, the embedding operator from $H^l(\overline{\Omega})$ into $H^r(\overline{\Omega})$ is completely continuous for $l > r$.

Theorem 2. (Schauder) *Let K be a bounded, closed, convex subset of the Banach space $\mathbf{X}$. If Ψ is a completely continuous operator that maps K into K, then there exists at least one fixed point $u \in K$ of Ψ.*

2.5. Properties of solutions of differential equations

We now recall a few theorems on the existence and qualitative properties of solutions to linear parabolic equations of second order (cf., [137]).

Let $\Omega \subset \mathbb{R}^n$ be a bounded domain with boundary S in the class H^{l+2} (defined in Section 6 of this chapter together with the class O^2). In the domain $\Omega_T = \Omega \times (0, T)$, a solution $u(x, t)$ is to be found to the equation

$$Lu \equiv D_t u - \sum_{i,j=1}^{n} a_{ij}(x,t) D_i D_j u + \sum_{i=1}^{n} a_i(x,t) D_i u + a(x,t)u = f(x,t), \quad (2.5)$$

subject to either of the boundary conditions

$$u|_{S_T} = \Phi(x,t), \qquad (2.6)$$

$$\left(\sum_{i=1}^{n} b_i(x,t) D_i u + b(x,t)u \right)\Big|_{S_T} \equiv Bu|_{S_T} = \Phi(x,t) \qquad (2.7)$$

on $S_T = S \times (0, T)$ and satisfying the initial condition

$$u\,|_{t=0} = \varphi(x). \qquad (2.8)$$

We shall assume that the operator L is uniformly parabolic in Ω_T,

$$\mu|\xi|^2 \leq \sum_{i,j=1}^{n} a_{ij} \xi_i \xi_j \leq \nu|\xi|^2, \quad \mu, \nu = \text{const.} > 0,$$

and that the functions $b_i(x,t)$ satisfy the condition

$$\left| \sum_{i=1}^{n} b_i(x,t) \nu_i(x) \right| \geq \delta > 0$$

on S_T, where $\nu = (\nu_1, ..., \nu_n)$ is the normal to S at $x \in S$.

We shall say that the functions f, φ, Φ satisfy *compatibility conditions of order m at $x \in S$ and $t = 0$*, if all the time derivatives of u up to order m appearing in equation (2.5) and the initial condition (2.8), fulfil the appropriate boundary condition (2.6) or (2.7) for $x \in S$.

Theorem 3. *Let $l > 0$ be non-integer, the coefficients of the operator L be in the class $H^{l,l/2}(\bar{\Omega}_T)$ and let the boundary S belong to H^{l+2}. Then, for any $f \in H^{l,l/2}(\bar{\Omega}_T)$, $\varphi \in H^{l+2}(\Omega)$ and $\Phi \in H^{l+2,(l+2)/2}(\bar{S}_T)$ (see Section 6 of this chapter for the appropriate definitions) satisfying compatibility conditions of order $[l/2] + 1$, there exists a unique solution $u \in H^{l+2,(l+2)/2}(\bar{\Omega}_T)$ of the problem (2.5),(2.6),(2.8). This solution satisfies the a priori estimate*

$$|u|_{\Omega_T}^{(l+2)} \leq c\left(|f|_{\Omega_T}^{(l)} + |\varphi|_{\Omega}^{(l+2)} + |\Phi|_{S_T}^{(l+2)}\right).$$

Theorem 4. *Let S, L, f, u and φ be as in Theorem 3, and let $b_i, b \in H^{l+1,(l+1)/2}(\bar{S}_T)$. Then, for any $\Phi \in H^{l+1,(l+1)/2}(\bar{S}_T)$ that fulfils compatibility conditions of order $[(l + 1)/2]$, the problem (2.5),(2.7),(2.8) has a unique solution $u(x, t)$. This solution satisfies the a priori estimate*

$$|u|_{\Omega_T}^{(l+2)} \leq c\left(|f|_{\Omega_T}^{(l)} + |\varphi|_{\Omega}^{(l+2)} + |\Phi|_{S_T}^{(l+1)}\right).$$

Theorem 5. *Let $S \in O^2$, $1 < q$, $q \neq 3/2$. Assume that the coefficients a_{ij} of the operator L are continuous functions on $\bar{\Omega}_T$ and the coefficients a_i, a have finite norms in $L_r(\Omega_T)$ and $L_s(\Omega_T)$, respectively, with*

$$r = \begin{cases} \max(q, n + 2) & \text{if} \quad q \neq n + 2 \\ n + 2 + \varepsilon & \text{if} \quad q = n + 2 \end{cases}$$

$$s = \begin{cases} \max(q, (n + 2)/2) & \text{if} \quad q \neq (n + 2)/2 \\ (n + 2)/2 + \varepsilon & \text{if} \quad q = (n + 2)/2 \end{cases}$$

where $\varepsilon > 0$ is arbitrarily small. Moreover, let $\|a_i\|_{r,\Omega_{t+\tau}\setminus\Omega_t}$, $\|a\|_{s,\Omega_{t+\tau}\setminus\Omega_t}$ converge to 0 as $\tau \to 0$. Then, for any $f \in L_q(\Omega_T)$, the problem (2.5),(2.6),(2.8) with $\Phi = 0, \varphi = 0$ has a unique solution $u \in W_q^{2,1}(\Omega_T)$. For this solution we have

$$\|u\|_{q,\Omega_T}^{(2)} \leq c\|f\|_{q,\Omega_T}.$$

The only reason for assuming $\Phi = 0, \varphi = 0$ in Theorem 5 was to avoid the necessity of introducing Sobolev spaces of fractional order. In the sequel we refer to Theorem 5 also in the case of non-vanishing φ and Φ. This is justified by the postulated regularity of these functions and by the character of the estimates used. The problem with non-homogeneous conditions on the boundary S and at $t = 0$ can be reduced to a problem with homogeneous conditions.

The following two theorems provide local estimates on the solutions. Let $\Omega' \subset \Omega'' \subset \Omega$, where the distances from Ω' to S and to $\Omega\setminus\Omega''$ are positive. Define $Q' = \Omega' \times (T_1, T_2)$ and $Q'' = \Omega'' \times (T_0, T_2)$, with $T_0 = 0$ if $T_1 = 0$.

Theorem 6. *Let* $u \in H^{l+2,(l+2)/2}(\bar{Q}'')$ *and let*

$$Lu = f$$

be satisfied in Q''. *Moreover, if* $T_1 = 0$, *then let the initial condition* (2.8) *be fulfilled. If the coefficients of the operator* L *belong to* $H^{l+2,(l+2)/2}(\bar{Q}'')$, *then*

$$|u|_{Q'}^{(l+2)} \le c_1(|f|_{Q''}^{(l)} + |\varphi|_{\Omega''}^{(l+2)}) + c_2|u|_{Q''}^{(0)}.$$

Theorem 7. *Let* $T_1 > T_0 > 0$, *the coefficients of the operator* L *satisfy the hypotheses of Theorem 5,* $u \in W_q^{2,1}(Q'')$ *and let* $Lu = f$ *in* Q''. *Then*

$$\|u\|_{q,Q'}^{(2)} \le c_3\|f\|_{q,Q''} + c_4\|u\|_{q,Q''},$$

where $1 \le p \le q$.

2.6. The Cauchy problem for the heat equation over smooth unbounded manifolds in the classes $H^{l+2,(l+2)/2}(S_T)$

Let S be the surface that bounds the domain Ω. Assume there exists $d > 0$ such that, for any $\xi \in S$, the surface S can be described in the ball with radius d and center at ξ by the equation

$$y_n = F(y'), \quad y' = (y_1, \ldots, y_{n-1}) \tag{2.9}$$

in a local coordinate system.

We shall say that $S \in H^l$, if $F \in H^l(\sigma)$ for each $\xi \in S$, where σ is the ball $|y'| < d/2$ and if the norms $|F|_\sigma^{(l)}$ are uniformly bounded.

S belongs to the class O^l if the derivatives of order $l - 1$ of the functions F are continuous, their first differentials exist and the appropriate derivatives of order l are uniformly bounded.

Let S admit a decomposition into subsurfaces $S^{(1)}, \ldots, S^{(k)}, \ldots$ such that the representation (2.9) holds for all $S^{(k)}$. Denote $S_T^{(k)} = S^{(k)} \times (0, T)$ and let $u^{(k)}(y', t)$ be the image of $u(x, t)$ on transforming $S_T^{(k)}$ onto σ_T, $\sigma_T = \sigma \times (0, T)$. The space $H^{l,l/2}(S_T)$ is defined as the set of functions $u(x, t)$ with the finite norm

$$\{u\}_{S_T}^{(l)} = \sup_h |u^{(k)}|_{\sigma_T}^{(l)}.$$

The norms corresponding to different decompositions of S into $S^{(k)}$ are equivalent.

The surface S can be treated as a Riemannian manifold with a metric tensor $\{g_{ij}\}$ such that

$$g_{ij} = D_i x \, D_j x, \quad i, j = 1, \ldots, n - 1$$

in local coordinates (y'), where $x = x(y')$. For any local coordinate system, the *Laplace-Beltrami operator* $\Delta_S u = \text{div}(\nabla u)$ can be represented in the form

$$\Delta_S u = |g|^{-1/2} \sum_{i,j=1}^{n-1} D_i(|g|^{1/2} g^{ij} D_j u^{(k)}),$$

in terms of the contravariant components g^{ij} of the metric tensor, where $|g| = \det(g_{ij})$.

Theorem 8. *Let* $S \in H^{l+2}, l > 0$, $u_0 \in H^{l+2}(S)$, $f \in H^{l,l/2}(S_T)$. *Then the Cauchy problem*

$$D_t u - \Delta_S u = f, \qquad \text{for} \quad (x, t) \in S_T;$$
$$u \,|_{t=0} = u_0, \qquad \text{for} \quad x \in S ,$$

has a unique solution $u \in H^{l+2,(l+2)/2}(S_T)$. *This solution satisfies the a priori bound*

$$|u|_{S_T}^{(l+2)} \le c\Big(|f|_{S_T}^{(l)} + |u_0|_S^{(l+2)}\Big).$$

For details of the results in Section 6 see [151].

3. Existence and uniqueness of the generalized solution to the Stefan problem

Let $Q_T = \Omega \times (0, T)$ be a cylindrical domain, with $\Omega \subset \mathbb{R}^n$ a bounded domain with piecewise regular boundary S, $S_T = S \times (0, T)$. The *Stefan problem* consists of determining a function $U(x, t)$ which satisfies the equation

$$U_t = \Delta \vartheta + f(U, x, t) \quad \text{in } Q_T, \tag{3.1}$$

with the conditions

$$\vartheta(s, t) = \vartheta^0(s, t), \qquad \text{for} \quad (s, t) \in S_T \tag{3.2}$$

and at the initial time

$$U(x, 0) = U_0(x), \qquad \text{for} \quad x \in \Omega. \tag{3.3}$$

We shall assume that the temperature ϑ is a non-decreasing function of the specific internal energy (enthalpy), $\vartheta = \chi(U)$, in the class $C^2(-\infty, -L] \cap C^2[0, \infty)$ and such that

$$A > \frac{\partial \chi}{\partial U} > a_0 > 0 \quad \text{for } U \notin (-L, 0), \qquad \chi(U) = 0 \quad \text{for } U \in [-L, 0].$$

Equation (3.1) will be understood in the weak sense of an appropriate integral identity.

Definition 1. A bounded measurable function U is called the *generalized solution* of the Stefan problem (3.1) – (3.3), if it satisfies the integral identity

$$\int_{Q_T} \{U\varphi_t + \chi(U)\Delta\varphi + f\varphi\}\, dxdt - \int_{S_T} \vartheta^0(s,t)\frac{\partial\varphi}{\partial\nu}\, dsdt$$

$$+ \int_\Omega U_0(x)\varphi(x,0)\, dx = 0 \qquad (3.4)$$

for any function $\varphi \in W_2^{2,1}(Q_T)$, vanishing at $t = T$ and on the boundary S_T, where ν is the outward normal to S_T.

Theorem 9. *Let the boundary temperature $\vartheta^0(s,t)$ and the initial enthalpy $U_0(x)$ be given by bounded measurable functions and let f be uniformly continuous with respect to its first variable in $\mathbb{R}$, so that*

$$|f(s_1,x,t) - f(s_2,x,t)| \le f_1\,|s_1 - s_2| \quad \text{for all } s_1, s_2 \in \mathbb{R}, \ (x,t) \in Q_T. \qquad (3.5)$$

Then there exists a unique generalized solution of the Stefan problem (3.1) – (3.3) such that $U \in L_\infty(Q_T)$. The solution U is stable in the space $L_1(\Omega)$ with respect to the initial data: if $\tilde{U}(x,t)$ is the solution of equation (3.1) with the initial-boundary conditions

$$\tilde\vartheta\,|_{S_T} = \vartheta^0, \qquad \tilde{U}\,|_{t=0} = \tilde{U}_0(x) \in L_\infty(\Omega), \qquad (3.6)$$

then

$$\int_\Omega |U(x,t) - \tilde{U}(x,t)|\, dx \le e^{f_1 t}\int_\Omega |U_0(x) - \tilde{U}_0(x)|\, dx. \qquad (3.7)$$

Proof. Without loss of generality, we can assume f to be a function of the single variable U.

Let us regularize the problems (3.1) – (3.3) and (3.1), (3.6), approximating the function χ by $\chi_\varepsilon \in C^{2+\alpha_1}(\mathbb{R})$ such that

$$A \ge \chi'_\varepsilon(s) \ge \varepsilon, \quad |\chi_\varepsilon(s) - \chi(s)| \le C_1\varepsilon.$$

Let the boundary and initial data be approximated by

$$\vartheta^0_\varepsilon \in C^{2+\alpha_1,(2+\alpha_1)/2}(S_T) \quad \text{and} \quad U_0^\alpha, \ \tilde{U}_0^\alpha \in C^{2+\alpha_1}(\bar\Omega),$$

respectively, so that

$$\sup_{\alpha,\varepsilon\in(0,1]} \left\{\|\vartheta^0_\varepsilon\|_{L_\infty(S_T)}, \|U_0^\alpha, \ \tilde{U}_0^\alpha\|_{L_\infty(\Omega)}\right\} \le \max\left\{\|\vartheta^0\|_{L_\infty(S_T)}, \|U_0, \ \tilde{U}_0\|_{L_\infty(\Omega)}\right\},$$

where $\vartheta^0_\varepsilon \to \vartheta^0$ strongly in $L_1(S_T)$ and $U_0^\alpha \to U_0$, $\tilde{U}_0^\alpha \to \tilde{U}_0$ strongly in $L_1(\Omega)$ as $\varepsilon \to 0$ and $\alpha \to 0$, respectively. The source term f will be approximated by functions $f_\varepsilon \in C^{1+\alpha}(\mathbb{R})$ such that $\sup_{\varepsilon,\mathbb{R}} |f'_\varepsilon| < f_1$ and $f_\varepsilon \to f$ in the norm of $C(\mathbb{R})$ as $\varepsilon \to 0$.
The initial and boundary conditions are to be regularized so as to preserve the first-order compatibility conditions for the regularized problems (3.1) – (3.3) and (3.1), (3.6). The

existence of solutions $U^{\varepsilon,\alpha}(x,t)$, $\tilde{U}^{\varepsilon,\alpha}(x,t) \in C^{2+\alpha_1,(2+\alpha_1)/2}(\bar{Q}_T)$ to the regularized problems follows in view of the appropriate results of [137].

By the uniform Lipschitz continuity of the functions $f_\varepsilon(s)$ we have

$$\sup_{\varepsilon,\alpha,Q_T} \left\{ |U^{\varepsilon,\alpha}(x,t)|, |\tilde{U}^{\varepsilon,\alpha}(x,t)| \right\} < C_2. \tag{3.8}$$

Indeed, consider the function $V(x,t) = \left(U^{\varepsilon,\alpha} - f_\varepsilon(0) \right) e^{-f_1 t}$. If V attains its maximum at (x_0,t_0), with $U^{\varepsilon,\alpha}(x_0,t_0) \le 0$, or at $(x_0,t_0) \in S_T$ or at $t_0 = 0$, then $U^{\varepsilon,\alpha}$ admits a finite upper bound independent of ε and α. Assume now that $(x_0,t_0) \in Q_T$ and $U^{\varepsilon,\alpha}(x_0,t_0) > 0$. Then $\Delta V = (\chi'_\varepsilon)^{-1} \Delta \vartheta^{\varepsilon,\alpha} e^{-f_1 t} - \chi''_\varepsilon (\nabla V)^2 e^{f_1 t}$, and therefore $\Delta \vartheta^{\varepsilon,\alpha}(x_0,t_0) < 0$. . The function $V(x,t)$ satisfies the equation

$$V_t = \Delta \vartheta^{\varepsilon,\alpha} e^{-f_1 t} + \left\{ \left(f_\varepsilon(U^{\varepsilon,\alpha}) - f_\varepsilon(0) \right) - f_1 U^{\varepsilon,\alpha} \right\} e^{-f_1 t}$$

in Q_T, hence $\dfrac{\partial V}{\partial t}(x_0,t_0) < 0$, in contradiction to the inequality $U^{\varepsilon,\alpha}(x_0,t_0) > 0$ which holds at the point where $V(x,t)$ attains its maximum over Q_T. Similarly, a finite lower bound independent of ε and α can be derived for $U^{\varepsilon,\alpha}(x,t)$ and the analogous estimates can be shown for $\tilde{U}^{\varepsilon,\alpha}(x,t)$.

The function $\bar{U}(x,t) := U^{\varepsilon,\alpha_1}(x,t) - U^{\varepsilon,\alpha_2}(x,t)$ satisfies the equation

$$\bar{U}_t = \Delta \bar{\vartheta} + \bar{f}$$

in Q_T, where $\bar{\vartheta} = \vartheta^{\varepsilon,\alpha_1} - \vartheta^{\varepsilon,\alpha_2}$ and $\bar{f} = f_\varepsilon(U^{\varepsilon,\alpha_1}) - f_\varepsilon(U^{\varepsilon,\alpha_2})$. After multiplying the above equation by the function $\mathrm{sgn}^\gamma \bar{U} = \bar{U}(\bar{U}^2 + \gamma)^{-1/2}$ and then integrating the result over Q_T, we obtain

$$\int_\Omega \left[\bar{U} \mathrm{sgn}^\gamma \bar{U} \right]_{t=0}^{t=\tau} dx - \int_{Q_\tau} \frac{\bar{U} \bar{U}_t \gamma}{(\bar{U}^2 + \gamma)^{3/2}} \, dx dt$$

$$= -\int_{Q_\tau} \left\{ \frac{\nabla \bar{\vartheta} \nabla \bar{U} \gamma}{(\bar{U}^2 + \gamma)^{3/2}} - \bar{f} \mathrm{sgn}^\gamma \bar{U} \right\} dx dt.$$

The second term on the left-hand side of this equality is bounded uniformly with respect to γ,

$$|\bar{U}\gamma|(\bar{U}^2 + \gamma)^{-3/2} \le |\bar{U}\gamma| \{ \max(|\bar{U}|, \sqrt{\gamma}) \}^{-3} \le 1,$$

hence, by the Lebesgue theorem,

$$\lim_{\gamma \to 0} \int_{Q_\tau} \gamma \bar{U} U_t (\bar{U}^2 + \gamma)^{-3/2} \, dx dt = 0.$$

Let us rewrite the first term on the right-hand side in the form

$$\int_{Q_\tau} \left\{ \bar{\vartheta} \gamma \Delta \bar{U} (\bar{U}^2 + \gamma)^{-3/2} - \bar{\vartheta} \bar{U} (\nabla \bar{U})^2 \gamma (\bar{U}^2 + \gamma)^{-5/2} \right\} dx dt.$$

Since $0 \leq \bar{\vartheta}/\bar{U} \leq A$, the upper limit of the first term above as $\gamma \to 0$ is negative. Passing to the limit as $\gamma \to 0$, we eventually obtain

$$\int_{\Omega} |U^{\varepsilon,\alpha_1} - U^{\varepsilon,\alpha_2}| \, |_{t=\tau} \, dx \leq \int_{\Omega} |U_0^{\alpha_1} - U_0^{\alpha_2}| \, dx + f_1 \int_{Q_\tau} |U^{\varepsilon,\alpha_1} - U^{\varepsilon,\alpha_2}| \, dxdt,$$

for all $\tau \in [0, T]$. By Gronwall's lemma (see Lemma 1 of Section 2), we thus conclude the estimate

$$\int_{\Omega} |U^{\varepsilon,\alpha_1} - U^{\varepsilon,\alpha_2}| \, |_{t=\tau} \, dx \leq e^{f_1 \tau} \int_{\Omega} |U_0^{\alpha_1} - U_0^{\alpha_2}| \, dx. \tag{3.9}$$

By repeating the same arguments, we obtain

$$\int_{\Omega} |\tilde{U}^{\varepsilon,\alpha_1} - \tilde{U}^{\varepsilon,\alpha_2}| \, |_{t=\tau} \, dx \leq e^{f_1 \tau} \int_{\Omega} |\tilde{U}_0^{\alpha_1} - \tilde{U}_0^{\alpha_2}| \, dx, \tag{3.10}$$

$$\int_{\Omega} |\tilde{U}^{\varepsilon,\alpha} - U^{\varepsilon,\alpha}| \, |_{t=\tau} \, dx \leq e^{f_1 \tau} \int_{\Omega} |\tilde{U}_0^{\alpha} - U_0^{\alpha}| \, dx. \tag{3.11}$$

To avoid redundancy, we refer to the proof of Theorem 10 on the comparison of solutions, which as a byproduct implies the compactness of the families $(U^{\varepsilon,\alpha})_{\varepsilon>0}$ and $(\tilde{U}^{\varepsilon,\alpha})_{\varepsilon>0}$ in the space $L_2(Q_T)$. It is thus possible to select subsequences $U^{\tilde{\varepsilon},\alpha}$ and $\tilde{U}^{\tilde{\varepsilon},\alpha}$ such that $U^{\tilde{\varepsilon},\alpha}(x,t) \to U^{\alpha}(x,t)$ and $\tilde{U}^{\tilde{\varepsilon},\alpha}(x,t) \to \tilde{U}^{\alpha}(x,t)$ in the space $L_2(Q_T)$ as $\tilde{\varepsilon} \to 0$. Since the function χ is continuous and $\chi_{\varepsilon} \to \chi$ in $C(\mathbb{R})$, the limits of subsequences $\vartheta^{\tilde{\varepsilon},\alpha}$ and $\tilde{\vartheta}^{\tilde{\varepsilon},\alpha}$ in $L_2(Q_T)$ as $\tilde{\varepsilon} \to 0$ are equal to $\vartheta^{\alpha} = \chi(U^{\alpha})$ and $\tilde{\vartheta}^{\alpha} = \chi(\tilde{U}^{\alpha})$, respectively. Passing to the limit in the integral identity

$$\int_{Q_T} \{U^{\tilde{\varepsilon},\alpha} \varphi_t + \chi_{\tilde{\varepsilon}}(U^{\tilde{\varepsilon},\alpha}) \Delta\varphi + f_{\tilde{\varepsilon}}(U^{\tilde{\varepsilon},\alpha})\varphi\} \, dxdt - \int_{S_T} \vartheta^0(s,t) \frac{\partial\varphi}{\partial n} \, dsdt$$

$$+ \int_{\Omega} U_0^{\alpha}(x)\varphi(x,0) \, dx = 0,$$

where $\varphi(x,t)$ is a test function in the sense of Definition 1, we conclude that $U^{\alpha}(x,t)$ is a solution of the Stefan problem with nonlinear source term f and smooth initial data $\tilde{U}_0^{\alpha}(x)$.

Passing to the limit as $\tilde{\varepsilon} \to 0$ in the bounds (3.9) – (3.11), we infer the stability of solutions with respect to smooth initial data,

$$\int_{\Omega} |U^{\alpha_1}(x,t) - U^{\alpha_2}(x,t)| \, dx \leq e^{f_1 t} \int_{\Omega} |U_0^{\alpha_1}(x) - U_0^{\alpha_2}(x)| \, dx, \tag{3.12}$$

$$\int_{\Omega} |\tilde{U}^{\alpha_1}(x,t) - \tilde{U}^{\alpha_2}(x,t)| \, dx \leq e^{f_1 t} \int_{\Omega} |\tilde{U}_0^{\alpha_1}(x) - \tilde{U}_0^{\alpha_2}(x)| \, dx, \tag{3.13}$$

$$\int_{\Omega} |\tilde{U}^{\alpha}(x,t) - U^{\alpha}(x,t)| \, dx \leq e^{f_1 t} \int_{\Omega} |\tilde{U}_0^{\alpha}(x) - U_0^{\alpha}(x)| \, dx. \tag{3.14}$$

The bounds (3.12) – (3.14) and the fact that the functions U^{α}, $\tilde{U}^{\alpha}$ are bounded uniformly in α allow the possibility of passing to the limit as $\alpha \to 0$ and thus the removal of the smoothness hypotheses on the initial data. Passing to the limit as $\alpha \to 0$ in (3.14), we obtain the estimate (3.7).   □

The uniqueness of the generalized solution to the Stefan problem (3.1) – (3.3) follows by the comparison theorem.

Theorem 10. (On the comparison of solutions) *Let the functions $f^1(U)$ and $f^2(U)$ be Lipschitz continuous in $\mathbb{R}$,*

$$U_0^1(x) \le U_0^2(x), \quad \vartheta_1^0(s,t) \le \vartheta_2^0(s,t), \quad f^1(U) \le f^2(U).$$

If U_1 and U_2 are the generalized solutions of the problems (3.1) – (3.3) with the appropriate initial-boundary conditions and right-hand sides of the differential equations, then

$$U_1(x,t) \le U_2(x,t), \quad for \quad (x,t) \in Q_T.$$

The domain Ω may be unbounded here.

Proof. Take the difference of the integral identities satisfied by U_1 and U_2. Denoting $U = U_2 - U_1$, $U_0 = U_0^2 - U_0^1$ and $\vartheta^0 = \vartheta_2^0 - \vartheta_1^0$, we obtain the equality

$$\int_\Omega U(x,\tau)\varphi(x,\tau)\,dx - \int_{Q_\tau} U\,(\varphi_t + \mu\Delta\varphi)\,dxdt$$

$$= \int_\Omega U_0(x)\varphi(x,0)\,dx - \int_{S_T} \vartheta^0 \frac{\partial\varphi}{\partial n}\,dsdt$$

$$+ \int_{Q_\tau} \varphi\big(f^2(U^2) - f^1(U^2)\big)\,dxdt + \int_{Q_\tau} \varphi\Delta f^1 U\,dxdt. \tag{3.15}$$

In (3.15),

$$\mu = \frac{\chi(U_2) - \chi(U_1)}{U_2 - U_1}, \quad \Delta f^1 = \frac{f^1(U_2) - f^1(U_1)}{U_2 - U_1},$$

with $0 \le \mu \le A$ and $|\Delta f^1| \le f_1$. The function $\varphi(x,t)$ vanishes on S_T.

Let us define $\nu_\varepsilon(x) = \varepsilon \exp(-|x|)$ for any fixed $\varepsilon > 0$ and consider the initial-boundary value problem

$$D_t\varphi^\varepsilon + (\mu + \nu_\varepsilon)\,\Delta\varphi^\varepsilon = \alpha\varphi^\varepsilon \quad \text{in} \quad Q_\tau,$$
$$\varphi^\varepsilon(x,\tau) = \varphi_0(x) \ge 0 \quad \text{for} \quad x \in \Omega, \tag{3.16}$$
$$\varphi^\varepsilon(s,t) = 0 \quad \text{for} \quad (s,t) \in S_\tau,$$

formulated for the function φ^ε with arbitrary positive data in $C^3(\Omega)$ prescribed at $t = \tau > 0$. To use the results of Section 2, let us consider a sequence of smooth non-negative bounded functions $\mu_n(x,t)$ strongly convergent to $\mu(x,t)$ in $L_2(Q_\tau)$. The domain Ω is to be approximated by a family of bounded domains $\Omega^{(n)}$ with smooth boundaries $S^{(n)}$ such that $\Omega^{(n)} \subset \Omega^{(n+1)} \subset \Omega$.

We construct approximate solutions $\varphi_n(x,t)$ of problem (3.16) as functions that satisfy the equation

$$D_t \varphi_n^\varepsilon + (\mu_n + \nu_\varepsilon)\,\Delta\varphi_n^\varepsilon = \alpha\varphi_n^\varepsilon$$

in $Q_\tau^{(n)}$, vanish on the boundary $S_T^{(n)}$ and coincide with $\varphi_0(x)$ at $t = \tau$.

To derive an estimate independent of the index n, multiply the equation for $\varphi_n^\varepsilon(x,t)$ by $\Delta\varphi_n^\varepsilon$ and then integrate over $\Omega^{(n)}$ (for any $t = \text{const.} > 0$). By simple computation it follows that

$$\int_{\Omega^{(n)}} (\mu_n + \nu_\varepsilon)|\Delta\varphi_n^\varepsilon|^2 \, dx - \frac{1}{2}\frac{d}{dt}\int_{\Omega^{(n)}} |\nabla\varphi_n^\varepsilon|^2 \, dx = -\int_{\Omega^{(n)}} \alpha|\nabla\varphi_n^\varepsilon| \, dx,$$

which by Hölder's inequality and Lemma 1 of Section 2 implies that

$$\max_{t\in(0,\tau)} \int_{\Omega^{(n)}} |\nabla\varphi_n^\varepsilon(x,t)|^2 \, dx + \int_{Q_\tau^{(n)}} (\mu + \nu_\varepsilon)|\Delta\varphi_n^\varepsilon|^2 \, dxdt \le C. \tag{3.17}$$

Moreover, by the maximum principle,

$$0 \le \varphi_n^\varepsilon(x,t) \le \max\varphi_0(x) \qquad \text{for} \quad (x,t) \in Q_\tau^{(n)}. \tag{3.18}$$

(3.17) and (3.18) together with the equation imply the estimates

$$\int_{Q_\tau^{(n)}} (D_t\varphi_n^\varepsilon)^2 \, dxdt + \int_{\Omega^{(n)}} (\nabla\varphi_n^\varepsilon)^2 \, dx \le C,$$

$$\int_{Q_\tau^{(n)}} \nu_\varepsilon(x)|\Delta\varphi_n^\varepsilon|^2 \, dxdt \le C. \tag{3.19}$$

According to (3.18) and the boundary conditions for φ_n^ε,

$$\frac{\partial\varphi_n^\varepsilon}{\partial n}(s,t) \le 0, \qquad \text{for} \quad (s,t) \in S_T^{(n)}.$$

From (3.19), by standard diagonalization we can select a subsequence strongly convergent in $L_2(Q_\tau^{(n)})$ and weakly convergent in $W_2^{2,1}(Q_\tau^{(n)})$ to a function $\varphi^\varepsilon(x,t) \in W_2^{2,1}(Q_\tau^{(n)})$ such that

$$0 \le \varphi^\varepsilon(x,t) \le \max\varphi_0(x),$$

$$\int_{Q_\tau} \nu_\varepsilon(x)|\Delta\varphi^\varepsilon(x,t)|^2 \, dxdt + \int_{Q_\tau} ((\varphi_t^\varepsilon)^2 + (\nabla\varphi^\varepsilon)^2) \, dxdt \le C. \tag{3.20}$$

We substitute the resulting function into identity (3.15) to obtain

$$\int_\Omega U(x,\tau)\varphi^\varepsilon(x,t) \, dx + \int_{Q_\tau} U\nu_\varepsilon(x)\Delta\varphi^\varepsilon(x,t) \, dxdt \ge \int_{Q_\tau} (\varphi\Delta f^1 U + \alpha\varphi) \, dxdt.$$

Without loss of generality, we can assume that $\varphi^\varepsilon(x,t) \to \varphi(x,t)$ strongly in $L_2(Q_\tau)$ and weakly in $W_2^{1,1}(Q_\tau)$. It follows from (3.20) that

$$\int_{Q_\tau} U\nu_\varepsilon(x)\Delta\varphi^\varepsilon \, dxdt \to 0 \qquad \text{as} \quad \varepsilon \to 0.$$

Thus after passing to the limit as $\varepsilon \to 0$ we obtain

$$\int_\Omega U(x,\tau)\varphi_0(x)\,dx \geq \int_{Q_\tau} \varphi(\alpha + \Delta f^1)U\,dxdt.$$

Let

$$\alpha - \max|\Delta f^1| \geq C_1 > 0 \quad \text{and} \quad U(x,t) = U^+(x,t) - U^-(x,t),$$

where

$$U^+(x,t) = \frac{1}{2}\big(|U(x,t)| + U(x,t)\big) \quad \text{and} \quad U^-(x,t) = \frac{1}{2}\big(|U(x,t)| - U(x,t)\big).$$

Then,

$$\int_\Omega U(x,\tau)\varphi_0(x)\,dx \geq -C_2 \int_{Q_\tau} U^-(x,t)\,dxdt.$$

This inequality is true for each function $\varphi_0(x)$ such that $\varphi_0 \in C^3(\bar\Omega)$, $1 \geq \varphi_0(x) \geq 0$. Since the constant C_2 is independent of φ_0, this inequality also holds for each function $\varphi_0 \in L_2(\Omega)$ such that $0 \leq \varphi_0(x) \leq 1$. Let us take $\varphi_0(x) = \operatorname{sgn} U^-(x,\tau)$ (i.e., equal to 1 whenever $U^-(x,\tau) > 0$ and vanishing at $U^-(x,\tau) = 0$). Then

$$-\int_\Omega U^-(x,\tau)\,dx > -C_2 \int_{Q_\tau} U^-(x,t)\,dxdt.$$

Hence, due to the arbitrariness of $\tau \in [0,T]$ by Lemma 1 of Section 2 we infer that

$$\int_\Omega U^-(x,\tau)\,dx = 0, \qquad \text{for} \quad \tau \in [0,T],$$

and, eventually, that

$$U(x,t) = U^+(x,t) \geq 0.$$

We shall now generalize a result due to Berger & Rogers [224] which states that the measure of the mushy phase in a Stefan problem is non-increasing.

In physical variables, by the mushy phase we shall mean the set

$$M = \big\{(x,t) \,:\, x \in \Omega,\ t \in (0,T),\ U(x,t) \in (-L,0)\big\}, \quad \text{with } M(t_0) = M \cap \{t = t_0\}.$$

Theorem 11. (cf., [225]) *Let f be uniformly Lipschitz continuous in the sense of (1.5). Then the following assertions are equivalent:*
1° $f(0) \geq 0$, $f(-L) \leq 0$;
2° *for all $\vartheta^0 \in L_\infty(S_T)$, $U_0 \in L_\infty(\Omega)$, the mushy phase corresponding to the generalized solution of problem (3.1) – (3.3) satisfies:*

$$M(t_2) \subset M(t_1), \qquad \text{for all} \quad t_2 > t_1,$$

with the measure of $M(t_2) \setminus M(t_1)$ equal to zero.

Proof. The implication $2^\circ \Rightarrow 1^\circ$ follows from results of [154], [189] on mushy phase formation in the presence of heat sources.

The main part of the proof uses the following Stefan problem, semi-discretized in time, with smooth initial data:

$$\frac{1}{\tau}(U^{n+1} - U^n) = \kappa \Delta \vartheta^{n+1} + f(U^{n+1}), \quad \text{for } n = 0, \ldots, N-1; \ \tau = \frac{T}{N}, \ (3.21)$$

$$\vartheta^n = \frac{1}{\tau} \int_{\tau n}^{\tau(n+1)} \vartheta^0(s,t)\, dt, \qquad \text{for } s \in S, \tag{3.22}$$

$$U^0 = U_0(x) \in C^{2+\alpha}(\Omega), \qquad \text{for } x \in \Omega. \tag{3.23}$$

We regularize the problem (2.1) – (2.3), by approximating χ with χ_ε and f with f_ε, defined earlier in this section. The existence of a bounded solution $U_\varepsilon(x,t) = U_\varepsilon^n(x)$ to the regularized problem for $t \in [\tau n, \tau(n+1)]$, $n = 0, \ldots, N-1$, under the condition $\tau f_1 < 1$, follows from Theorem 5.4 of [138] and the a priori bound

$$|U_\varepsilon^n(x,t)| \leq \max\left\{ \|\vartheta^0\|_{L_\infty(S_T)} + L, (1 - \tau f_1)^{-1} \|U_\varepsilon^n\|_{L_\infty(\Omega)} + \tau f(0) \right\}$$

$$\leq \max\left\{ \|\vartheta^0\|_{L_\infty(S_T)}, e^{f_1 T} \|U_0\|_{L_\infty(\Omega)} + f_\varepsilon(0)T \right\} \leq C_1. \tag{3.24}$$

By the regularity of the initial data and χ_ε, the functions ϑ_ε^n are three times and U_ε^n twice continuously differentiable for all n in every inner subdomain of Ω. By multiplying equation (3.21) by $\vartheta_\varepsilon^n \varphi(x)$, where $\varphi(x)$ is a smooth function equal to 1 in $\Omega^\gamma = \{x \in \Omega : \rho(x,S) > \gamma\}$ and vanishing outside of $\Omega^{\gamma/2}$, then integrating the resulting identity over Ω and taking the sum over n, by (2.4) we get

$$\|\nabla \vartheta_\varepsilon\|_{L_2(0,T;L_2(\Omega^\gamma))} \leq C_2(\gamma) \qquad \text{for all} \quad \gamma > 0. \tag{3.25}$$

If $V = \frac{\partial}{\partial x_k} U_\varepsilon$, then by differentiating the regularized equation (3.21) with respect to x_k we obtain

$$\frac{1}{\tau}(V^{n+1} - V^n) - \kappa \Delta(\chi_\varepsilon' V^{n+1}) + f_\varepsilon' V^{n+1} = 0.$$

We now multiply this equation by $\varphi \operatorname{sgn}^\delta V^{n+1} = \varphi V^{n+1}((V^{n+1})^2 + \delta)^{-1/2}$, where the non-negative function $\varphi(x)$ belongs to $\mathring{C}^2(\bar{\Omega})$, is equal to 1 in Ω^γ and vanishes outside of $\Omega^{\gamma/2}$. Then integrating over Ω and taking the sum over n, we get

$$\int_\Omega \varphi \sum_{n=0}^m (V^{n+1} - V^n) \frac{V^{n+1}}{\{(V^{n+1})^2 + \delta\}^{1/2}}\, dx + \kappa \tau \sum_{n=0}^m \int_\Omega \nabla(\chi_\varepsilon' V^{n+1}) \nabla(\operatorname{sgn}^\delta V^{n+1} \varphi)\, dx$$

$$= \tau \int_\Omega \sum_{n=0}^m f_\varepsilon' V^{n+1} \operatorname{sgn}^\delta V^{n+1} \varphi\, dx.$$

The first term on the left-hand side, which we denote by I_1^δ, admits the representation

$$I_1^\delta = \int_\Omega \varphi\left(V^{m+1} \operatorname{sgn}^\delta V^{m+1} - V^0 \operatorname{sgn}^\delta V^1 + \sum_{n=1}^m V^n(\operatorname{sgn}^\delta V^n - \operatorname{sgn}^\delta V^{n+1})\right) dx.$$

Passing to the limit as $\delta \to 0$, gives us

$$\lim_{\delta \to 0} I_1^\delta = \int_\Omega \varphi\big(|V^{m+1}| - V^0 \mathrm{sgn}\, V^1 + \sum_{n=1}^m |V^n|(1 - \mathrm{sgn}\, V^n \mathrm{sgn}\, V^{n+1})\big)\, dx$$

$$\geq \int_\Omega \varphi\big(|V^{m+1}| - |V^0|\big)\, dx.$$

Now we integrate the second term I_2^δ by parts,

$$I_2^\delta = \kappa\tau \sum_{n=0}^m \int_\Omega \big(V^{n+1}(\nabla \mathrm{sgn}^\delta V^{n+1} \nabla \chi'_\varepsilon)\varphi + \chi'_\varepsilon \nabla V^{n+1} \nabla \mathrm{sgn}^\delta V^{n+1} \varphi$$

$$- \chi'_\varepsilon V^{n+1} \mathrm{div}\,(\mathrm{sgn}^\delta V^{n+1} \nabla\varphi)\big)\, dx = J_1^\delta + J_2^\delta + J_3^\delta.$$

From the equality $\nabla \mathrm{sgn}^\delta V^{n+1} = \delta V^{n+1} \nabla V^{n+1}\big((V^{n+1})^2 + \delta\big)^{-3/2}$, we infer that $\lim_{\delta \to 0} J_1^\delta = 0$ (see also the bound (3.9)) and $J_2^\delta \geq 0$. For the third term J_3^δ, we introduce the representation

$$J_3^\delta = -\tau \int_\Omega \big(\sum_{n=0}^m \chi'_\varepsilon V^{n+1} \nabla \mathrm{sgn}^\delta V^{n+1} \nabla\varphi + \chi'_\varepsilon V^{n+1} \mathrm{sgn}^\delta V^{n+1} \Delta\varphi\big)\, dx.$$

Passing to the limit as $\delta \to 0$, we get the inequality

$$\lim_{\delta \to 0} J_3^\delta \geq -\sum_{n=0}^m \tau \int_\Omega |\vartheta^{n+1}|\, |\Delta\varphi|\, dx \geq -C_3(\gamma).$$

Hence,

$$\int_\Omega |V^{m+1}|\varphi\, dx \leq C_3(\gamma) + \sum_{n=0}^m \int_\Omega \tau f_1 |V^{n+1}|\varphi\, dx$$

and, eventually,

$$\int_{\Omega^\gamma} |\nabla U_\varepsilon^n|\, dx \leq C_4(\gamma) \qquad \text{for all} \quad \gamma > 0.$$

These estimates are sufficient for obtaining as $\varepsilon \to 0$ the solution U_τ of problem (3.1) – (3.3) satisfying estimates analogous to (3.4), (3.5) and

$$\max_{i=0,\ldots,N} \|U^i\|_{BV(\Omega^\gamma)} \leq C_5(\gamma) \qquad \text{for all} \quad \gamma > 0. \tag{3.26}$$

We now prove the implication $2^o \Rightarrow 1^o$ for problem (3.21) – (3.23). The function $\chi(U^{n+1})$ satisfies Poisson's equation with a bounded right-hand side, hence $\chi(U^{n+1}) \in W^2_{2,\mathrm{loc}}(\Omega)$. Therefore, $\Delta\chi(U^{n+1}) = 0$ and $U^{n+1} = U^n + \tau f(U^{n+1})$ almost everywhere whenever $U^{n+1} \in (-L, 0)$.

Let $M^{n+1} = \{x \in \Omega : U^{n+1}(x) \in (-L, 0), \delta\chi(U^{n+1}) = 0\}$. Let $x \in M^{n+1}$ and suppose that $U^n(x) \notin (-L, 0)$, for instance, $U^n(x) \geq 0$. Then

$$U^{n+1}(x) \geq \tau\big(f(U^{n+1}(x)) - f(0)\big) \geq -\tau f_1 |U^{n+1}(x)|.$$

Thus,

$$|U^{n+1}(x)| \leq \tau f_1 |U^{n+1}(x)| \quad \text{and hence} \quad \tau f_1 \geq 1.$$

We have arrived at a contradiction to the assumption $\tau < 1/f_1$, i.e., $U^n(x) \in (-L, 0)$. This implies that the measure of the set $M_{n+1} = \{x \in \Omega : U^{n+1}(x) \in (-L, 0)\} \setminus M_n$ is equal to zero, which completes the proof. $\square$

Let us now consider the piecewise constant interpolation, linear in t, of the function U_τ,

$$u_\tau(x, t) = U^{n+1}(x) \frac{t - n\tau}{\tau} + U^n(x) \frac{(n+1)\tau - t}{\tau}, \quad \text{for} \quad t \in [n\tau, (n+1)\tau].$$

In the sequel we shall use the following modified version of Lemma 5 of [226], formulated in [227].

Lemma. *Let $u(x, t)$ be a measurable function in the cylinder $\Omega^\delta \times [0, T]$, with*

$$|u(x, t)| \leq M_1, \quad \sup_{0 \leq t \leq T} \|u\|_{BV(\Omega^\delta)} \leq M_2.$$

Suppose that, for all t, $t + \Delta t \in [0, T]$, $\Delta t > 0$, and for each function $g(x)$ continuously differentiable in $\Omega^{2\delta}$,

$$\int_{\Omega^{2\delta}} g(x) \left\{ u(x, t + \Delta t) - u(x, t) \right\} dx \leq C(\delta) \, \Delta t \max_{x \in \bar{\Omega}^{2\delta}} \{ g + |\nabla g| \}.$$

Then

$$\int_{\Omega^{2\delta}} |u(x, t + \Delta t) - u(x, t)| \, dx \leq C_1(\delta) \min_{0 < 2h < \delta} \left\{ h + \frac{\Delta t}{h} \right\}, \quad \text{for } 0 \leq t \leq t + \Delta t \leq T.$$

By multiplying equation (3.21) by any finite function $g(x) \in \mathring{C}^1(\bar{\Omega}^{2\delta})$ and then integrating over $\Omega^{2\gamma} \times [t, t + \Delta t]$, we get the bound

$$\left| \int_{\Omega^{2\gamma}} g(x) \left\{ u_\tau(x, t + \Delta t) - u_\tau(x, t) \right\} dx \right| = \left| \int_t^{t+\Delta t} (\chi' \nabla U_\tau \nabla g + f(U_\tau)g) \, dx dt \right|$$
$$\leq C_6(\gamma) \max_{x \in \bar{\Omega}^{2\gamma}} \{ |g| + |\nabla g| \} \, \Delta t.$$

By the above estimate, the function $u_\tau(x, t)$ fulfils the hypotheses of the lemma, thus

$$\int_{\Omega^\gamma} |u_\tau(x, t + \Delta t) - u_\tau(x, t)| \, dx \leq C_7(\gamma) \min_{0 < 2h < \gamma} \left\{ h + \frac{\Delta t}{h} \right\} \leq C_8(\gamma) \, (\Delta t)^{1/2}. \quad (3.27)$$

By the estimates (3.26) and (3.27), there exists a subsequence $u_{\tilde{\tau}}$ strongly convergent in $L_2(Q_T)$ such that

$$\lim_{\tilde{\tau} \to 0} u_{\tilde{\tau}} = \lim_{\tilde{\tau} \to 0} U_{\tilde{\tau}} = U(x, t).$$

Passing to the limit in the integral identity

$$\int_{Q_T} \{u_{\tilde{\tau}}\varphi_t + \chi(U_{\tilde{\tau}})\Delta\varphi + f(U_{\tilde{\tau}})\varphi\}\, dxdt = -\int_{\Omega} U_0\varphi(x,0)\, dx + \int_{S_T} \vartheta^0 \frac{\partial\varphi}{\partial n}\, dsdt$$

satisfied for all $\varphi(x,t) \in W_2^{2,1}(Q_T)$ such that $\varphi(x,T) = 0$ and $\varphi\,|_{S_T} = 0$, we conclude that U is the generalized solution of the Stefan problem (3.1) – (3.3) with nonlinear source term.

With the strong convergence $U_{\tilde{\tau}} \to U$ shown, we have completed the proof of the main assertion in the case of smooth initial data. By virtue of the stability of solutions to problem (3.1) – (3.3) with respect to the initial data (see Theorem 1), the smoothness assumptions can be relaxed.   □

Corollary 1. *Let $\vartheta^0 \in L_\infty(S_T)$, $U_0 \in L_\infty(\Omega)$ and the function $f(s)$ satisfy the hypotheses of the theorem and assertion 1. Then the generalized solution $U(x,t)$ of problem (3.1) – (3.3) admits the following characterization of the corresponding mushy phase: there is a non-negative function $G : \Omega \to \mathbb{R} \cup \{+\infty\}$ such that*

$$M = \{(x,t) \ : \ x \in M(0),\ 0 \le t < G(x)\}.$$

Moreover, for each point $x \in M(0)$, the function U satisfies the Cauchy problem

$$U_t = f(U), \qquad U(x,0) = U_0(x)$$

on the interval $[0, G(x))$.

The first assertion of Corollary 1 follows directly from Theorem 2. The second assertion can be proved by introducing the time-discretized Stefan problem (3.21) – (3.23), as was done with Theorem 11. Due to the strong convergence $U_\tau \to U$ as $\tau \to 0$ we have proved that the conclusions apply also to the limit problem (3.1) – (3.3).

In some situations, Corollary 1 implies an upper bound on the function G; then the mushy phase disappears in finite time.

Corollary 2. *Let all the hypotheses of Corollary 1 be satisfied. Suppose there exist $s_0 \in (-L, 0)$ and $\alpha > 0$ such that $f(s) > \alpha$ for all $s \in (s_0, 0)$. Moreover, let $U_0(x) > s_0$ for all $x \in M(0)$. Then*

$$G(x) \le -\frac{s_0}{\alpha}.$$

Chapter II

Classical solution of the multidimensional Stefan problem

---

1. The one-phase Stefan problem. Main result

Let us recall the setting of the one-phase Stefan problem. Let $\Omega_T \subset \mathbb{R}_T^n = \mathbb{R}^n \times (0,T)$ be a domain located between a given surface $\mathcal{F}_T = \mathcal{F} \times (0,T)$, $\mathcal{F} \subset \mathbb{R}^n$, and an unknown surface $\Gamma_T = \{(x,t) \; : \; x \in \Gamma(t) \subset \mathbb{R}^n, \, t \in (0,T)\}$ such that $\Omega_T = \{(x,t) \; : \; x \in \Omega(t) \subset \mathbb{R}^n, \, t \in (0,T)\}$. The problem consists of determining the domain Ω_T and a function ϑ (temperature) which satisfies the nonlinear heat equation

$$a(\vartheta)D_t\vartheta = \sum_{i=1}^n D_i^2\vartheta + f(x,t). \tag{1.1}$$

in Ω_T.

On the known boundary $\mathcal{F}_T$,

$$\vartheta = \vartheta^1(x,t), \qquad \left(\text{ or } \sum_{i=1}^n b_i(x,t)D_i\vartheta + b(x,t)\vartheta = \vartheta^2 \right). \tag{1.2}$$

On the free boundary Γ_T (unknown), the temperature assumes the melting value,

$$\vartheta(x,t) = 0 \;, \tag{1.3}$$

and the Stefan condition

$$V_\nu + \sum_{i=1}^n \nu_i D_i\vartheta = 0 \tag{1.4}$$

holds, with V_ν the normal velocity of surface $\Gamma(t)$ where ν is the normal vector to $\Gamma(t)$. If $\nu = \nabla\vartheta|\nabla\vartheta|^{-1}$ (such a definition of the normal vector is admissible as long as the zero level set $\Gamma(t)$ of the temperature is regular, i.e. $|\nabla\vartheta| \neq 0$), then

$$V_\nu = -D_t\vartheta|D\vartheta|^{-1} \qquad \text{for} \quad (x,t) \in \Gamma_T. \tag{1.5}$$

To complete the problem formulation it is necessary to prescribe the initial position of the free boundary and the initial temperature distribution,

$$\Gamma(0) = S, \quad \vartheta(x,0) = \vartheta_0(x), \quad x \in \Omega(0) = G. \tag{1.6}$$

In order to give a more precise formulation of the main result, we shall modify the form of the Stefan condition. Let S be a C^2-surface and assume that $|D\vartheta_0(x_0)| \neq 0$

for all $x_0 \in S$. Then

$$\nu(x_0) = D\vartheta_0(x_0)|D\vartheta_0(x_0)|^{-1}.$$

There exists $d > 0$, dependent only on the differentiation characteristics of the surface, such that a scalar function $R(x_0, t)$ defines $\Gamma(t)$ by the equality

$$x = x_0 + R(x_0, t)\nu(x_0), \qquad (1.7)$$

provided

$$|R(x_0, t)| \leq 2d. \qquad (1.8)$$

The constant d can be chosen small enough to ensure that $\Gamma(t)$ defined by (1.7) does not intersect the surface $\mathcal{F}$, if at the initial time instant $\Gamma(0) = S$ does not intersect $\mathcal{F}$.

By differentiating the identity

$$\vartheta(x_0 + R(x_0, t)\nu(x_0), t) = 0$$

with respect to t and recalling (1.5), we obtain

$$V_\nu = |D\vartheta|^{-1}(D\vartheta\nu(x_0))D_t R(x_0, t), \qquad (1.9)$$

where $D\vartheta$ is computed at (x, t) and the point x is given by (1.7). According to (1.9), the Stefan condition can be written in the form

$$D_t R(x_0, t) = X(x_0, D\vartheta) \equiv -|D\vartheta|^2 (D\vartheta\nu(x_0))^{-1}, \qquad (1.10)$$

where $D\vartheta$, as in (1.9), is computed at (x, t) with x given by (1.7).

Theorem 1. *Assume the following hypotheses:*
(I) *Closed surfaces $\mathcal{F}$ and S bound the sets $U_\mathcal{F}$ and U_S respectively such that $U_\mathcal{F} \subset U_S$, $\mathcal{F} \cap S = \emptyset$ and $\mathcal{F}$, S belong to the class H^{2l}, $[l] = m + 1 = n + 5$.*
(II) *$a \in C^{m+2}[0, \infty)$, the functions $\vartheta^1, \vartheta^2, b, b_i$ $(i = 1, ..., n)$ are in the space $H^{2l,l}(\mathcal{F}_\infty)$ and*

$$a \geq M_0^{-1} = \text{const.} > 0, \qquad \sum_{i=1}^{n} b_i q_i \geq M_0^{-1},$$

where $q = (q_1, ..., q_n)$ is the normal to the surface $\mathcal{F}$ at the point x.
(III) *$f \in H^{2l,l}(\mathbb{R}^n_\infty)$, $\vartheta_0 \in H^{2l}(\bar{G})$, on the surfaces $\mathcal{F}$ and S the compatibility conditions up to order $m + 1$ that follow from (1.1) – (1.3), (1.10) hold; moreover*

$$\vartheta_0(x) > 0, \quad x \in G; \quad \left|\log|D\vartheta_0(x)|\right| \leq N_0, \quad x \in S. \qquad (1.11)$$

(IV) *The norms of the functions $f, \vartheta_0, \vartheta^1, \vartheta^2$ in appropriate spaces are bounded by a common constant N_0, and the norms of the coefficients a, b, b_i $(i = 1, ..., n)$ and the norms of functions defined in local coordinates of surfaces $\mathcal{F}$ and S are bounded by the constant M_0.*

Then there exists a sufficiently small $T^* > 0$, dependent only on M_0, N_0 and d, such that the problem (1.1) – (1.6) admits a unique solution $\{R, \vartheta\}$ with

$$R \in C^{2,1}(S_{T_*}), \quad \vartheta \in C^{2,1}(\bar{\Omega}_{T_*}). \qquad \square$$

Remark 1. The assertion of Theorem 1 remains valid for the more general equation

$$D_t \vartheta = \sum_{j,i=1}^{n} D_i(a_{ij}(x,t,\vartheta)D_j\vartheta) + f(x,t,\vartheta,D\vartheta), \tag{1.12}$$

as long as the strong parabolicity of this equation and sufficient regularity of the functions a_{ij} and f are guaranteed. The Stefan condition (1.4) is then replaced by

$$V_\nu + \sum_{i,j=1}^{n} a_{ij}\nu_i D_j\vartheta = 0.$$

A proof of this result is given in [151].

Remark 2. The solution of problem (1.1) – (1.6) is also a solution of the original Stefan problem if the temperature $\vartheta(x,t)$ is strictly positive in Ω_T. If $f(x,t) = 0$, it then follows from (1.11) and the maximum principle that $\vartheta(x,t)$ is positive at least on a sufficiently small time interval $(0,T)$. If $f \neq 0$, then the first of the conditions (1.11) implies local in time positivity of the solution outside a small neighbourhood of the free boundary Γ_T. On Γ_T the solution $\vartheta(x,t)$ is identically zero, but, due to its regularity and by the second of the conditions (1.11), the normal derivative of $\vartheta(x,t)$ is different from zero at least on a small time interval. The positivity of the derivative along an inward normal ensures the positivity of the solution in a sufficiently small neighbourhood of Γ_T. This obviously implies the possibility of choosing a time interval $(0,T_*)$ small enough to guarantee that the solution $\vartheta(x,t)$ is positive everywhere on Ω_{T_*}.

2. The simplest problem setting

The basic step in the proof of the classical solvability of the multidimensional Stefan problem consists of establishing a priori estimates for the solution in a neighbourhood of the free boundary Γ_T. For this, the condition (1.11) is used to construct local coordinates (von Mises variables) in which a part of the plane corresponds to the free boundary. The new unknown (the coordinate in x−variables co-linear with the normal to the surface S) in the von Mises coordinates satisfies a nonlinear second-order parabolic equation and a nonlinear boundary condition on the image of the free boundary, which involves the time derivative of the unknown solution. After linearizing the equation and boundary condition, the resulting linear boundary value problem is no longer "parabolic", i.e. the boundary condition does not satisfy the complementarity condition. Since there is no complete theory of such problems, a detailed study of the nonlinear problem so constructed is rather complicated. Fortunately, on a small time interval, energy estimates can be established for this problem by differentiating the equation and the boundary condition with respect to t and the tangent variables, then multiplying the result with the appropriate derivative and, eventually, by integrating it.

As an illustration we shall perform the above procedure for the problem in its simplest form.

Theorem 2. *Let the hypotheses of Theorem 1 hold with $a = 1$, $\mathcal{F}$ being the plane $\{x_n = 1\}$, S the plane $\{x_n = 0\}$, $f = 0$, $\vartheta^1 = 1$ (we consider the problem with temperature prescribed on the surface $\mathcal{F}_T$) and $\vartheta_0(x)$ be a 1-periodic (i.e., periodic with period 1) function with respect to $x' = (x_1, \ldots, x_{n-1})$. Moreover, let $\vartheta_0(x)$ be dependent only on x_n in a neighbourhood of the plane $\mathcal{F}$ and let*

$$\left|\log|D_n\vartheta_0(x)|\right| \leq N_0 \quad \text{for} \quad x \in G. \tag{2.1}$$

Then, on a sufficiently small time interval $(0, T_*)$ dependent only on the constant N_0, there exists a unique solution $R(x', t)$, 1-periodic with respect to x', of the Stefan problem in the space $C^{2,1}(S_{T_*})$ that defines the domain $\Omega_{T_*} = \{(x, t) : |x'| < \infty, R(x', t) < x_n < 1, 0 < t < T_*\}$ and $\vartheta(x, t)$ in the space $C^{2,1}(\bar{\Omega}_{T_*})$.

Proof. We shall split the proof into several steps. First, it follows directly from the uniqueness theorem that the solution is 1-periodic with respect to x'. Thus, we can confine ourselves to considering the intersections of the sets $\mathcal{F}, G, S, \Omega_T$ with the half-strip $\Pi = \{x \in \mathbb{R}^n : 0 < x_i < 1, i = 1, \ldots, n-1, -\infty < x_n < 1\}$, where the former notation has been maintained.

Due to the postulated regularity of the solution $\vartheta(x, t)$ it follows from the inequality (2.1) that on a sufficiently small time interval $(0, T_1)$

$$\left|\log|D_n\vartheta|\right|_{\Omega_{T_1}}^{(0)} < \infty. \tag{2.2}$$

In view of the above inequality it is useful to introduce new independent variables

$$\tau = t, \quad y' = x', \quad y_n = \vartheta(x, t), \tag{2.3}$$

in which $G_T = G \times (0, T)$ corresponds to the domain Ω_T and the surface $S_T = S \times (0, T)$ corresponds to the free boundary Γ_T. The surface $\mathcal{F}_T$ is invariant under the mapping $(x, t) \to (y, \tau)$.

The new unknown function $u(y, \tau) = x_n$, 1-periodic with respect to y', satisfies the nonlinear parabolic equation

$$D_\tau u - \Delta' u + D_n\left\{\frac{1}{p_n}\left(1 + \sum_{i=1}^{n-1} p_i^2\right)\right\} = 0 \tag{2.4}$$

in G_T, subject to the conditions

$$u = 1 \quad \text{when} \quad (y, \tau) \in \mathcal{F}_T, \qquad u\,|_{\tau=0} = u_0 \qquad \text{for} \quad y \in G, \tag{2.5}$$

$$D_\tau u + \frac{1}{p_n}\left(1 + \sum_{i=1}^{n-1} p_i^2\right) = 0 \qquad \text{for} \quad (y, \tau) \in S_T. \tag{2.6}$$

In (2.4) and (2.6) we have set $p_i = D_i u$, $\Delta' u = \sum\limits_{i=1}^{n-1} D_i^2 u$.

Equation (2.4) corresponds to the heat equation for $\vartheta(x,t)$ and (2.6) is the counterpart of the Stefan condition

$$D_t \vartheta = |D\vartheta|^2.$$

The first of the conditions (2.5) is just an expression of the equation $x_n = 1$ for the surface $\mathcal{F}$, and the function $u_0(y)$ in the second of conditions (2.5) can be specified by the equation

$$y_n = \vartheta_0(y', u_0(y)).$$

Due to (2.1),the last equation always admits a solution.

In deriving (2.4), (2.6) we have used the relations

$$D_i \vartheta = -\frac{p_i}{p_n}, \quad \frac{\partial}{\partial x_i} = \frac{\partial}{\partial y_i} - \frac{p_i}{p_n}\frac{\partial}{\partial y_n}, \quad i = 1,\ldots n-1, \tag{2.7}$$

$$D_t \vartheta = -\frac{D_\tau u}{p_n}, \quad D_n \vartheta = \frac{1}{p_n}, \quad \frac{\partial}{\partial x_n} = \frac{1}{p_n}\frac{\partial}{\partial y_n}. \tag{2.8}$$

The resulting initial-boundary value problem (2.4)–(2.6) (we shall refer to it as *Problem (B)*) is equivalent to the original Stefan problem (further referred to as *Problem (A)*) in terms of (x,t), if

$$\big| \log |p_n| \big|_{G_T}^{(0)} < \infty.$$

For (2.6) the complementarity condition does not hold, hence it is impossible to apply known methods of constructing the solution (for instance, if it is virtually one-dimensional). On the other hand, as already mentioned, the solution of Problem (B) satisfies the simplest energy estimates and hence its norm in the space $C^{2,1}(G_T)$ can be estimated on a sufficiently small time interval $(0,T)$. The problem arises from constructing a sequence of "regular" (or "parabolic") problems approximating Problem (B) and preserving formerly established energy estimates which are independent of the approximation index.

Such a sequence of functions can easily be found in case of Problem (B). This is the Problem (B_ε) which consists of determining functions u^ε which satisfy equation (2.4), conditions (2.5) and the boundary condition

$$D_\tau u^\varepsilon + \frac{1}{p_n^\varepsilon}\left(1 + \sum_{i=1}^{n-1}(p_i^\varepsilon)^2\right) = \varepsilon\Delta' u^\varepsilon \tag{2.9}$$

on the boundary S_T.

Condition (2.9) which approximates (2.6) is "regular", i.e. it satisfies the compatibility condition, and the corresponding initial-boundary value problem is uniquely soluble.

Lemma 1. *Under the hypotheses of Theorem 2 there exists a unique solution u^ε of Problem (B_ε) in $H^{2l,l}(\bar{G}_T)$, $T < T_\varepsilon$, on a sufficiently small time interval $(0,T_\varepsilon)$ specified by* *the relations*

$$\mathcal{I}(T, u^{\varepsilon}) < \infty \quad when \quad T < T_{\varepsilon}, \quad \mathcal{I}(T, u^{\varepsilon}) = \infty, \tag{2.10}$$

where $\mathcal{I}(T, u) = |p|_{G_T}^{(\gamma)} + |\log |p||_{G_T}^{(0)}$ *for some* $\gamma \in (0, 1)$, $p = Du = (p_1, \dots, p_n)$.

Proof. We begin by recalling how the compatibility conditions on the surface S can be derived from equation (2.4) and boundary condition (2.6). By comparing the derivative $D_{\tau} u$ computed from equation (2.4) with its value resulting from condition (2.6), we get the following identity holding on the boundary S_T :

$$\Delta' u - D_n \left\{ \frac{1}{p_n} \left(1 + \sum_{i=1}^{n-1} p_i^2 \right) \right\} = -\frac{1}{p_n} \left(1 + \sum_{i=1}^{n-1} p_i^2 \right) . \tag{2.11}$$

By the postulated regularity of the solution $u(y, \tau)$, the identity (2.11) remains valid also for the initial function $u_0(y)$,

$$\Delta' u_0 - D_n \left\{ \frac{1}{D_n u_0} \left(1 + \sum_{i=1}^{n-1} (D_i u_0)^2 \right) \right\} = -\frac{1}{D_n u_0} \left(1 + \sum_{i=1}^{n-1} (D_i u_0)^2 \right) , \quad y \in S .$$

The above relation is the compatibility condition of the first order which follows from (2.4), (2.6) for the initial function $u_0(y)$. If the solution we look for should be more regular, an appropriate compatibility condition of higher order can be obtained by successively differentiating the identity (2.11) with respect to time and then substituting the derivatives $D_{\tau} u$ which result from equation (2.4). The total number m of such conditions is equal to the number of time derivatives $D_{\tau}^m u$ that should exist.

Having "improved" the boundary condition (2.6), we shall also modify appropriately the compatibility conditions to be fulfilled by the initial function $u_0(y)$. Correspondingly, the approximate solution u^{ε} should satisfy the "improved" initial condition (2.5) involving $u_0^{\varepsilon}(y)$. Such a correction of the initial condition is always possible. There exists a sequence u_0^{ε} convergent to u_0 as $\varepsilon \to 0$ in the norm of $H^{2l}(\bar{G})$ and such that the compatibility conditions up to order m, following from conditions (2.4) and (2.9), hold for the functions u_0^{ε}. In the special case we consider, the construction turns out to be simpler since u_0 depends only on y_n in a neighbourhood of the surface S, and the compatibility conditions which follow from (2.4), (2.6) for such a function coincide with those corresponding to (2.4), (2.9). It remains to note that the compatibility conditions for the function $u_0(y)$ on the surface S result as a consequence of those holding for $\vartheta_0(x)$ on S in the original Problem (A).

First we show the existence of solutions to Problem (B_{ε}) in the space $H^{r, r/2}(\bar{G}_T)$, $r = 2 + \gamma$ with some $\gamma \in (0, 1)$.

Let us take an arbitrary

$$N > |Du_0|_G^{(\gamma)} + \left| \log |D_n u_0|_G^{(0)} \right|.$$

By $\mathcal{M}$ we denote a closed convex set of vector functions $Du_0 = (D_1 u_0, \dots, D_n u_0)$ such that

$$|\omega|_{G_T}^{(\gamma)} \leq N, \quad e^{-N} \leq |\omega_n| \leq e^N . \tag{2.12}$$

For a given vector function $\omega \in \mathcal{M}$ we shall consider the linear boundary value problem

$$D_\tau v = \Delta' v - \sum_{i=1}^{n-1} \frac{2\omega_i}{\omega_n} D_i D_n v + \frac{1}{\omega_n^2}\left(1 + \sum_{i=1}^{n-1} \omega_i^2\right) D_n^2 v, \quad (y,\tau) \in G_T, \quad (2.13)$$

$$v(y', 1, \tau) = 1; \quad v(y, 0) = u_0(y), \quad y \in G, \tag{2.14}$$

$$D_\tau v - \varepsilon \Delta' v = -\frac{1}{\omega_n}\left(1 + \sum_{i=1}^{n-1} \omega_i^2\right), \quad (y,\tau) \in S_T, \tag{2.15}$$

formulated in terms of $v(y,\tau) \in H^{r,r/2}(\bar{G}_T)$. Its solvability follows from known results [137]. Indeed, let us treat the condition (2.15) on S_T expressed in terms of $v(y', 0, \tau)$ as a non-homogeneous heat equation with the right-hand side

$$f = -\left(1 + \sum_{i=1}^{n-1} \omega_i^2\right)\omega_n^{-1},$$

subject to the initial condition $v(y', 0, 0) = u_0(y', 0)$. By Theorem 3 of Section 2, Chapter I, the function $v(y', 0, \tau)$ is uniquely defined and satisfies the bound

$$|v|_{S_T}^{(2+\gamma)} \le c(\varepsilon)\left(|u_0|_G^{(2+\gamma)} + |f|_{S_T}^{(\gamma)}\right) \le N_1(\varepsilon, N_0, N).$$

It follows from the same theorem that there exists in $H^{r,r/2}(\bar{G}_T)$ a unique solution $v(y,\tau)$ of equation (2.13) which satisfies the conditions (2.14) and coincides with the solution of equation (2.15) on the boundary S_T. This solution satisfies the a priori bound

$$|v|_{G_T}^{(r)} \le c(N)\left(|u_0|_G^{(r)} + |v|_{S_T}^{(r)}\right) \le N_2(\varepsilon, N_0, N). \tag{2.16}$$

We can easily note that if $\omega_i = D_i v$, then equation (2.13) coincides with (2.4) and the boundary condition (2.15) with (2.9). Hence, if $Dv = \Psi(\omega)$ is an operator which associates with any vector function $\omega \in \mathcal{M}$ the vector function Dv, where $v(y,\tau)$ is the solution of the problem (2.13)–(2.15), then the fixed point of this operator is a solution of Problem (B_ε).

The operator Ψ from the set $\mathcal{M}$ into the space of vector functions that belong to $H^{\gamma,\gamma/2}(\bar{G}_T)$ is continuous. This follows from the continuous dependence of solutions to linear parabolic equations of the second order on their coefficients [137]. Moreover, Ψ is completely continuous since it maps the set $\mathcal{M}$ into a bounded set in the space $H^{1+\gamma,(1+\gamma)/2}(\bar{G}_T)$ (see estimate (2.16)). In order to justify the use of Schauder's fixed point theorem, it remains to show that the operator Ψ maps $\mathcal{M}$ into itself. This property can be shown provided that the time interval on which we consider the boundary value problem (2.13)–(2.15) is short enough.

Because $Dv(y, 0) = Du_0(y)$ we can conclude that

$$|Dv(y,\tau) - Du_0(y)| \le |Dv - Du_0|_{G_T}^{(1+\gamma)} \tau^{(1+\gamma)/2} \le 2N_2 T^{(1+\gamma)/2}.$$

Returning to Lemma 4 of Section 2, Chapter I, we can see that

$$|Dv - Du_0|_{G_T}^{(\gamma)} \le \mu |Dv - Du_0|_{G_T}^{(1+\gamma)} + \frac{c}{\mu\gamma}|Dv - Du_0|_{G_T}^{(0)}$$

for arbitrary μ, $0 < \mu < T^{1/2}$. Thus,

$$|Dv|_{G_T}^{(\gamma)} \le |Du_0|_G^{(\gamma)} + |Dv - Du_0|_{G_T}^{(\gamma)} \le N + N_3\{\mu + \mu^{-\gamma}T^{(1+\gamma)/2}\}.$$

It follows in a similar way that

$$|\log(D_n v)|_{G_T}^{(0)} \le \left|\log|D_n u_0|\right|_G^{(0)} + N_4 T^{(1+\gamma)/2}.$$

Assuming $T < 1$ and taking $\mu = T$, we obtain the estimate

$$|Dv|_G^{(\gamma)} < N + N_4 T^{(1-\gamma)/2}.$$

All the hypotheses of Schauder's theorem will be satisfied as soon as

$$T \le (N - |Du_0|_G^{(\gamma)} - \left|\log|D_n u_0|\right|_G^{(0)})^{2/(1-\gamma)} \{N_5(\varepsilon, N_0, n)\}^{2/(1-\gamma)}.$$

In order to show the desired regularity of the solution to Problem (B_ε) (see Lemma 1), let us recall condition (2.9). This condition "lifts" the regularity of the solution one order. Let $Du \in H^{\beta,\beta/2}(\bar{G}_T)$. By treating (2.9) as the heat equation with the right-hand side

$$f = -(D_n u)^{-1}\left(1 + \sum_{i=1}^{n-1} |D_i u|^2\right)$$

in the space $H^{\beta,\beta/2}(\bar{S}_T)$, in view of Theorem 3, Section 2 of Chapter I, we can conclude that $u \in H^{2+\beta,(2+\beta)/2}(\bar{S}_T)$. By the same theorem, $u(y,\tau)$, as the solution of equation (2.4) that satisfies the conditions (2.5) and coincides with a function belonging to $H^{2+\beta,(2+\beta)/2}(\bar{S}_T)$ on the boundary S_T, is itself an element of the space $H^{2+\beta,(2+\beta)/2}(\bar{G}_T)$, provided that the appropriate compatibility conditions are satisfied at the initial time instant. Thus, $Du \in H^{1+\beta,(1+\beta)/2}(\bar{G}_T)$ with the value of β dependent only on the regularity of the initial function $u_0(y)$ and the order of the compatibility condition holding at the initial time instant. The solution can be extended onto a maximal time interval $(0, T_\varepsilon)$ specified by conditions (2.10).

We shall now derive uniform energy estimates for the solutions of Problem (B_ε), independent of ε. To this end, let us apply the operator of tangent differentiation

$$D_0^{\bar{m}} = D_\tau^{m_0} D_1^{m_1} \dots D_{n-1}^{m_{n-1}}, \quad |\bar{m}| = m,$$

to equation (2.4) and boundary condition (2.9), multiply the result by $D_0^{\bar{m}}u$ (whenever possible, we shall omit the index ε in the functions $u^\varepsilon(y,\tau)$), then integrate over G_T, $T < T_\varepsilon$ and S_T , respectively, and take the sum of the expressions obtained. The boundary integrals over $\{x_i = 0, 1\}$, $i = 1, \dots, n-1$, reduce due to the periodicity of the solution in x'. By elementary transformations, we get

$$\frac{1}{2}\|D_0^{\bar{m}}u(T)\|_{2,G}^2 + \frac{1}{2}\|D_0^{\bar{m}}u(T)\|_{2,S}^2 + \varepsilon\sum_{i=1}^{n-1}\|D_0^{\bar{m}}p_i\|_{2,S_T}^2$$

$$+\|\frac{1}{p_n}D_0^{\bar{m}}p_n\|_{2,G_T}^2 + \sum_{i=1}^{n-1}\|D_0^{\bar{m}}p_i - \frac{p_i}{p_n}D_0^{\bar{m}}p_n\|_{2,G_T}^2$$

$$= \frac{1}{2}\|D_0^{\bar{m}}u_0\|_{2,G}^2 + \frac{1}{2}\|D_0^{\bar{m}}u_0\|_{2,S}^2 + \int_{G_T}\Psi D_0^{\bar{m}}p_n\,dy\,d\tau, \qquad (2.17)$$

where

$$\Psi = \sum_{|\bar{k}|=1}^{[m/2]}\Psi_{\bar{k}}D_0^{\bar{m}-\bar{k}}p + \Psi_0\,.$$

$\Psi_{\bar{k}}$, Ψ_0 include derivatives of the functions p up to order $[m/2]$, and Ψ_0, Ψ_k are continuous functions which are bounded if their arguments and the quantity $|\log|p_n||$ are bounded. Let N be arbitrary,

$$N > N_0 \geq |u_0|_G^{(2l)} + |\log|D_n u_0||_G^{(0)}\,.$$

For each fixed $\varepsilon > 0$, there exists a maximal time interval $(0, T_*^\varepsilon)$ (with T_*^ε dependent on ε, in general) where

$$|\hat{D}^{\bar{k}}p^\varepsilon|_{G_{T_*^\varepsilon}}^{(0)} < N, \quad k \leq \left[\frac{m}{2}\right]; \quad |\log|p_n^\varepsilon||_{G_{T_*^\varepsilon}}^{(0)} < N, \qquad (2.18)$$

where $\hat{D}^{\bar{k}} = D_0^{k_0}D_1^{k_1}\ldots D_n^{k_n}$, $|\bar{k}| = k_0 + k_1 + \ldots + k_n = k$.

We shall derive all further estimates on the interval $(0, T)$ where $T \leq T_*^\varepsilon$. On this interval, the coefficients of the highest derivatives (of order $[m/2] + 1$ and higher) in the expression for Ψ in equality (2.17) are bounded by a constant dependent only on N. Let us estimate the last term on the right-hand side of (2.17) by applying the Hölder inequality and the bound (see Lemma 2 of Section 2, Chapter I)

$$\|D_0^{\bar{r}}p\|_{2,G_T}^2 \leq \mu\|\hat{D}^{\bar{m}}p\|_{2,G_T}^2 + c_1(\mu)\|p\|_{2,G_T}^2\,,$$

with $|\bar{r}| < m$ and any constant $\mu > 0$. At the same time, we estimate the left-hand side of (2.17) from below to get

$$\|D_0^{\bar{m}}p\|_{2,G_T}^2 \leq N_1\left\{\mu\|\hat{D}^{\bar{m}}p\|_{2,G_T}^2 + c_1(\mu)\right\}, \quad N_1 = N_1(N_0, N). \qquad (2.19)$$

There are no terms in (2.19) containing ε, hence all the constants in this inequality and also in the sequel are independent of ε.

It follows directly from equation (2.4) that the derivatives $D_n^i D_0^{\bar{k}}p$, $i + |\bar{k}| \leq m$ can be estimated by $D_0^{\bar{m}}p$. Indeed, by applying the operator $D_0^{\bar{r}}$, $|\bar{r}| = m - 1$, to (2.4) and then using the identity $D_n D_0^{\bar{r}}p_i = D_i D_0^{\bar{r}}p_n$, $i = 1, \ldots, n - 1$, we determine the derivative $D_n D_0^{\bar{r}}p_n$ in terms of $D_0^{\bar{m}}p$. Further, we apply the operator $D_n^{\bar{k}-1}D_0^{\bar{m}-\bar{k}}$ to equation (2.4) to determine successively the derivatives $D_n^{\bar{k}}D_0^{\bar{m}-\bar{k}}p$, $|\bar{k}| \leq m$, in terms of $D_n^{\bar{j}}D_0^{\bar{m}-\bar{j}}p$, $0 \leq |\bar{k}| \leq m$ and hence also in terms of $D_0^{\bar{m}}p$. We also use the inequality

$$\|\hat{D}^{\bar{m}}p\|_{2,G_T}^2 \leq N_2(N)\|D_0^{\bar{m}}p\|_{2,G_T}^2\,.$$

By applying (2.19) to the right-hand side of the above inequality, we get the bound

$$\|\hat{D}^{\bar{m}}p\|_{2,G_T}^2 \le N_3\left\{\mu\|\hat{D}^{\bar{m}}p\|_{2,G_T}^2 + c_3(\mu)\right\}, \quad N_3 = N_3(N_0, N) .$$

For μ sufficiently small this yields

$$\|\hat{D}^{\bar{m}}p\|_{2,G_T}^2 \le N_4(N_0, N) . \tag{2.20}$$

Due to inequality (2.20), all derivatives of the functions p up to order $[m/2]$ can be estimated with the use of the simplest embedding theorems. In fact, for each section $\{t = \text{const.}\}$ we can estimate the integrand in the representation

$$\hat{D}^{\bar{r}}p(y,\tau) = \hat{D}^{\bar{r}}p(y,0) + \int_0^\tau D_t\hat{D}^{\bar{r}}p(y,t)\,dt, \quad |\bar{r}| \le \left[\frac{m}{2}\right],$$

by using Theorem 1 of Section 2, Chapter I to conclude

$$|\hat{D}^{\bar{q}}p(t)|_G^{(0)} \le c\|\hat{D}^{\bar{m}}p(t)\|_{2,G}, \quad |\bar{q}| = |\bar{r}| + 1 . \tag{2.21}$$

Here the constant c depends only on the geometric characteristics of the domain G and t is a parameter. A necessary condition for (2.21) to hold is that

$$n < 2(m - 1 - \max|\bar{r}|) = 2(m - 1 - [m/2]) .$$

This relation is always true, as can easily be shown.

Therefore, for $|\bar{r}| \le [m/2]$,

$$|\hat{D}^{\bar{r}}p|_{G_T}^{(0)} \le N_0 + T^{\frac{1}{2}}\left(\int_0^T \{|\hat{D}^{\bar{q}}p(t)|_G^{(0)}\}^2\,dt\right)^{\frac{1}{2}}$$

$$\le N_0 + T^{\frac{1}{2}}c\left(\int_0^T \|\hat{D}^{\bar{m}}p(t)\|_{2,G}^2\,dt\right)^{1/2} \le N_0 + T^{\frac{1}{2}}cN_4^{\frac{1}{2}}. \tag{2.22}$$

From the representation

$$\log|p_n| = \log|D_nu_0| + \int_0^\tau \frac{1}{p_n}D_tp_n(y,t)\,dt$$

and inequality (2.18) we get

$$\left|\log|p_n|\right|_{G_T}^{(0)} \le N_0 + N_5(N)T . \tag{2.23}$$

Let

$$T_* = \min\left\{\frac{N - N_0}{N_5}, \frac{(N - N_0)^2}{cN_4}\right\}.$$

Then the right-hand sides of (2.22) and (2.23) do not exceed N for $T \le T_*$. It can be proved that all intervals $(0, T_*^\varepsilon)$ contain the interval $(0, T_*)$, i.e. $T_* < T_*^\varepsilon$.

Indeed, suppose the contrary. Then there would exist $\varepsilon > 0$ such that $T_*^\varepsilon \le T_*$. The corresponding solution p^ε satisfies the estimates (2.22) and (2.23) :

$$|\hat{D}^{\bar{r}}p^\varepsilon|_{G_{T_*^\varepsilon}}^{(0)} \le N_0 + c(T_*^\varepsilon N_4)^{\frac{1}{2}} < N, \quad |\bar{r}| \le [m/2], \tag{2.24}$$

$$\left|\log|p_n^\varepsilon|\right|_{G_{T_*^\varepsilon}}^{(0)} \le N_0 + N_5 T_*^\varepsilon < N. \tag{2.25}$$

Since $[m/2] \ge 3$ for $n \ge 2$, we have

$$\mathcal{I}(T_*^\varepsilon, u^\varepsilon) \equiv |p^\varepsilon|_{G_{T_*^\varepsilon}}^{(\gamma)} + \left|\log|p_n^\varepsilon|\right|_{G_{T_*^\varepsilon}}^{(0)} \le 2N < \infty.$$

Hence, the interval $(0, T_\varepsilon)$ where Problem (B_ε) admits a smooth solution u^ε contains the interval $(0, T_*^\varepsilon)$, $T_\varepsilon > T_*^\varepsilon$. But this implies that the inequalities (2.24), (2.25) are satisfied on an interval larger than $(0, T_*^\varepsilon)$, hence leading to a contradiction to the assumption of the maximality of the latter.

As already mentioned, $[m/2] \ge 3$ and, by (2.18), all derivatives of $u^\varepsilon(y, \tau)$ up to third order are bounded on the interval $(0, T_*)$:

$$|\hat{D}^{\bar{r}} u^\varepsilon|_{G_{T_*}}^{(0)} \le N, \quad |\bar{r}| \le 3.$$

Thus there exists a subsequence $u^{\varepsilon k}(y, \tau)$, uniformly convergent in $C^{2,1}(\bar{G}_{T_*})$ to a function $u(y, \tau)$ that obviously is the desired solution of Problem (B).

With $t = \tau$ and $x' = y'$, we determine the classical solution of the Stefan Problem (A) on the whole interval $(0, T_*)$ by

$$x_n = u(x', \vartheta(x, t), t).$$

The free boundary equation $x_n = R(x', t)$ is explicitly given by the function $u(y, \tau)$:

$$R(x', t) \equiv u(x', 0, t) . \qquad \square$$

3. Construction of approximate solutions to the one-phase Stefan problem over a small time interval

In general, for an arbitrary surface S, the construction of approximate solutions is more complicated, since no unique parametrization of the free boundary Γ_T in the form $x_n = R(x', t)$ is available and one is unable to "improve" the boundary condition (1.5) by applying the Laplace operator along the tangent coordinates x'. It turns out to be possible, however, to regularize the Stefan condition by transforming it to the form (1.10) equivalent to (1.4) on sufficiently small time intervals. The Laplace-Beltrami operator defined on the initial surface S (see Section 2, Chapter I for its definition) plays here the role of the tangent Laplace operator.

In this connection, we look for a function $R^\varepsilon(x_0, t)$ defined on the surface $S_T = S \times (0, T)$ which determines the surface Γ_T^ε by

$$x = x_0 + R^\varepsilon(x_0, t)\nu(x_0) \qquad \text{for} \quad x_0 \in S , \tag{3.1}$$

where $\nu(x_0)$ is the normal to the surface S at the point x_0, and a function $\vartheta^\varepsilon(x, t)$ defined over Ω_T^ε (this is the domain bounded by surfaces $\mathcal{F}_T, \Gamma_T^\varepsilon$ and the planes $\{t = 0\}$

and $\{t = T\}$) by the system

$$a(\vartheta^\varepsilon) D_t \vartheta^\varepsilon = \sum_{i=1}^n D_i^2 \vartheta^\varepsilon + f \qquad \text{for} \quad (x,t) \in \Omega_T^\varepsilon; \tag{3.2}$$

$$\vartheta^\varepsilon = \vartheta^1 \quad \text{or} \quad \sum_{i=1}^n b_i D_i \vartheta^\varepsilon + b \vartheta^\varepsilon = \vartheta^2 \qquad \text{for} \quad (x,t) \in \mathcal{F}_T; \tag{3.3}$$

$$\vartheta^\varepsilon(x,t) = 0 \qquad \text{for} \quad (x,t) \in \Gamma_T^\varepsilon; \tag{3.4}$$

$$\vartheta^\varepsilon(x,0) = \vartheta_0^\varepsilon(x) \qquad \text{for} \quad x \in G, \qquad R^\varepsilon \mid_{t=0} = 0; \tag{3.5}$$

$$D_t R^\varepsilon - \varepsilon \Delta_S R^\varepsilon = X(x_0, D\vartheta^\varepsilon) \qquad \text{for} \quad (x_0, t) \in S_T. \tag{3.6}$$

In (3.6), Δ_S is the Laplace-Beltrami operator on the surface S,

$$X(x_0, D\vartheta) = -|D\vartheta|^2 (D\vartheta \cdot \nu(x_0))^{-1}.$$

The derivative $D\vartheta$ in the last expression is computed at (x,t), with x given by (3.1).

At $\varepsilon = 0$, the problem (3.2)–(3.6) coincides with the original Stefan problem and it can be expected that the solutions of problem (3.2)–(3.6) for $\varepsilon > 0$ (we shall refer to it as to problem (A_ε)) approximate the solution of the Stefan problem and converge to this solution as $\varepsilon \to 0$.

As already mentioned in the preceding section, by changing the boundary condition (1.5) we have also modified the compatibility conditions to be satisfied by function $\vartheta_0(x)$. Thus, the initial condition for the approximate solution $\vartheta^\varepsilon(x,t)$ has to be changed in order to fulfil the compatibility conditions that follow from (3.2), (3.4), (3.6).

Lemma 2. *Let $\vartheta_0 \in H^{2l}(\bar{G})$ and assume that the compatibility conditions up to order $[l] = m + 1$ that follow from (1.1), (1.4), (1.3) hold on the surface S. Then for each $\varepsilon > 0$ there exists a function $\vartheta_0^\varepsilon(x)$ in $H^{2l}(\bar{G})$ coinciding with $\vartheta_0(x)$ outside a small neighbourhood of S and such that*

$$\lim_{\varepsilon \to 0} |\vartheta_0 - \vartheta_0^\varepsilon|_G^{(2l)} = 0 \ ;$$

on S the function $\vartheta_0^\varepsilon(x)$ satisfies the compatibility conditions up to order $m + 1$ that follow from (3.2), (3.4), (3.6).

A proof of this lemma is given in [151]. The function ϑ_0^ε in condition (3.5) is everywhere in the sequel defined as in Lemma 2.

Theorem 3. *Let the surfaces $\mathcal{F}, S$, the functions $\vartheta^1, \vartheta^2, f, b, b_i$, $i = 1, \ldots, n$, be the same as in Theorem 1 and the functions $\vartheta_0^\varepsilon(x)$ be defined according to Lemma 2. Then for each $\varepsilon > 0$ there exists a solution of Problem (A_ε) that consists of $R^\varepsilon \in H^{2l,l}(S_T)$ defining the domain Ω_T^ε and $\vartheta^\varepsilon \in H^{2l,l}(\bar{\Omega}_T^\varepsilon)$ for $T < T_\varepsilon$. The value $T_\varepsilon > 0$ can be found from the conditions*

$$J(T, \vartheta^\varepsilon) < \infty, \quad T < T_\varepsilon, \quad J(T_\varepsilon, \vartheta^\varepsilon) = \infty, \tag{3.7}$$

where

$$J(T, \vartheta) = |D\vartheta|_{S_T}^{(\gamma)} + \left|\log|(D\vartheta\nu(x_0))|\right|_{S_T}^{(0)} + |\log(2d - R)|_{S_T}^{(0)},$$

for arbitrary $\gamma \in (0, 1)$, $\nu(x_0)$ *is the normal to the surface* S *at the point* $x_0 \in S$ *and* $D\vartheta$ *in the definition of* J *is computed at the point* x *given by equation (3.1).*

Proof. In principle, we shall repeat the proof of Lemma 1 in the preceding section. Let $\mathcal{M}$ be the closed convex set of vector functions ω defined on the surface S_T, such that

$$\omega(x_0, 0) = \mathring{\omega}(x_0), \quad |\omega|_{S_T}^{(\gamma)} \leq N$$

$$\left|\log|(\omega(x_0, t)\nu(x_0))|\right|_{S_T}^{(0)} \leq N ;$$

$$\mathring{\omega}(x_0) = D\vartheta_0^\varepsilon(x_0) \qquad \text{for} \quad x_0 \in S ;$$

$$N > N_0 \geq \max\left\{|\mathring{\omega}|_S^{(\gamma)}, \left|\log|(\mathring{\omega}\nu)|\right|_S^{(0)}\right\} .$$

It follows from Theorem 8 of Section 2, Chapter I, that the Cauchy problem

$$\begin{aligned} D_t \tilde{R} - \varepsilon \Delta_S \tilde{R} &= X(x_0, \omega), \\ \tilde{R}\,|_{t=0} &= 0 \end{aligned} \tag{3.8}$$

admits a unique solution for all $\omega \in \mathcal{M}$ and

$$|\tilde{R}|_{S_T}^{(2+\gamma)} \leq N_1(\varepsilon, N) . \tag{3.9}$$

The operator Ψ_1 from the set $\mathcal{M}$ into $H^{r, r/2}(S_T), r = 2 + \gamma$, which associates with any $\omega \in \mathcal{M}$ the corresponding solution of problem (3.8) is continuous.

Since $|\tilde{R}(x_0, t)| \leq \max|D_t\tilde{R}|t \leq N_1 T$, for $T < dN_1^{-1}$ the function $\tilde{R}$ determines the surface $\tilde{\Gamma}_T$ and the domain $\tilde{\Omega}_T$ bounded by the surfaces $\mathcal{F}_T, \tilde{\Gamma}_T$ and the planes $\{t = 0\}$, $\{t = T\}$. For each function $\tilde{R} = \Psi_1(\omega)$ the initial-boundary value problem (3.2)–(3.5) has a unique solution $\tilde{\vartheta} \in H^{r, r/2}(\tilde{\Omega}_T)$. It follows by standard arguments (cf., [137], [151]) that the operator Ψ_2 from the set of functions $\tilde{R} \in H^{r, r/2}(S_T), |\tilde{R}|_{S_T}^{(0)} \leq d$, into the space $H^{r, r/2}(\tilde{\Omega}_T)$, associating with each such function $\tilde{R}$ the solution $\tilde{\vartheta}$ of problem (3.2)–(3.5), is continuous and

$$|\tilde{\vartheta}|_{\tilde{\Omega}_T}^{(r)} \leq N_2(N_1) . \tag{3.10}$$

The superposition $\Psi_2 \circ \Psi_1$ of the operators Ψ_1 and Ψ_2 is also continuous from $\mathcal{M}$ into $H^{r, r/2}(\tilde{\Omega}_T)$, and the operator Ψ that associates with $\omega \in \mathcal{M}$ the vector function

$$D\tilde{\vartheta}(x_0 + \tilde{R}(x_0, t)\nu(x_0), t) = \Psi(\omega), \quad x_0 \in S,$$

is completely continuous on $\mathcal{M}$, since $|D\tilde{\vartheta}|_{S_T}^{(1+\gamma)} \leq N_2$.

As in Lemma 1 of Section 2, we shall take the interval $(0, T)$ sufficiently small to ensure that the operator Ψ maps the set $\mathcal{M}$ into itself. For this, we use the inequalities

$$|\tilde{R}(x_0, t)| \leq |D_t\tilde{R}|t \leq N_1 T,$$

$$|D\tilde{\vartheta}(x_0 + \tilde{R}\nu(x_0), t) - \mathring{\omega}(x_0)| \leq |D\tilde{\vartheta} - \mathring{\omega}|_{\Omega_T}^{(1+\gamma)}(|\tilde{R}(x_0, t)|^{1+\gamma} + t^{(1+\gamma)/2})$$

$$\leq 2N_2 \left\{(|\tilde{R}|_{S_T}^{(0)})^{1+\gamma} + T^{(1+\gamma)/2}\right\} \leq N_3 T^{(1+\gamma)/2}.$$

Lemma 4 of Section 2, Chapter I implies

$$|D\tilde{\vartheta} - \mathring{\omega}|_{S_T}^{(\gamma)} \leq \mu |D\tilde{\vartheta} - \mathring{\omega}|_{S_T}^{(1+\gamma)} + \frac{c}{\mu^\gamma}|D\tilde{\vartheta} - \mathring{\omega}|_{S_T}^{(0)}$$

with arbitrarily small μ, $0 < \mu < T^{1/2}$. Together there yield the estimate

$$|D\tilde{\vartheta}|_{S_T}^{(\gamma)} \leq |\mathring{\omega}|_S^{(\gamma)} + \mu(N_0 + N_2) + \frac{c}{\mu^\gamma}N_3 T^{(1+\gamma)/2}.$$

In an analogous way we obtain the bound

$$\max_{(x_0,t)\in S_T} \left|\log|(D\tilde{\vartheta}(x_0 + \tilde{R}\nu(x_0), t) \cdot \nu(x_0)|\right| \leq \left|\log|(\mathring{\omega}\nu)|\right|_S^{(0)} + N_4 T^{(1+\gamma)/2}.$$

By taking $\mu = T$ and choosing T sufficiently small, we can easily infer the inequalities

$$|D\tilde{\vartheta}|_{S_T}^{(\gamma)} \leq N, \quad \left|\log|(D\tilde{\vartheta}\nu)|\right|_{S_T}^{(0)} \leq N.$$

These relations, together with the equality $D\tilde{\vartheta}(x_0, 0) = \mathring{\omega}(x_0)$, $x_0 \in S$, ensure that the operator Ψ transforms the set $\mathcal{M}$ into itself. Thus, by Schauder's fixed point theorem, Ψ has a fixed point which is at the same time a solution of Problem (A_ε).

Obviously, the solution of Problem (A_ε) can be extended in time as long as $J(T, \vartheta^\varepsilon) < \infty$. The maximal regularity of this solution can be concluded as in Lemma 1.   □

The same scheme applies to constructing the approximate solution of the more general equation (1.12). The only difference is the definition of $\Psi_2(\tilde{R})$ as a solution of the appropriate initial-boundary value problem. At this stage, we refer to [203] for results on the local in time solvability of standard boundary value problems for arbitrary nonlinear parabolic equations.

4. A lower bound on the existence interval of the solution. Passage to the limit

In contrast to the simplest case, the global transformation that fixes the free boundary is not applicable to the problem in its general form. But because the problem is considered on a small time interval, the global change of variables is not necessary. Thus, by the partition of unity and using local estimates it is enough to restrict our attention to a small neighbourhood (in Ω_T^ε) of the standard set $s \subset S$.

Let G_d be the d−neighbourhood of the surface S in G :

$$G_d = \{x \in G : x = x_0 + R\nu(x_0), \quad x_0 \in S, \ |R| < d\},$$

where $\nu(x_0)$ is the normal to the surface S at x_0. By $G_{d,T}$ we denote the intersection of the cylinder $G_d \times (0, T)$ with Ω_T^ε. We also set $G_T^* = \Omega^\varepsilon \setminus \bar{G}_{2d,T}$.

We shall assume, changing the constants N_0 and d if necessary, that the inequality (1.11) for the functions $\vartheta_0^\varepsilon(x)$ holds not only on S, but also on a larger set G_{2d} :

$$\left|\log|D\vartheta_0^\varepsilon|\right|_{G_{2d}}^{(0)} \leq N_0 \ . \tag{4.1}$$

Let $N > N_0$ be an arbitrary number. The functions ϑ^ε clearly satisfy an inequality similar to (4.1), with the constant N, on a small time interval $(0, T)$ and the $2d$−neighbourhood of the free boundary Γ_T^ε. Furthermore, we shall assume that on the chosen time interval the distance between the free boundary Γ_T^ε and the initial surface S does not exceed d, and the norms of the derivatives of $D\vartheta^\varepsilon$ up to order $[m/2]$ in the space $C(\bar{G}_{2d,T})$ are bounded by the constant N. Therefore, for each $\varepsilon > 0$ there exists a maximal time interval $(0, T_*^\varepsilon)$ such that for $T < T_*^\varepsilon$

$$|R^\varepsilon|_{S_T}^{(0)} \leq d, \quad \left|\log|D\vartheta^\varepsilon|\right|_{G_{2d,T}}^{(0)} \leq N \ , \tag{4.2}$$

$$|\hat{D}^{\bar{k}}(D\vartheta^\varepsilon)|_{G_{2d,T}}^{(0)} \leq N, \quad |\bar{k}| \leq [m/2] \ , \tag{4.3}$$

where $R^\varepsilon, \vartheta^\varepsilon$ is the solution of Problem (A_ε). Due to the regularity of the solution, as already proved, and by hypothesis (4.1), T_*^ε are strictly positive. The interval $(0, T_*^\varepsilon)$ is specified by the inequalities (4.2) and (4.3); we are going to show that T_*^ε does not converge to 0 as $\varepsilon \to 0$.

Since $|R^\varepsilon| < d$ on the interval $(0, T_*^\varepsilon)$, the distance between the domain G_T^* and the free boundary Γ_T^ε is larger than d. From local estimates on the solutions of second-order parabolic equations (see Theorem 6 of Section 2, Chapter I), we have

$$|D\vartheta^\varepsilon|_{G_T^*}^{(2l)} \leq N_1(N_0, M_0, d), \quad T \leq T_*^\varepsilon \ . \tag{4.4}$$

Let us consider the systems of open sets $\{s^{(k)}\} = \mathcal{N}_S^\lambda$ and $\{S^{(k)}\} = \mathcal{M}_S^\lambda$ on S, such that

1° $s^{(k)} \subset S^{(k)} \subset S$, $\bigcup_k s^{(k)} = \bigcup_k S^{(k)} = S$;

2° in the local coordinate system obtained from the original Cartesian system of coordinates by an orthogonal affine transformation $z = \mathcal{A}_k(x)$, the sets $s^{(k)}$ and $S^{(k)}$ are given by the equations

$$s^{(k)} = \{z \ : \ z_n = Z^{(k)}(z'), z' \in V^\lambda\}, \ Z^{(k)}(0) = 0 \ ,$$
$$S^{(k)} = \{z \ : \ z_n = Z^{(k)}(z'), z' \in V^{2\lambda}\}, \ DZ^{(k)}(0) = 0 \ ,$$
$$V^\lambda = \{z' \in R^{n-1} \ : \ |z'| < \gamma\lambda\}, \ \gamma = \text{const.} > 0 \ ;$$

3° there exists a number k_0 independent of λ, such that the intersection of any k_0 sets from $\mathcal{N}_S^\lambda$ (the same is clearly true for the sets from $\mathcal{M}_S^\lambda$) is empty.

Let us define the sets

$$\Omega_T^{(k)} = \{(x, t) \in \Omega_T^\varepsilon \ : \ x = x_0 + R\nu(x_0), \ x_0 \in S^{(k)}, \ |R| < 2d\},$$
$$\omega_T^{(k)} = \{(x, t) \in \Omega_T^\varepsilon \ : \ x = x_0 + R\nu(x_0), \ x_0 \in s^{(k)}, \ |R| < 2d\}.$$

In the domain $\Omega_T^{(k)}$ we can apply a coordinate transformation similar to (2.3). We first briefly describe the further procedure:

Step 1. Formulation of the heat equation and the regularized Stefan condition in the local coordinates $z = A_k(x)$.

Step 2. Transformation to new variables

$$\tau = t, \quad y' = z', \quad y_n = v(z,t) \equiv \vartheta^\varepsilon \left(A_k^{-1}(z), t \right) \tag{4.5}$$

in each of the domains $\Omega_T^{(k)}$ and formulation of the equivalent boundary value problem for the function $u(y, \tau) = z_n$.

Step 3. Derivation of energy estimates in the norm of W_2^m for the functions expressed in (y, τ)–coordinates. These estimates will not be complete, with the norms of Du in $W_2^m(Q_T^1)$ bounded by the analogous norms in $W_2^m(Q_T^2)$ (Q_T^1 and Q_T^2 are the images of the sets $\omega_T^{(k)}$ and $\Omega_T^{(k)}$ under the mapping (4.5), respectively) with small multiplier $\mu > 0$ and by known quantities.

Step 4. Closure of the energy estimates and resulting variables; choice of the interval $(0, T_*)$ contained in all intervals $(0, T_*^\varepsilon)$.

We now proceed to a detailed exposition of the above procedure.

Step 1. Let $z = A_k(x)$, $\hat{f}(z,t) = f(A_k^{-1}(z), t)$. In the domain $\Omega_T^{(k)}$ the function $v(z,t) = \vartheta^\varepsilon(A_k^{-1}(z), t)$ satisfies the nonlinear heat equation

$$a(v)D_t v = \sum_{i=1}^n D_i^2 v + \hat{f} . \tag{4.6}$$

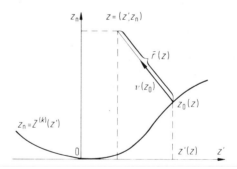

Figure 4

We shall rewrite the regularized Stefan condition in the following form useful in our further considerations:

$$-D_t v + |Dv|^2 = \varepsilon \frac{(Dv \cdot Dv_0(z_0))}{|Dv_0(z_0)|} \Delta_S R^\varepsilon, \tag{4.7}$$

where

$$v_0 = v|_{t=0}, \quad \nu(x_0) = \frac{Dv_0(z_0)}{|Dv_0(z_0)|}, \quad z_0 = \mathcal{A}_k(x_0).$$

Because we already know how the left-hand side of (4.7) is transformed by (4.5) (see Section 2), let us confine our further attention to the right-hand side of condition (4.7). An inconvenient feature of this condition is that its left-hand side is expressed in terms of (z, t) while the right-hand side is given in terms of (z_0, t) belonging to surface S_T. There is a one-to-one correspondence between these points,

$$z = z_0(z) + \bar{r}(z)\nu(z_0) , \tag{4.8}$$

if the quantity λ defining the size of set $S^{(k)}$ and the number d are small enough. To remove the above inconvenience, we shall rewrite the right-hand side of condition (4.7) in terms of variables (z, t).

Since $\bar{r} = z_n$ at $z' = 0$ (the normal ν at $(0, z_n)$ is co-linear with axis z_n), by reducing the values of λ and d if necessary, we obtain

$$\left|\log|D_n\bar{r}|\right| \le M_0 \tag{4.9}$$

in the $2d$−neighbourhood of $S^{(k)}$.

If the equation $z_n = Z_\varepsilon^{(k)}(z', t)$ describes the free boundary Γ_T^ε in local coordinates, then

$$R^\varepsilon = \bar{r}(z', Z_\varepsilon^{(k)}).$$

Let $\{g_{ij} = D_i x(z) \, D_j x(z)\}$, $i, j = 1, \ldots, n-1$, be the metric tensor of the surface S in local coordinates, $|g| = \det(g_{ij})$ and let $\{g^{ij}\}$ be the tensor corresponding to $\{g_{ij}\}$, defined by

$$\sum_{j=1}^{n-1} g^{ij} g_{ik} = \delta_k^i .$$

Then

$$\Delta_S R^\varepsilon = \frac{1}{|g|^{1/2}} \sum_{i,j=1}^{n-1} \frac{d}{dz_i}\left\{|g|^{1/2} g^{ij} \frac{d}{dz_j}\bar{r}(z', Z_\varepsilon^{(k)})\right\}.$$

Further, because

$$\frac{(Dv \cdot Dv_0)}{|Dv_0|} = \sum_{i=1}^{n} D_i v \mathring{v}_i(z)$$

where $\mathring{v}_i(z) = D_i v_0(z_0(z)) \, |Dv_0(z_0(z))|^{-1}$, we obtain

$$-D_t v + |Dv|^2 = \frac{\varepsilon}{|g|^{1/2}}\left\{\sum_{i=1}^{n} \mathring{v}_i D_i v\right\}\left(\sum_{i,j=1}^{n-1} \frac{d}{dz_i}\left\{|g|^{1/2} g^{ij} \frac{d}{dz_j}\bar{r}(z', Z_\varepsilon^{(k)})\right\}\right). \tag{4.10}$$

Because the surface S is sufficiently regular, the functions $\mathring{v}_i$ and $\bar{r}$ are at least $(m+3)$-times differentiable and all their derivatives up to this order are bounded by the constant M_0. Moreover, since

$$\sum_{i=1}^{n} \mathring{v}_i D_i v|_{t=0} = |Dv_0| \geq \exp(-N_0) \,,$$

in addition to the inequalities (4.2) and (4.3) we can choose the interval $(0, T_*^\varepsilon)$ small enough to provide

$$\Lambda_0 \;=\; \frac{1}{|g|^{1/2}} \sum_{i=1}^{n} \mathring{v}_i D_i v \;\geq\; N^{-1} > 0 \,. \tag{4.11}$$

Step 2. As already mentioned, due to the second inequality of (4.2), the transformation (4.5) is well-defined. In new variables (y, τ), the part of the free boundary Γ_*^ε contained in $\Omega_T^{(k)}$ is transformed into the cylinder

$$\Pi_T^2 = \Pi^2 \times (0, T), \quad \Pi^2 = \{(y', y_n) \,:\, |y'| < 2\gamma\lambda, \; y_n = 0\},$$

and the surface

$$\{(z, t) \in \bar{\Omega}_T^{(k)} \,:\, z = z_0 + 2\nu(z_0)d\}$$

is transformed onto

$$\Pi_T^0 = \{(y, \tau) \,:\, |y'| < \gamma\lambda, \; y_n = Y^0(y', \tau), \; \tau \in (0, T)\}.$$

Let Q_T^1 and Q_T^2 be the images under the transformation (4.5) of the domains $\omega_T^{(k)}$ and $\Omega_T^{(k)}$, respectively, and let Π_T^1 be the intersection of Π_T^2 with $\bar{Q}_T^1$. In Q_T^2 the function $u(y, \tau) = z_n$ satisfies the quasilinear parabolic equation

$$a(y_n)D_\tau u - \Delta' u + D_n \left\{ \frac{1}{p_n}\left(1 + \sum_{i=1}^{n-1} p_i^2\right) \right\} = F \,, \tag{4.12}$$

where $p_i = D_i u$, $i = 1, \ldots, n$, $F(y, \tau, u, p) = -p_n \hat{f}(y', u, \tau)$.

We now derive the boundary condition for the function u on Π_T^2. The left-hand side $J = -D_t v + |Dv|^2$ of (4.10) can be given the form

$$J = \frac{1}{p_n}\left\{ D_\tau u + \frac{1}{p_n}\left(1 + \sum_{i=1}^{n-1} p_i^2\right) \right\}.$$

Since $D_i v = -(D_i u)/p_n$, $i = 1, \ldots, n-1$, $D_n v = 1/p_n$ and $\bar{r}(z', Z^{(k)}\epsilon) = \bar{r}(y', u(y', 0, \tau))$, we have $\Lambda_0 = p_n^{-1}\Lambda(y', u, p')$, $p' = (p_1, \ldots, p_{n-1})$, and the condition (4.10) in coordinates (y, τ) is equivalent to

$$D_\tau u + \frac{1}{p_n}\left\{1 + \sum_{i=1}^{n-1} p_i^2\right\} = \varepsilon\left\{\Lambda(y', u, p') \sum_{i,j=1}^{n-1} \frac{d}{dy_i}\left(a^{ij}(y', u)\frac{d}{dy_j}\bar{r}(y', u)\right)\right\}, \tag{4.13}$$

where $\quad a^{ij} = |g|^{1/2}g^{ij}$.

The choice of the initial surface S ensures that the quadratic form $\{a^{ij}\}$ is strictly positive definite:

$$\sum_{i,j=1}^{n-1} a^{ij}\xi_i\xi_j \geq M_0^{-1}|\xi|^2. \tag{4.14}$$

Eventually, adding the bounds (4.2), (4.9) and (4.11) we get

$$\Lambda \geq N_2(M_0, N), \quad \left|\log\left|\frac{d\bar{r}}{du}(y', u)\right|\right| \leq N_2 . \tag{4.15}$$

Inequality (4.4) implies that $u \in H^{2l,l}$ on the boundary Π_T^0 which itself belongs to $H^{2l,l}$, with the appropriate norms bounded by a constant dependent only on N_1. The function Y^0 which defines the surface Π_T^0 is strictly positive there,

$$Y^0(y', \tau) \geq N_3(M_0, N, d) > 0. \tag{4.16}$$

Remark. If the solution ϑ^ε of the problem considered satisfies the more general equation

$$D_t\vartheta - \sum_{i,j=1}^{n} D_i(b_{ji}(x,t,\vartheta)D_j\vartheta) = f(x,t,\vartheta,D\vartheta)$$

and the Stefan condition

$$-D_t\vartheta + \sum_{i,j=1}^{n} b_{ij}D_i\vartheta D_j\vartheta = \varepsilon(D\vartheta \cdot \nu(x_0))\Delta_S R ,$$

then in the coordinates (y, τ) the function $u(y, \tau) = z_n$ satisfies the equation

$$D_\tau u - \sum_{i=1}^{n-1} D_i A_i + D_n\left\{\frac{1}{p_n}\left(\sum_{i=1}^{n-1} p_i A_i - A_n\right)\right\} = F \quad \text{in } Q_T^2$$

and the boundary condition

$$D_\tau u + \frac{1}{p_n}\left(\sum_{i=1}^{n-1} p_i A_i - A_n\right) = \varepsilon\Lambda \sum_{i=1}^{n-1}\frac{d}{dz_i}\left\{a^{ij}\frac{d}{dz_j}\bar{r}(z', u)\right\} \quad \text{on } \Pi_T^2 .$$

Here

$$A_i = \sum_{j=1}^{n-1} c_{ij}p_j - b_{in} , \qquad c_{ij} = \sum_{q,l=1}^{n} A_{il}A_{jq}b_{ql}(A_k^{-1}(z),t,v) ,$$

where A_{ij} are constant elements of the matrix $(\partial z/\partial x)$,

$$F(y,\tau,u,p) = -p_n f(A_k^{-1}(z),t,v,Dv), \quad p = Du.$$

Step 3. Except for technical details, the energy estimates in the W_2^m−norm can be derived in nearly the same way as described in Section 2 of the present chapter. The first of these bounds requires estimating integrals on the lateral boundaries of Q_T^2 different from both Π_T^0 and Π_T^2. To avoid the necessity of considering the resulting integrals, it is necessary to multiply the equation obtained by differentiation of (4.12) by $\xi^2(y')D_0^{\bar{m}}u$ where the function $\xi(y')$ is equal to 1 on Q_T^1 and equal to 0 outside Q_T^2.

After simple transformations, the left-hand side $\mathcal{J}_1$ of the identity

$$\mathcal{J}_1 = \mathcal{J}_2 + \mathcal{J}_3 + \mathcal{J}_4 ,\tag{4.17}$$

analogous to (2.17), reduces to

$$\mathcal{J}_1(T) = \frac{1}{2}\left(\|\xi a D_0^{\bar{m}}u\|_{2,Q^2(T)}^2 + \|\xi D_0^{\bar{m}}u\|_{2,\Pi^2}^2\right) + \sum_{i=1}^{n-1}\left\|\xi\left(D_0^{\bar{m}}p_i - \frac{p_i}{p_n}D_0^{\bar{m}}p_n\right)\right\|_{2,Q_T^2}^2$$

$$+\varepsilon\left\|\xi^2\sum_{i,j=1}^{n-1}\frac{\partial\bar{r}}{\partial u}a^{ij}D_0^{\bar{m}}p_iD_0^{\bar{m}}p_j\right\|_{1,\Pi_T^2} + \left\|\frac{\xi}{p_n}D_0^{\bar{m}}p_n\right\|_{2,Q_T^2}^2$$

where $Q^2(T)$ is the intersection of $Q_{T\varepsilon}^2$ and the plane $\{t=T\}$.

Since $\xi=1$ in Q_T^1 and on Π_T^1, and the quadratic form $\{\frac{\partial\bar{r}}{\partial u}a^{ij}\}$ is strictly positive definite, $\mathcal{J}_1$ admits the following obvious lower bound

$$\mathcal{J}_1 \geq N_4\left\{\|D_0^{\bar{m}}p\|_{2,Q_T^1}^2 + \varepsilon\|D_0^{\bar{m}}p'\|_{2,\Pi_T^1} + \|D_T^m u\|_{2,Q^1(T)}^2 + \|D_T^m u(T)\|_{2,\Pi^1}^2\right\} .$$

In this inequality N_4 depends on N_0, M_0, N; everywhere below the constants N_k depend on the indicated constants.

The second difference arises in the way of estimating the integral

$$\mathcal{J}_2 = \|\xi^2 D_0^{\bar{m}}u D_0^{\bar{m}}F\|_{1,Q_T^2}$$

which appears on the right-hand side of (4.17). In the latter integral, except for the main term

$$\mathcal{J}_{02} = \|\xi^2 D_0^{\bar{m}}u\frac{\partial F}{\partial p}D_0^{\bar{m}}p\|_{1,Q_T^2},$$

all the other terms exhibit the same structure as in the integral

$$\mathcal{J}_3 = \|D_0^{\bar{m}}(\xi^2 p_n)\Psi\|_{1,Q_T^2}.$$

This can be obtained by differentiating the product

$$\frac{1}{p_n}\left(1+\sum_{i=1}^{n-1}p_i^2\right)$$

and almost completely reproduces the corresponding expression in (2.17). Hence, $\mathcal{J}_2$ up to the component $\mathcal{J}_{02}$ admits the same upper estimates as $\mathcal{J}_3$,

$$\mathcal{J}_2 - \mathcal{J}_{02} + \mathcal{J}_3 \leq \bar{\mu}\|\hat{D}^m p\|_{2,Q_T^2}^2 + N_5(\mu),\tag{4.18}$$

with arbitrarily small number μ. If $D_0^{\bar{m}}u \neq D_T^m u$, then $D_0^{\bar{m}}u = D_0^{\bar{q}}p_i, |\bar{q}| = m-1$, and for some i the expression

$$\left\|\xi^2\frac{\partial F}{\partial p}D_0^{\bar{m}}u D_0^{\bar{m}}p\right\|_{1,Q_T^2}$$

can be estimated from above by the right-hand side of (4.18). We shall bound the term

$$\mathcal{J}_{12} = \left\| \xi^2 \frac{\partial F}{\partial p} D_\tau^m u D_0^{\bar m} p \right\|_{1,Q_T^2}$$

from above by using the Hölder inequality

$$\mathcal{J}_{12} \leq \mu \left\| \hat{D}^{\bar m} p \right\|_{2,Q_T}^2 + N_6(\mu) \left\| D_\tau^m u \right\|_{2,Q_T^2}^2.$$

And, finally, the integral $\mathcal{J}_4$ was obtained by differentiating the right-hand side of condition (4.13). The main term obtained by differentiation, multiplication by $\xi^2 D_0^{\bar m} u$ and integration on Π_T^2, entered the left-hand side of (4.17). The "maximal" terms of the integral $\mathcal{J}_4$ are components of the form $\varepsilon D_0^{\bar m} u D_0^{\bar m} p'$. If $D_0^{\bar m} u \neq D_\tau^m u$, then $D_0^{\bar m} u = D_0^{\bar q} p_i$ for some i and $\bar q$, $|\bar q| = m - 1$, and

$$\varepsilon \| D_0^{\bar m} u D_0^{\bar m} p' \|_{1,\Pi_T^2} \leq \varepsilon \mu \| D_0^{\bar m} p' \|_{2,\Pi_T^2}^2 + N_7(\mu).$$

Otherwise,

$$\varepsilon \| D_\tau^m u D_0^{\bar m} p' \|_{1,\Pi_T^2} \leq \varepsilon \mu \| D_0^{\bar m} p' \|_{2,\Pi_T^2}^2 + N_8(\mu) \| D_\tau^m u \|_{2,\Pi_T^2}^2.$$

The other terms can be estimated as in $\mathcal{J}_3$:

$$\mathcal{J}_4 \leq \varepsilon \mu \| D_0^{\bar m} p' \|_{2,\Pi_T^2}^2 + N_9(\mu) \| D_\tau^m u \|_{2,\Pi_T^2}^2 + N_{10}(\mu).$$

After adding all the above estimates, we eventually get the bound

$$\| D_\tau^m u \|_{2,Q^1(T)}^2 + \| D_\tau^m u(T) \|_{2,\Pi_1}^2 + \| \hat{D}^{\bar m} p \|_{2,Q_T^1}^2 + \varepsilon \| D_0^{\bar m} p' \|_{2,\Pi_T^1}^2$$

$$\leq \mu \| \hat{D}^{\bar m} p \|_{2,Q_T^2}^2 + \varepsilon \mu \| D_0^{\bar m} p' \|_{2,\Pi_T^2}^2 + N_{11}(\mu) \left\{ \| D_\tau^m u \|_{2,\Pi_T^2}^2 + \| D_\tau^m u \|_{2,Q_T}^2 \right\} + N_{12}(\mu),$$
$$(4.19)$$

where the bounds of the derivatives $\hat{D}^{\bar m} p$ by $D_0^{\bar m} p$ follow directly from equation (4.12), as in Section 2 of this chapter.

Step 4. Passing to the original physical variables (z, t) is especially simple for the derivatives of the function R^ε. Indeed, let $R^\varepsilon(z_0, t) = R^{(k)}(z', t)$ in the local coordinates $z = \mathcal{A}_k(x)$, then by definition

$$R^{(k)}(z', t) = \bar{r}(z', u(z', 0, t))$$

and

$$N_{13}^{-1} \left\{ -1 + \| D_0^{\bar m} R^{(k)}(T) \|_{2,S^{(k)}}^2 \right\} \leq \| D_0^{\bar m} u(T) \|_{2,\Pi^2}^2 \leq N_{13} \left\{ 1 + \| D_0^{\bar m} R^{(k)}(T) \|_{2,S^{(k)}}^2 \right\},$$

$$N_{13}^{-1} \left\{ \| D_0^{\bar m} (DR^{(k)}) \|_{2,S_T^{(k)}}^2 - 1 \right\} \leq \| D_0^{\bar m} p' \|_{2,\Pi_T^2}^2 \leq N_{13} \left\{ \| D_0^{\bar m} (DR^{(k)}) \|_{2,S_T^{(k)}}^2 + 1 \right\}.$$

The derivatives of $u(y, \tau)$ can be expressed in terms of the derivatives of $v(z, t)$ according to formulae analogous to (2.7), (2.8) :

$$\hat{D}^{\bar m} p_i = \sum_{j=1}^n l_{ij} \hat{D}^{\bar m}(D_j v) + l_{i0}, \quad D_\tau^m u = l_0 D_t^m v + l_{00},$$

where l_{ij}, l_0 depend on z, t, v, Dv and $l_{i0}, i = 0, 1, \ldots, n,$ depend on the derivatives of Dv of order lower than m and their structure is similar to that of the multiplier Ψ in the identity (2.17). Let us estimate the L_2-norms of $\hat{D}^{\bar{q}}(Dv), |\bar{q}| < m,$ by the L_2-norm of the highest derivatives $\hat{D}^{\bar{m}}(Dv)$ with a small coefficient in the appropriate term, and by the lower order norms with taking into account the boundedness of the derivatives of Dv up to order $[m/2]$ on the interval $(0, T),$ where $T \leq T_*^\varepsilon,$ to obtain the inequality

$$N_{13}^{-1}\left\{\|\hat{D}^{\bar{m}}(Dv)\|_{2,\Omega_T^{(k)}}^2 - 1\right\} \leq \|\hat{D}^{\bar{m}}p\|_{2,\Omega_T^{(k)}}^2 \leq N_{13}\left\{\|\hat{D}^{\bar{m}}(Dv)\|_{2,\Omega_T^{(k)}}^2 + 1\right\}.$$

We can transform the derivatives $D_\tau^m u$ analogously. In this case one only has to remember that the components of the form $D_\tau^{m-|\bar{q}|}pD_\tau^{|\bar{q}|}u, |\bar{q}| \leq [m/2],$ in the representation

$$D_t^m v = -\frac{1}{p_n}D_\tau^m u + \frac{D_\tau u}{p_n^2}D_\tau^{\bar{q}}p_n + \ldots, \qquad |\bar{q}| = m - 1,$$

can be bounded by using the inequalities (4.3) and the representation

$$D_\tau^{m-q}p(T) = D_\tau^{m-q}p(0) + \int_0^T D_\tau^{m-q+1}p(t)\, dt,$$

which implies that

$$\|D_\tau^{m-q}p(T)\|_{2,\Omega^2(T)}^2 \leq \mu\|D_\tau^{m-q+1}p\|_{2,Q_T^2}^2 + N_{14}(\mu)\ .$$

Hence

$$N_{15}^{-1}\left\{\|D_t^m v(T)\|_{2,\omega^{(k)}(T)}^2 - 1\right\} - \mu\|\hat{D}^{\bar{m}}(Dv)\|_{2,\Omega_T^{(k)}}^2$$

$$\leq \|D_\tau^m u(T)\|_{2,Q^1(T)}^2 \leq N_{15}(\mu)\left\{\|D_t^m v(T)\|_{2,\omega^{(k)}(T)}^2 + 1\right\} + \mu\|\hat{D}^{\bar{m}}(Dv)\|_{2,\Omega_T^{(k)}}^2\ .$$

Using the above inequalities in estimating the left-hand side of (4.19) from below, the right-hand side from above and summing the results over all the sets $\omega_T^{(k)},$ due to the property 3° of sets $\mathcal{N}_S^\lambda$ we obtain the inequality

$$\|D_t^m \vartheta^\varepsilon(T)\|_{2,\Omega^\varepsilon(T)}^2 + \|\hat{D}^{\bar{m}}R^\varepsilon(T)\|_{2,S}^2 + \varepsilon\|\hat{D}^{\bar{m}}(DR^\varepsilon)\|_{2,S_T}^2 + \|\hat{D}^{\bar{m}}(D\vartheta^\varepsilon)\|_{2,\Omega_T^\varepsilon}^2$$

$$\leq N_{16}(k_0)\left\{\mu\|\hat{D}^{\bar{m}}(D\vartheta^\varepsilon)\|_{2,\Omega_T^\varepsilon}^2 + \|D_t^m\vartheta^\varepsilon\|_{2,\Omega_T^\varepsilon}^2 + \varepsilon\mu\|\hat{D}^{\bar{m}}(DR^\varepsilon)\|_{2,S_T}^2 + \|\hat{D}^{\bar{m}}R^\varepsilon\|_{2,S_T}^2\right\}$$

$$+N_{17}(\mu).$$

Here we have used the bound (4.4) outside $2d$-neighbourhoods of the surface S. Selecting the parameter μ small enough and applying Gronwall's lemma (cf., Lemma 1, Section I.2),

$$\|D\vartheta^\varepsilon\|_{W_2^m(\Omega_T^\varepsilon)}^2 \leq N_{18}(k_0, M_0, N_0, N, d), \qquad T \leq T_*^\varepsilon. \qquad (4.20)$$

As can be seen, the parameter ε has not entered the bound (4.20). Thus all the remaining considerations do not differ from those in Section 2 of this chapter. More precisely, the estimate (4.20) and the simplest embedding theorems give rise to bounds of the left-hand side of the inequalities (4.2) and (4.3) from above by constant N. At the same time, they contribute to estimating from below the left-hand side of the inequality (4.11) by N^{-1} on a certain sufficiently small time interval $(0, T_*)$ dependent only on the constants k_0, M_0, N_0, N, d.

In contrast to Section 2 where the functions R^ϑ were not introduced, in each local coordinate system the derivatives of the functions R^ε up to order $[m/2]$ can be expressed in terms of the appropriate derivatives of ϑ^ε up to the same order, and are bounded on the time interval $(0, T_*)$ due to the estimate (4.3). Since $[m/2] \geq 3$ for $n \geq 2$, according to (4.3) there are subsequences of $\{R^\varepsilon\}$ and $\{\vartheta^\varepsilon\}$ convergent in $C^{2,1}$. Clearly, the appropriate limits of these subsequences, $R(x_0, t)$ and $\vartheta(x, t)$, represent the solution of the Stefan problem (1.1) – (1.5), (1.7) on a small time interval $(0, T_*)$.   □

Remark. The case of unbounded surfaces $\mathcal{F}$ and S can be studied similarly. Small changes must be made in the last step, in obtaining the bounds in the W_2^m-norm, where local estimates of the solutions of parabolic equations have to be used.

Let us now consider a special one-dimensional problem. In this case, the requirements on the differential properties of the problem data are minimal, and the existence of a classical solution follows from Theorem 3, since every approximate solution coincides with the exact one.

To be determined is a pair of functions: $R(t)$ which defines the domain $\Omega_T = \{(x, t) : 0 < R(t), \ 0 < t < T\}$ and $\vartheta(x, t)$ which satisfies the nonlinear heat equation

$$a(\vartheta)\frac{\partial\vartheta}{\partial t} = \frac{\partial^2\vartheta}{\partial x^2} + f(x, t) \tag{4.21}$$

in Ω_T. On the fixed boundary $x = 0$,

$$\vartheta = \vartheta^0(t) \qquad \text{or} \qquad \frac{\partial\vartheta}{\partial x} + b(t)\vartheta = \vartheta^1(t). \tag{4.22}$$

On the free boundary $x = R(t)$,

$$\vartheta = 0, \qquad \frac{dR}{dt} = -\frac{\partial\vartheta}{\partial x}. \tag{4.23}$$

In deriving the Stefan condition (4.23) we have assumed that $\vartheta(x, t)$ is strictly positive throughout Ω_T. Finally, at the initial time instant we have

$$R(0) = x_0 > 0, \quad \vartheta(x, 0) = \vartheta_0(x) \qquad \text{for} \quad x \in (0, x_0). \tag{4.24}$$

Theorem 4. *Assume that* $a(\vartheta) \geq a_0 = \text{const.} > 0$ *,* $\vartheta_0(x) \geq 0$ *for* $x \in G = (0, x_0)$, $a \in C^1[0, \infty)$, $f \in H^{\gamma, \gamma/2}(\mathbb{R}_\infty^+)$, $\mathbb{R}_\infty^+ = \{(x, t) : 0 < x, \ t < \infty\}$, *and* $b, \vartheta^1 \in H^{(1+\gamma)/2}[0, \infty)$, $\vartheta_0 \in H^{2+\gamma}(\bar{G})$.

Further, suppose that, at the points $x = 0$ *and* $x = x_0$, *the first-order compatibility conditions hold either in the form*

$$\vartheta^0(0) = \vartheta_0(0), \quad a(\vartheta^0(0))\frac{d\vartheta^0}{dt}(0) = \frac{d^2\vartheta_0}{dx^2}(0) + f(0, 0)$$

or

$$\frac{d\vartheta_0}{dx}(0) + b(0)\vartheta_0(0) = \vartheta^1(0);$$

$$\vartheta_0(x_0) = 0, \quad \frac{d^2\vartheta_0}{dx^2}(x_0) + f(x_0, 0) = a(0)\left|\frac{d\vartheta_0}{dx}(x_0)\right|^2.$$

Furthermore, let $\dfrac{d\vartheta_0}{dx}(x_0) < 0$ *if* $f \not\equiv 0$.

Then, on a sufficiently small time interval $(0, T_*)$, *there exists a unique solution of problem (4.21)–(4.24), with*

$$\vartheta \in H^{2+\gamma,(2+\gamma)/2}(\bar{\Omega}_T), \quad R \in H^{(3+\gamma)/2}[0, T], \quad T < T_*.$$

The above interval $(0, T_*)$ *is characterized by the conditions*

$$J_0(T) \equiv |\vartheta|_{\Omega_T}^{(1+\gamma)} < \infty \text{ if } T < T_*, \quad J_0(T_*) = \infty; \quad \vartheta(x, t) > 0 \quad \text{in} \quad \Omega_{T_*}.$$

As already mentioned, each approximate solution (the solution of problem (A_ε)) in the one-dimensional case coincides with the exact solution of the Stefan problem. In defining the maximal existence interval of the solutions in Theorem 3 three factors were taken into account:

1) the finiteness of $J_0(T)$;
2) the regularity of the domain Ω_T;
3) the boundedness both from below and above of the modulus of the gradient of the solution on the free boundary.

In the case of one space variable the domain Ω_T is always regular, because the function $R(t)$ is then strictly increasing. This follows from the positiveness of the solution $\vartheta(x, t)$ in the domain Ω_T and the conditions (4.23). The boundedness both from below and above of the modulus of the gradient of the solution on the free boundary is not used in the proof of Theorem 3. This property is necessary for obtaining uniform estimates (independent of the approximation index) on the approximate solutions in a vicinity of the free boundary. Thus, the finiteness of the value $J_0(T)$ remains the only condition for the existence of a positive solution on the interval $(0, T)$.

5. The two-phase Stefan problem

All results on the solvability of the one-phase Stefan problem apply without any change to the problem in the two-phase setting. We shall consider the simplest single-front problem.

Let the closed surfaces $\mathcal{F}^-, S$ and $\mathcal{F}^+$ that bound the domains U^-, U, U^+, respectively, be given with $U^- \subset U \subset U^+$ and $\mathcal{F}^\pm \cap S = \emptyset$. To be determined are a surface Γ_T that initially coincides with S and satisfies the condition $\mathcal{F}^\pm_T \cap \Gamma_T = \emptyset$, $\mathcal{F}^\pm_T = F^\pm \times (0, T)$, as well as a continuous function $\vartheta(x, t)$ which in each of the domains $\Omega_T^\pm$ ($\Omega_T^\pm$ is bounded by the surfaces $\mathcal{F}^\pm_T, \Gamma_T$ and the planes $t = 0$ and $t = T$) satisfies the nonlinear heat equation

$$a(\vartheta)D_t\vartheta = \sum_{i=1}^{n} D_i^2\vartheta + f(x,t). \tag{5.1}$$

The function $\vartheta(x,t)$ coincides on the given surfaces $\mathcal{F}_T^\pm$ with the prescribed function $\vartheta^1(x,t)$,

$$\vartheta(x,t) = \vartheta^1(x,t) \qquad \text{for} \quad (x,t) \in \mathcal{F}_T^\pm, \tag{5.2}$$

or

$$\sum_{i=1}^{n} b_i(x,t)D_i\vartheta + b(x,t)\vartheta = \vartheta^2(x,t) \qquad \text{for} \quad (x,t) \in \mathcal{F}_T^\pm. \tag{5.3}$$

On the free surface Γ_T,

$$\vartheta(x,t) = 0 \tag{5.4}$$

and

$$D_t R(x_0,t) = X^-(x_0, D\vartheta) - X^+(x_0, D\vartheta) \qquad \text{for} \quad (x_0,t) \in S_T. \tag{5.5}$$

Here, $R(x_0,t)$ is a smooth function defined on the surface S_T, determining the surface Γ_T by

$$x = x_0 + \nu(x_0)R(x_0,t),$$

where $\nu(x_0)$ is the normal vector to the surface S at the point $x_0 \in S$;

$$X^\pm = \lim_{r\to 0} |D\vartheta(x_r^\pm, t)|^2 \left(D\vartheta(x_r^\pm, t) \cdot \nu(x_0)\right)^{-1},$$

$$x_r^\pm = x_0 + \nu(x_0)(R(x_0,t) \pm r) \in \Omega_T^\pm.$$

In deriving the Stefan condition (5.5) we have assumed that $\vartheta(x,t)$ is positive in the domain Ω_T^+ and negative in Ω_T^-. Finally, at the initial time instant

$$\vartheta(x,0) = \vartheta_0(x), \quad x \in G = \Omega(0); \qquad R(x_0,0) = 0, \quad x_0 \in S. \tag{5.6}$$

Theorem 5. *Let the surfaces $\mathcal{F}^\pm, S$ and the functions $\vartheta_0, \vartheta^1, \vartheta^2, a, f, b, b_i (i = 1,\ldots,n)$ fulfil all hypotheses of Theorem 1 (except that ϑ_0 is regular not in G but in each of the domains $G^\pm = \Omega^\pm(0)$, and $a(s)$ is regular for $s \in (-\infty, 0]$ and $s \in [0, +\infty)$). Assume that the compatibility conditions up to order $m+1$ are fulfilled on the surfaces $\mathcal{F}^\pm$ and S, the function $\vartheta_0(x)$ is strictly positive in G^+ and strictly negative in G^-, as well as*

$$\lim_{r\to 0} \big|\log |D\vartheta_0(x \pm r\nu(x))|\big| < \infty \qquad \text{for} \quad x \in S.$$

Then there exists a unique solution $R \in C^{2,1}(S_{T_*})$, $\vartheta \in C^{2,1}(\bar\Omega_{T_*}^\pm)$ of the problem (5.1), (5.2) (or (5.3)), (5.4)-(5.6), where the value $T_* > 0$ (sufficiently small) is determined only by the problem data.

Remark. The assertions of Theorem 5 remain true also for unbounded surfaces $\mathcal{F}^{\pm}$ and S. In particular, the proof of Theorem 5 extends without any changes to the case where the surfaces $\mathcal{F}^{\pm}, S$ are given by $x_n = F(x')$ and all the problem data (the function F included) are periodic with respect to x'.

The only difference in the proof of Theorem 5 compared to that of Theorem 1 consists of the way of estimating the W_2^m-norm of the solution in the vicinity of the free boundary. We shall explain it by an example.

Let $\mathcal{F}^{\pm}$ denote the planes $\{x_n = \pm 1\}$ and let S be the plane $\{x_n = 0\}$. Moreover, let all data of the problem be periodic functions of x', $\vartheta^1(x', \pm 1, t) = \pm 1$ and

$$\left| \log |D_n \vartheta_0(x)| \, \right| < \infty \qquad \text{for} \quad x \in \bar{G}^{\pm}.$$

The last condition allows the possibility of introducing the variables (2.3) on the whole domain Ω_T for sufficiently small T. Under this transformation, the fixed cylinder $G_T^{\pm} = G \times (0, T)$ corresponds to the domain $\Omega_T^{\pm}$. The surfaces $\mathcal{F}_T^{\pm}$ are transformed onto themselves and the free boundary Γ_T is mapped onto the known surface S_T. In each of the domains $G_T^{\pm}$, the new unknown variable $u(y, \tau) = x_n$ fulfils equation (2.4), is constant on the given boundaries $\mathcal{F}_T^{\pm}$ and is continuous across the surface S_T,

$$u(y', +0, \tau) = u(y', -0, \tau). \tag{5.7}$$

Equation (5.7) also means that the derivatives of the function $u(y, \tau)$ with respect to the tangential variables y' and with respect to time are continuous everywhere in G_T. The Stefan condition (5.5) leads thus to

$$-D_\tau u = \left(1 + \sum_{i=1}^{n-1} p_i^2\right) \left\{ \frac{1}{p_n(y', +0, \tau)} - \frac{1}{p_n(y', -0, \tau)} \right\}, \tag{5.8}$$

where

$$p_i = D_i u, \ i = 1, \ldots, n, \qquad p_n(y', \pm 0, \tau) = \lim_{y_n \to \pm 0} p_n(y', y_n, \tau).$$

In order to derive the estimate of the solution $u(y, \tau)$ in the $W_2^m(G_T^{\pm})$-norm, we differentiate equation (2.4) m times with respect to y' and τ (i.e., we apply the operator $D_0^{\bar{m}}$ to it), multiply by $D_0^{\bar{m}} u$, integrate over each of the domains $G_T^{\pm}$ and finally take the sum of all the obtained relations. As in Section 2 (equality (2.17)), we transform $\mathcal{J} = \mathcal{J}^+ + \mathcal{J}^-$, where

$$\mathcal{J}^{\pm} = \int_{G_T^{\pm}} D_0^{\bar{m}} u D_n \left\{ D_0^{\bar{m}} \left[\frac{1}{p_n} \left(1 + \sum_{i=1}^{n-1} p_i^2\right) \right] \right\} \, dy d\tau \ ,$$

into

$$\mathcal{J} = \int_{G_T} D_0^{\bar{m}} p_n \, D_0^{\bar{m}} \left[\frac{1}{p_n} \left(1 + \sum_{i=1}^{n-1} p_i^2\right) \right] \, dy d\tau$$

$$- \int_{S_T} D_0^{\bar{m}} u D_0^{\bar{m}} \left\{ \left(1 + \sum_{i=1}^{n-1} p_i^2 \right) \left(\frac{1}{p_n(y', +0, \tau)} - \frac{1}{p_n(y', -0, \tau)} \right) \right\} dy' d\tau$$

and we take advantage of condition (5.8) in the integral over the boundary S. After differentiating m times the first term in $\mathcal{J}$ and integrating

$$\mathcal{J}_0^{\pm} = \int_{G_T^{\pm}} D_0^{\bar{m}} u \Delta' D_0^{\bar{m}} u \, dy d\tau$$

by parts, we obtain the same equality as in (2.17). We then proceed as in the case of the one-phase problem.

As in Section 4, we shall consider the one-dimensional Stefan problem separately. To be determined are the functions $\vartheta(x, t)$, continuous on $\Omega_T = \Omega \times (0, T)$, $\Omega = \{x : |x| < 1\}$, and $R(t)$, defining the domains

$$\Omega_T^{\pm} = \{(x, t) \in \Omega_T : (\pm 1)(x - R(t)) > 0\} .$$

These functions satisfy the quasilinear parabolic equation

$$a(\vartheta) \frac{\partial \vartheta}{\partial t} = \frac{\partial^2 \vartheta}{\partial x^2} + f(x, t) \qquad \text{for} \quad (x, t) \in \Omega_T^{\pm} \tag{5.9}$$

and the conditions

$$\vartheta(\pm 1, t) = \vartheta^{\pm}(t) \text{ or } \frac{\partial \vartheta}{\partial x} + b^{\pm} \vartheta = \vartheta_0^{\pm}(t) \qquad \text{for} \quad x = \pm 1, \tag{5.10}$$

$$R(0) = x_0 \in \Omega, \ \vartheta(x, 0) = \vartheta_0(x) \qquad \text{for} \quad x \in \Omega, \tag{5.11}$$

$$\vartheta(R(t), t) = 0, \qquad\qquad t \in (0, T), \tag{5.12}$$

$$\frac{dR(t)}{dt} = \frac{\partial \vartheta}{\partial x}(R(t) + 0, t) - \frac{\partial \vartheta}{\partial x}(R(t) - 0, t), \quad t \in (0, T). \tag{5.13}$$

Theorem 6. *Let $a \geq a_0 = \text{const.} > 0$, $\vartheta_0(x) < 0$ for $x \in \Omega^-(0)$, $\vartheta_0(x) > 0$ for $x \in \Omega^+(0)$; $a \in C^1(-\infty, 0] \cap C^1[0, +\infty)$, $f \in H^{\gamma, \gamma/2}(\bar{\Omega}_\infty)$; $b^{\pm}, \vartheta_0^{\pm} \in H^{(1+\gamma)/2}[0, +\infty)$, $\vartheta^{\pm} \in H^{(2+\gamma)/2}[0, +\infty), \vartheta_0 \in H^{2+\gamma}(\Omega_{\pm}(0))$.*

Suppose that at the points $x = \pm 1$ and $x = x_0$, compatibility conditions of the first order are satisfied either in the form

$$\vartheta^{\pm}(0) = \vartheta_0(\pm 1), \qquad \vartheta_0(x_0 \pm 0) = 0,$$

$$a(\vartheta_0(\pm 1)) \frac{d\vartheta^{\pm}}{dt}(0) = \frac{d^2\vartheta_0}{dx^2}(\pm 1) + f(\pm 1, 0)$$

or

$$\frac{d\vartheta_0}{dx}(\pm 1) + b^{\pm}(0)\vartheta_0(\pm 1) = \vartheta_0^{\pm}(0),$$

$$\frac{d^2\vartheta_0}{dx^2}(x_0 \pm 0) + f(x_0, 0) = -a(0) \frac{d\vartheta_0}{dx}(x_0 \pm 0) \frac{dR}{dt}(0),$$

$$\frac{dR}{dt}(0) = \frac{d\vartheta_0}{dx}(x_0 + 0) - \frac{d\vartheta_0}{dx}(x_0 - 0).$$

Moreover, let $\dfrac{d\vartheta_0}{dx}(x_0 \pm 0) \neq 0$ *for* $f \neq 0$. *Then there exists a unique solution of the problem* (5.9)-(5.13) *on a sufficiently small time interval* $(0, T_*)$, *with*

$$\vartheta \in H^{2+\gamma,(2+\gamma)/2}(\bar{\Omega}_T^\pm), \quad R \in H^{(3+\gamma)/2}[0, T], \quad T < T_*.$$

Here T^* *is such that, for some* $\beta \in (0, 1)$,

$$J(T) \equiv |\vartheta|_{\Omega_T^-}^{(1+\beta)} + |\vartheta|_{\Omega_T^+}^{(1+\beta)} + \left|\log\left|1 - R^2(T)\right|\right| < \infty$$

for $T < T_*$, *and*

$$\lim_{T \to T_*} J(T) = \infty \quad \text{and} \quad (\pm 1)\vartheta(x, t) > 0 \text{ in } \Omega_T^\pm.$$

Chapter III

Existence of the classical solution to the multidimensional Stefan problem on an arbitrary time interval

---

1. The one-phase Stefan problem

Let us consider the reference problem introduced in Section II.2. We shall assume that the known surface $\mathcal{F}$ is the plane $\{x_n = 1\}$, the initial free boundary S coincides with the plane $\{x_n = 0\}$, the temperature prescribed on $\mathcal{F}$ is everywhere constant, and the initial temperature $\vartheta_0(x)$ is a periodic function of $x' = (x_1, \ldots, x_{n-1})$ with period 1. If

$$\Gamma_T = \{(x,t) \: : \: |x'| < \infty, \; x_n = R(x',t), \; t \in (0,T)\}$$

is an unknown surface to be determined, then

$$\Omega_T = \{(x,t) \: : \: R(x',t) < x_n < 1, \; |x'| < \infty, \; t \in (0,T)\}$$

and

$$D_t\theta = \sum_{i=1}^{n} D_i^2\theta \qquad \text{for} \quad (x,t) \in \Omega_T \, , \tag{1.1}$$

$$\theta(x,t) = 1 \qquad \text{for} \quad (x,t) \in \mathcal{F}_T, \tag{1.2}$$

$$\theta(x,t) = 0 \qquad \text{for} \quad (x,t) \in \Gamma_T, \tag{1.3}$$

$$D_t\theta = |D\theta|^2 \qquad \text{for} \quad (x,t) \in \Gamma_T, \tag{1.4}$$

$$\theta(x,0) = \theta_0(x) \qquad \text{for} \quad x \in \Omega(0) = G \, . \tag{1.5}$$

We have already shown that the unique classical solution

$$R \in C^{2,1}(\mathbb{R}_{T_*}^{n-1}), \quad \vartheta \in C^{2,1}(\overline{\Omega}_{T_*})$$

of problem (1.1)–(1.5) exists on a sufficiently small time interval $(0, T_*)$.

How far can this solution be extended in time? Let us recall that in order to construct a solution that is regular on the whole interval $(0, T_*,)$ as in Theorem 2 of Chapter II, the initial function $\vartheta_0(x)$ has to be in $H^{2\ell}(\overline{G})$, where $[\ell] = m + 1 = n + 5$. In other words, the solution $\vartheta(x,t)$ can be extended onto a larger time interval, provided that it belongs to $H^{2\ell}(\overline{\Omega(T_*)})$. It follows from the results of Chapter II that such regularity of the solution cannot be expected. Moreover, we have to remember inequality (2.2), Chapter II, which ensures the possibility of introducing von Mises variables.

We can thus formulate the following statement: the existence interval $(0, T_\infty)$ of the classical solution to problem (1.1)–(1.5) is characterized by the relations

$$
\begin{cases}
J_0(T) = \left| \log |D_n \vartheta| \right|_{\Omega_T}^{(0)} < \infty, \\[2mm]
J_1(T) = |\vartheta|_{\Omega(T)}^{(2l)} < \infty, \quad T < T_\infty, \\[2mm]
\lim\limits_{T \to T_\infty} \{J_0(T) + J_1(T)\} = \infty.
\end{cases}
\tag{1.6}
$$

It turns out that, under some restrictions on the problem data (more precisely, under hypotheses ensuring that the first of the inequalities (1.6) is satisfied on an infinite time interval), the classical solution of the reference Stefan problem exists for all positive times.

Crucial here are the Duvaut transformation, a result due to Caffarelli on the regularity of the free boundary in the problem

$$
\sum_{i=1}^{n} D_i^2 v = f, \quad x \in C; \qquad v|_{\partial C} = |Dv|\,|_{\partial C} = 0
\tag{1.7}
$$

and an equivalent formulation of the Stefan problem in Legendre variables, where a plane corresponds to the free boundary, and the new unknown function satisfies a uniformly parabolic equation and vanishes on that plane.

To simplify the exposition, we confine ourselves to the case of two space variables, since the proof of Caffarelli's theorem is then rather simple (the proof we give below is originally due to A. Friedman).

We split the considerations into several stages. First, in the von Mises variables we shall estimate the first derivatives of the solution in time and show the boundedness of $J_0(T)$ in (1.6) for all bounded T.

Further, by using the Duvaut transformation

$$
v(x, t) = \int_{g(x)}^{t} \vartheta(x, \tau) \, d\tau ,
\tag{1.8}
$$

where $g(x) = t$ is the equation of the free surface $\Gamma(t)$, the original problem (1.1)–(1.5) transforms to the equivalent setting (1.7) in terms of the function v to which Caffarelli's results apply.

And finally, for the Stefan problem formulated in the von Mises variables we shall show that its solution is infinitely differentiable for all positive t.

Lemma 1. *Let the hypotheses of Theorem 2, Chapter II hold, let $T < T_\infty$ and*

$$
\Delta \vartheta_0(x) \geq 0 \quad \text{for} \quad x \in G,
$$

$$
\Delta \vartheta_0(x) > 0 \quad \text{for} \quad x \in S.
$$

Then

$$
\left| D_t \vartheta, D\vartheta, \log |D_2 \vartheta| \right|_{\Omega_T}^{(0)} \leq M_1 ,
\tag{1.9}
$$

where the constant M_1 depends on the value

$$\kappa = |D\vartheta_0|_G^{(0)} + \left|\log|D_2\vartheta_0|\right|_G^{(0)} + |\Delta\vartheta_0|_G^{(0)} + \left\{\min_{x_2\in(0,2\delta)}|\Delta\vartheta_0(x)|\right\}^{-1}$$

(where $\delta > 0$ is a sufficiently small constant) and is bounded for all bounded $T > 0$, $\delta^{-1} > 0$, $\kappa > 0$.

Proof. We shall prove an inequality analogous to (1.9), but expressed in the von Mises variables (y, τ),

$$\tau = t, \quad y_1 = x_1, \quad y_2 = \vartheta(x, t),$$

where the reference Stefan problem (1.1)–(1.5) reduces to the following equivalent initial-boundary value problem: determine a solution $u(y, \tau) = x_2$ of the equation

$$D_\tau u = D_1^2 u - D_2\left\{\frac{1}{p_2}(1 + p_1^2)\right\} \quad \text{in} \quad G_T = G \times (0, T), \tag{1.10}$$

prescribed on the surface $\mathcal{F}_T$ by

$$u = 1 \quad \text{for} \quad (y, \tau) \in \mathcal{F}_T \tag{1.11}$$

and satisfying the boundary condition

$$D_\tau u + \frac{1}{p_2}(1 + p_1^2) = 0 \tag{1.12}$$

on the image $S_T = S \times (0, T)$ of the free boundary Γ_T under the transformation $(x, t) \rightarrow (y, \tau)$.

At the initial time instant,

$$u(y, 0) = u_0(y) \quad \text{for} \quad y \in G, \tag{1.13}$$

with the function $u_0(y)$ defined by the identity $y_2 = \vartheta_0(y_1, u_0(y))$.

We have taken $p_i = D_i u$, $i = 1, 2$, in (1.10) and (1.12). The desired estimate has the form

$$|p_1|_{G_T}^{(0)} + |D_\tau u|_{G_T}^{(0)} + \left|\log|p_2|\right|_{G_T}^{(0)} \le M_2$$

and in view of the relations

$$D_2\vartheta = \frac{1}{p_2}, \quad D_1\vartheta = -\frac{p_1}{p_2}, \quad D_t\vartheta = -\frac{1}{p_2}D_\tau u$$

is equivalent to the bound (1.9).

Equation (1.10) and boundary conditions (1.11), (1.12) can be differentiated in time and in y_1. The derivatives $D_1 u = p_1$ and $D_\tau u = r$ satisfy the initial-boundary value problem that consists of determining a solution $w(y, \tau)$ of the equation

$$D_\tau w = D_1^2 w + D_2\{bD_1 w + cD_2 w\} \quad \text{in} \quad G, \tag{1.14}$$

such that

$$w = 0 \quad \text{for} \quad (y, \tau) \in \mathcal{F}_T, \tag{1.15}$$

$$D_\tau \omega = b D_1 \omega + c D_2 \omega \qquad \text{for} \quad (y, \tau) \in S_T, \tag{1.16}$$

$$\omega(y, 0) = \omega_0(y) \qquad \text{for} \quad y \in G, \tag{1.17}$$

where $\omega_0 = D_1 u_0$ if $\omega = p_1$, and $\omega_0 = D_\tau u(y, 0) \equiv r_0(y)$, if $\omega = r$.
In (1.14) and (1.16) we have taken

$$b = -\frac{2 p_1}{p_2}, \qquad c = \frac{1}{p_2^2}(1 + p_1^2).$$

The maximal and minimal values of the function ω cannot be achieved in the interior of G_T (maximum principle). They also cannot be achieved on the boundary S_T, because, in particular, at the maximum point $D_2 \omega < 0$ (Hopf-Zaremba-Giraud principle, [92]), $D_\tau \omega \geq 0$ and $D_1 \omega = 0$, contradicting (1.16).

Since, by the hypotheses of the lemma, $D_\tau u$ is non-positive at the initial time, it remains non-positive everywhere in G_T and

$$|D_1 u, D_\tau u|_{G_T}^{(0)} \leq M_3, \qquad M_3 = \max |D_1 u_0, D_\tau u(0)|_G^{(0)}. \tag{1.18}$$

Equation (1.14) is also fulfilled by the function $p_2 = D_2 u$ which, consequently, achieves its minimal and maximal values only on the boundaries $\mathcal{F}_T, S_T$ or at the initial time instant. But equation (1.10) holds on the boundary $\mathcal{F}_T$, hence, according to the boundary condition (1.11),

$$D_2 p_2 = 0 \qquad \text{for} \quad (y, \tau) \in \mathcal{F}_T.$$

By the last equality, function p_2 cannot achieve its extremal values on the boundary $\mathcal{F}_T$ (Hopf-Zaremba-Giraud principle), i.e. the maximal and minimal values of p_2 can be achieved either on the boundary S_T or at the initial time instant. Since the bound (2.1) of Chapter II holds at the initial time, we only need to estimate p_2 on the boundary S_T.

Using boundary condition (1.12), we can express p_2 in terms of p_1 and r as $p_2 = -(1 + p_1^2)/r$.

The function $r(y, \tau)$ is non-positive and bounded in its absolute value from above, hence

$$p_2(y, \tau) \geq M_4 > 0 \qquad \text{for} \quad (y, \tau) \in G_T . \tag{1.19}$$

In order to estimate p_2 from above on the boundary S_T, it is necessary to get such an estimate for the non-positive function r. This does not follow directly from the maximum principle, because r vanishes on $\mathcal{F}_T$. A standard procedure in such a case consists of constructing a barrier φ which estimates the function $r(y, \tau)$ on G_T and does not vanish on S_T.

We are thus looking for a non-positive function $\varphi(y, \tau)$ such that the difference $q = r - \varphi$ is non-positive on $\mathcal{F}_T$ and at the initial time instant, and in addition does not achieve its maximum in the interior of the domain G_T or on the boundary S_T. The latter property can be ensured in the simplest way by postulating the inequalities

$$L q \leq 0 \qquad \text{for} \quad (y, \tau) \in G_T,$$

$$L_0 q \leq 0 \qquad \text{for} \quad (y, \tau) \in S_T,$$

where

$$Lq \equiv D_\tau q - D_1^2 q - bD_1 D_2 q - cD_2^2 q - D_2 bD_1 q - D_2 cD_2 q,$$

$$L_0 q \equiv D_\tau q - bD_1 q - cD_2 q.$$

There is a little difficulty to be overcome on this path. Whatever the function φ (non-constant), the expression for $Lq = -L\varphi$ will include derivatives of the functions b and c as coefficients of the terms $D_i\varphi$. In turn, these derivatives include second derivatives of the function u which are still to be estimated in the further stages. What can be done?

The simplest way is to apply local parabolic estimates on a domain Q "far" from the image S_T of the free boundary Γ_T under the mapping $(x,t) \to (y,\tau)$. More precisely, Theorem I.6 can be applied on the image of Q under the mapping $(x,t) \to (y,\tau)$, yielding bounds of the second derivatives of ϑ by the maximum of the C-norm of the solution, the norm of the initial function ϑ_0 and the distance from the boundary of Γ_T. Outside Q, φ is to be taken as a given function of time.

So let Q be a domain whose distance from the boundary S_T is larger than a given $\delta > 0$. The functions b, c and their derivatives will be bounded in the domain Q, if the distance of the image of Q in the domain Ω_T of variables (x,t) under the inverse transformation $(y,\tau) \to (x,t)$ from the boundary Γ_T be $\nu > 0$. Indeed, if this is so, then the second derivatives of ϑ will be bounded in Q by a constant dependent only on $\nu, |\vartheta|_{\Omega_T}^{(0)}$ and the corresponding norm of the initial function ϑ_0 (see Theorem 6 of Chapter I). Expressing the second derivatives of u in terms of the derivatives of ϑ and taking advantage of the inequalities (1.18) and (1.19), we obtain the required bounds on the maximal absolute value of the derivatives of the functions b and c.

To estimate the distance ν from Γ_T to the image of the domain Q under the transformation $(y,\tau) \to (x,t)$, let us consider the level set $\Pi = \{(x,t) \in \Omega_T : \vartheta(x,t) = \delta\}$. The distance ν coincides with the distance between the surface Π and the zero level set Γ_T of the function ϑ. The latter value is clearly proportional (with the proportionality coefficient dependent only on the constant M_3 in inequality (1.18)) to the value

$$\nu_0 = \min_{\substack{x_1 \in (0,1) \\ t \in (0,T)}} |x_2' - x_2''|,$$

where x_1', x_2'' are the coordinates of the intersection of the line $\{(x,t) \in \Omega_T : x_1 = \text{const.}, t = \text{const.}\}$ with the surfaces Π and Γ_T, respectively.

In turn,

$$\delta = \left| \vartheta|_{(x_1,x_2',t)\in\Pi} - \vartheta|_{(x_1,x_2'',t)\in\Gamma_T} \right|$$

$$\leq |D_2\vartheta| \, |x_2' - x_2''| \leq \frac{1}{|p_2|} |x_2' - x_2''| \leq M_4^{-1} |x_2' - x_2''| .$$

Therefore, $\nu_0 \geq \delta M_4$, as desired.

Let us now take

$$\varphi = \varphi(y_2, \tau) = \begin{cases} -\varepsilon \exp(-\alpha\tau) & \text{if } 0 \le y_2 < \delta \\ -\varepsilon \exp(-\alpha\tau)P(y_2) & \text{if } \delta \le y_2 < 2\delta \\ 0 & \text{if } 2\delta \le y_2 \le 1 \end{cases}$$

where α and ε are positive constants to be specified, and $P(y_2)$ is a 5th-order polynomial. To arrange that the function φ has continuous second derivatives with respect to y_2, it is sufficient to postulate that $P(\delta) = 1$, $P'(\delta) = P''(\delta) = 0$, $P(2\delta) = P'(2\delta) = P''(2\delta) = 0$. It is easy to show that that the polynomial

$$P(y_2) \equiv P_0(\xi) = 6\xi^5 - 15\xi^4 + 10\xi^3, \qquad \xi = (2\delta - y_2)/\delta$$

fulfils all the above requirements and, moreover, $P(y_2)$ is monotone decreasing in y_2 on the interval $(\delta, 2\delta)$.

By the hypothesis of the lemma, there exists a sufficiently small number $\varepsilon > 0$ such that

$$D_\tau u(y, 0) = r_0(y) < \varphi(y_2, 0).$$

Hence, it is enough to show that everywhere in G_T

$$L_q \equiv L(r - \varphi) = -L\varphi = -D_\tau\varphi + cD_2^2\varphi + D_2cD_2\varphi \le 0,$$

and on the boundary S_T

$$L_{0q} \equiv L_0(r - \varphi) = -D_\tau\varphi + bD_1\varphi + cD_2\varphi < 0.$$

The latter inequality is obvious since at the values y_2 close to zero, $D_1\varphi = D_2\varphi = 0$ and $-D_\tau\varphi = -\varepsilon\alpha\exp(-\alpha\tau) < 0$.

It is also obvious that $L\varphi$ remains non-negative in the domain G_T everywhere outside the strip $\{(y, \tau) \in G_T : \delta < y_2 < 2\delta\}$. In the remaining points of Q we have

$$-L\varphi = -\varepsilon\exp(-\alpha\tau)\{\alpha P(y_2) + cP''(y_2) + D_2cP'(y_2)\} \equiv -\varepsilon\exp(-\alpha\tau)P_1(y, \tau).$$

The first term in the expression $P_1(y, \tau)$ is positive everywhere except on the line $\{y_2 = 2\delta\}$. In turn, at the values y_2 close to 2δ the leading positive term in P_1 is its second term,

$$cP''(y_2) + D_2cP'(y_2) = \frac{60}{\delta}\frac{(2\delta - y_2)}{p_2^2}\left\{(1 + p_1^2) + P_2(y, \tau)\right\},$$

where $P_2 \to 0$ as $y_2 \to 2\delta$. Here we have used the explicit form of the function c and its derivative,

$$D_2c = \frac{1}{p_2^2}\left\{2p_1 D_2p_1 - 2\frac{1 + p_1^2}{p_2}D_2p_2\right\}$$

and the boundedness of the value $(p_2)^{-1}$ (estimate (1.19)).

Hence, by taking sufficiently large α it is easy to provide positivity of the function P_1 which completes the proof of the lemma. $\qquad\qquad\qquad\qquad\qquad\qquad\qquad\qquad\square$

The bound (1.9) implies that the function $R(x_1, t)$ defining the surface Γ_T is Lipschitz continuous in both variables, with Lipschitz constant independent of the interval $(0, T)$. More precisely, this constant is bounded as long as T remains finite. Moreover, the function $R(x_1, t)$ is decreasing in time (the derivative $D_\tau u$ is strictly negative on S_T). The latter feature ensures that the Duvaut transformation (1.8) is well-defined in the whole domain Ω_T. In the following it is enough to restrict the considerations to a subdomain C_T of Ω_T, containing all points $(x, t) \in \Omega_T$ such that $x_2 < 0$ and $t \in (T/2, T)$.

Let $g(x) = t$ be the equation describing the surface $\Gamma(t)$. By differentiating the identity

$$\vartheta(x_1, x_2, g(x_1, x_2)) = 0$$

with respect to x_1 and x_2, and using the Stefan condition (1.4), we get

$$\begin{aligned} D_i \vartheta &= -D_t \vartheta D_i g & \text{for} \quad (x, t) \in \Gamma_T, \quad i = 1, 2, \\ D_i g &= -D_i \vartheta |D\vartheta|^{-2} & \text{for} \quad (x, t) \in \Gamma_T, \quad i = 1, 2. \end{aligned}$$

By straightforward computation it follows that everywhere in the domain C_T

$$-D_t v + \sum_{i=1}^{n} D_i^2 v = 1, \tag{1.20}$$

and on the boundary Γ_T

$$v = D_1 v = D_2 v = 0. \tag{1.21}$$

Our next objective is to show that Γ_T is a Lyapunov surface (i.e., the first derivatives of $R(x_1, t)$ satisfy the Hölder condition). For this purpose we shall use the following result.

Theorem 1. (L. Caffarelli & A. Friedman [158], p. 141). *Let the function $v(x)$ of variables (x_1, x_2) defined on a domain $C \subset \mathbb{R}^2$ have second derivatives bounded by a constant N. Assume that $v(x)$ satisfies the boundary value problem (1.7), where the boundary Γ is the curve $\{x_2 = R(x_1) : |x_1| < \infty\}$, with a function R such that*

$$|R(s') - R(s'')| \le N|s' - s''|.$$

Suppose, moreover, that $f > 0$ in C, and let the first derivatives of the function f be bounded by the same constant N. Then the function $R(s)$ is continuously differentiable and

$$\left| \frac{dR}{ds}(s') - \frac{dR}{ds}(s'') \right| \le N_0 |s' - s''|^\alpha,$$

with constants $N_0 > 0$ and $\alpha \in (0, 1]$, dependent only on N.

We shall give a proof of this theorem at the end of this section.

Let us consider the solution $v(x, t)$ of the boundary value problem (1.20), (1.21) on the intersections $C(t)$ of the domain C_T with the planes $\{t = \text{const.}\}$, where $t \in (\frac{T}{2}, T)$.

This solution satisfies the boundary value problem (1.7) with

$$f = 1 + D_t v = 1 + \vartheta \geq 1.$$

By estimate (1.9), the first derivatives of f with respect to the space variables are uniformly bounded for all t (by a constant finite for all finite T). Thus, it is sufficient for providing the applicability of Theorem 1 that the second derivatives of v are bounded:

$$|D_i D_j v|_{C_T}^{(0)} \leq M_5, \quad i, j = 1, 2.$$

This follows from the representation

$$D_i D_j v = \frac{D_i \vartheta D_j \vartheta}{|D\vartheta|^2}, \quad (x, t) \in \Gamma_t, \tag{1.22}$$

the boundedness of these derivatives on the sets $\{t = T/2\}$ and $\{x_2 = 0\}$ (due to the local estimates on the solutions of linear parabolic equations), and the maximum principle.

The function $R(x_1, t)$ is thus continuously differentiable with respect to x_1, and its derivative $D_1 R$ fulfils the Hölder condition

$$|D_1 R(x_1', t) - D_1 R(x_1'', t)| \leq M_6 |x_1' - x_1''|^\alpha,$$

with constants M_6 and α dependent only on the constant M_1 as in (1.9).

Let us once more return to the boundary value problem (1.7) and consider the section $C(t)$ of the domain C_T. Let

$$\zeta = \psi(z), \quad \zeta = \zeta_1 + i\zeta_2, \quad z = x_1 + ix_2$$

be the conformal mapping (with time t as a parameter) of the strip $C(t)$ onto the strip $Q = \{\zeta : |\text{Re}\, \zeta| < \infty, \ 0 < \text{Im}\, \zeta < 1\}$ such that the boundary $\Gamma(t)$ is mapped onto the line $\{\text{Im}\, \zeta = 0\}$.

It follows from the general theory of conformal mappings (cf., [71]) that the derivative $d\psi(z)/dz$ is Hölder continuous with index α in the closure of $C(t)$. This lifts the regularity of the function v. More precisely, let $\tilde{v}(\zeta)$ be the same function v, but expressed in terms of ζ. The function $\tilde{f}$ is defined in an analogous way. The function $\tilde{v}$ in the new variables is the solution of the Poisson equation in Q, with the right-hand side

$$\tilde{F} = \tilde{f} \left| \frac{d\psi}{dz} \right|^2$$

which belongs to the space $H^\alpha(\overline{Q})$. Since $\tilde{v}$ vanishes on the boundary $\{\text{Im}\, \zeta = 0\}$ of Q, we have $\tilde{v} \in H^{2+\alpha}(\overline{Q})$. Therefore, for $i = 1, 2$,

$$|D_i D_j v(x', t) - D_i D_j v(x'', t)| \leq M_7 |x' - x''|^\alpha$$

uniformly for all t in the interval $(T/2, T)$.

Notice that the first-order derivatives $D_i v$, $i = 1, 2$, satisfy a Lipschitz condition with respect to t (estimate (1.9)):

$$|D_t D_i v|_{C_T}^{(0)} = |D_i \vartheta|_{C_T}^{(0)} \leq M_1.$$

The hypotheses of Lemma 6, Chapter I, are now satisfied, i.e., the second-order deriva-tives D_iD_jv of the function v are Hölder continuous also in t. Maintaining the former notation of the Hölder index α, we get

$$|v|^{(2+\alpha)}_{C_T} \leq M_8(M_1). \tag{1.23}$$

We now show that the function v is infinitely differentiable. It is actually enough to prove the boundedness of the norm

$$|v|^{(2\ell+1)}_{C_T} \leq M_9$$

for all $T < T_\infty$. The latter inequality contradicts the definition of interval $(0, T_\infty)$, if T_∞ is finite.

By the estimate (1.9), equalities (1.22) on the boundary Γ_T and the inequality (1.23), it follows that there exists a sufficiently small positive number δ such that the bound

$$\left|\log|D_2^2v|\right|^{(0)}_\Pi \leq M_{10}. \tag{1.24}$$

holds in the strip Π of width δ, contained in C_T and adjacent to the boundary Γ_T. In the strip Π, we apply the Legendre transformation

$$\tau = t, \quad \xi_1 = x_1, \quad \xi_2 = -D_2v. \tag{1.25}$$

Let Π^0 be the image of Π under the transformation $(x, t) \to (\xi, \tau)$. The function

$$w(\xi, \tau) = \xi_2 x_2 + v(x, t)$$

satisfies the equation

$$D_\tau w = D_1^2 w - (1 + |D_1 D_2 w|^2)(D_2^2 w)^{-1} - 1 \tag{1.26}$$

in the domain Π^0 and vanishes on the image $\Gamma^0 = \{\xi : \xi_2 = 0\}$ of the free boundary Γ_T.

In fact, by differentiating the identity

$$w(\xi_1, \xi_2, \tau) = \xi_2 x_2(\xi, \tau) + v(\xi_1, x(\xi, \tau), \tau)$$

with respect to ξ and τ, we obtain the equalities

$$D_2 w = x_2, \quad D_1 w = D_1 v, \quad D_\tau w = D_t v. \tag{1.27}$$

By differentiating the second equation of (1.27) with respect to x_1 and the first equation with respect to x_1 and x_2, we get

$$D_1^2 v = D_1^2 w + D_1 D_2 w D_1 \xi_2,$$
$$0 = D_1 D_2 w + D_2^2 w D_1 \xi_2,$$
$$1 = D_1^2 w D_2 \xi_2.$$

Finally we differentiate the last equation of (1.25) with respect to x_2 :

$$D_2^2 v = -D_2 \xi_2 = -(D_2^2 w)^{-1}. \tag{1.28}$$

Equality (1.28) ensures that the Jacobian of the mapping $(x, t) \to (\xi, \tau)$ does not vanish and, thus, the mapping (1.25) is a diffeomorphism of the domain Π onto Π^0.

The same equality together with estimate (1.24) provides parabolicity of equation
(1.26). More precisely, uniform parabolicity on Π^0 will hold for the equation in terms
of the derivatives q ($q = D_1 w$ or $q = D_\tau w$), which has been obtained from equation
(1.26) by differentiation with respect to ξ_1 or τ :

$$D_\tau q = D_1^2 q + b D_1 D_2 q + c D_2^2 q, \tag{1.29}$$

where $b = -2 D_1 D_2 w (D_2^2 w)^{-2}$, $c = (1 + |D_1 D_2 w|^2)(D_2^2 w)^{-2}$.

On the boundary Γ^0, being the image of the free boundary Γ_T, the function q is
identically equal to zero. Since the coefficients of equation (1.29) are Hölder continuous with index α, by Theorem 6 of Chapter I, the function q belongs to the space
$H^{2+\alpha,(2+\alpha)/2}(\overline{\Pi}^0)$. We do not use anywhere the infinite differentiability of the solution
w (or v) in every closed domain with positive distance from the boundary Γ^0 (the
boundary Γ_T) and the initial plane $\{\tau = 0\}$ (local bounds for solutions of the linear
parabolic equations). In the case of equation (1.26) we can conclude that the derivative
$D_2^2 w$ belongs to the space $H^{1+\alpha,(1+\alpha)/2}(\overline{\Pi}^0)$. Therefore, the coefficients of equation
(1.26) are of one order more regular than postulated. Hence also the function q is more
regular. By an infinite extension of the above process we can prove that the function q
as well as v is infinitely differentiable. This proves the following result.

Theorem 2. *Let the hypotheses of Lemma 1 in this chapter hold. Then the classical solution of the Stefan problem (1.1)–(1.5) is infinitely differentiable for $t > 0$ and extendable
onto an arbitrary finite time interval $(0, T)$.*

Remark. The conditions on the differential properties of the initial function $\vartheta_0(x)$ and
on the order of compatibility conditions can be relaxed. Actually, what one needs is that
inequality (2.1) is fulfilled under the hypotheses of Theorem 2, Chapter II, and that there
hold the inequalities in the hypotheses of Lemma 1 of this chapter. We shall use such a
scheme for studying the one-dimensional Stefan problem.

Proof of Theorem 1. All the following considerations have a local character and, as a
matter of fact, are reduced to a small neighbourhood of the boundary Γ with center in
any fixed point z_0 on Γ.

Without losing the generality of the construction, the function f can be considered
prescribed for all values of x and t (in our case, at $f = \vartheta + 1$, it can be extended by 1
across the boundary Γ_T).

Let $A(z)$ be a bounded solution of the Poisson equation $\Delta A = f$ on the unit ball
$B_1(z_0) = \{z : |z - z_0| < 1\}$ with the center $z_0 \in \Gamma$. According to [138, p. 145], for
any $\gamma \in (0, 1)$,

$$A \in H^{2+\gamma}(\overline{B_1(z_0)}).$$

Let us consider the complex function

$$a(z) = \frac{\partial A}{\partial z} \equiv (1/2)(D_1 A - i D_2 A).$$

As can be easily verified, $a(z)$ satisfies the equation

$$\frac{\partial a}{\partial z} \equiv (1/2)\,(D_1 a + i D_2 a) \;=\; f\,.$$

The function

$$w^*(z) \;=\; a(z) - a(z_0) - \frac{\partial a}{\partial z}(z_0)(z - z_0)$$

satisfies the same equation and, moreover, vanishes together with its derivative with respect to z at the point $z = z_0$:

$$\frac{\partial w^*}{\partial \bar z} = f, \quad w^*(z_0) = \frac{\partial w^*}{\partial z}(z_0) \;=\; 0. \tag{1.30}$$

Let $\varepsilon > 0$ be arbitrary. Then there exists $\delta > 0$ such that

$$\left| \frac{\partial w^*}{\partial z} \right| \left| \frac{\partial w^*}{\partial \bar z} \right|^{-1} \;<\; \varepsilon \;<\; 1\,. \tag{1.31}$$

in the ball $B_\delta(z_0)$ with center at z_0 and radius δ. This follows by the hypothesis $f \neq 0$ of the theorem and due to the smallness of the derivative $\dfrac{\partial w^*}{\partial z}(z)$, if the point z is close to z_0.

As a consequence of (1.31), the mapping $\xi = w^*(z) \; : \; z \to \xi$ of the ball $B_\delta(z_0)$ onto a neighbourhood of the point $\xi = 0$ is a diffeomorphism, with Jacobian

$$\mathcal{J} = \frac{1}{2}\left(\left| \frac{\partial w^*}{\partial z} \right|^2 - \left| \frac{\partial w^*}{\partial \bar z} \right|^2 \right)$$

non-vanishing on the ball $B_\delta(z_0)$.

The inverse mapping $z = \eta(\xi)$ of the neighbourhood of $\xi = 0$ onto the ball $B_\delta(z_0)$ is in the class $H^{1+\gamma}$.

Let v be the solution of problem (1.7). We shall take

$$w(z) \;=\; \frac{1}{2}\,(D_1 v - i D_2 v) \;\equiv\; \frac{\partial v}{\partial z}\,.$$

Then $\dfrac{\partial w}{\partial \bar z} = f$.

The function $h(z) = w^*(z) - w(z)$ is analytic in the domain $C \cap B_\delta(z_0)$, continuous in its closure and coincides with $w^*(z)$ on the boundary Γ :

$$h(z) = w^*(z), \qquad \text{for} \quad z \in \Gamma. \tag{1.32}$$

Under the hypotheses of Theorem 1 there exists an analytic function $\zeta = \psi(z)$ which conformally maps the domain C onto the strip $Q = \{\zeta \; : \; |\text{Re } \zeta| < \infty,\; 0 < \text{Im } \zeta < 1\}$, with the boundary Γ transformed onto the line $\{\zeta \; : \; \text{Im } \zeta = 0\}$. By $z = \varphi(\zeta)$ we shall denote the inverse mapping of the strip Q onto the domain C.

Let $B_\lambda(\zeta_0)$ be the ball in the plane ζ, with center at $\zeta_0 = \psi(z_0)$ and radius λ. We define in $B_\lambda(\zeta_0)$,

$$
\Psi(\zeta) = \begin{cases} \varphi(\zeta) & \text{if} \quad \text{Im } \zeta \geq 0 \\ \eta\big(h(\varphi(\bar{\zeta}))\big) & \text{if} \quad \text{Im}\zeta < 0 \end{cases}
$$

where λ has been chosen small enough to ensure that the image of the set $\{\zeta \in B_\lambda(\zeta_0) : \text{Im } \zeta \geq 0\}$ under the mapping $\varphi : \zeta \to 0$ is contained in the ball $B_\delta(z_0)$.

The mapping $\varphi(\zeta)$ can always be defined in such a way that each bounded subset Q' of Q is transformed onto a bounded subset C' of C. In turn, the measure of the set C' is represented by an integral over C' which by a transformation of variables becomes an integral of the square of the derivative of the function φ over Q'. Correspondingly, the function $\varphi(\zeta)$ has locally square-integrable derivatives. Since, by hypothesis, $h(z)$ is Lipschitz continuous, and $\eta(\zeta)$ is in the class $H^{1+\gamma}$, the derivatives of the function $\eta(h(\varphi(\bar{\zeta})))$ are also square-integrable in the part of $B_\lambda(\zeta_0)$ which does not intersect Q.

On the boundary $\{\zeta : \text{Im } \zeta = 0\}$, we have $\varphi(\zeta) = \varphi(\bar{\zeta})$ and $\eta(h(z)) = \eta(\omega^*(z)) = z$. The function $\Psi(\zeta)$ is thus continuous across the line $\{\text{Im } \zeta = 0\}$ and therefore belongs to the space $W_2^1(B_\lambda(\zeta_0))$.

A direct calculation shows that at $\text{Im } \zeta < 0$,

$$
\left|\frac{\partial\Psi}{\partial\bar{\zeta}}\right| \left|\frac{\partial\Psi}{\partial\zeta}\right|^{-1} = \left|\frac{\partial\omega^*}{\partial z}\right| \left|\frac{\partial\omega^*}{\partial\bar{z}}\right|^{-1} \equiv |\mu|
$$

where $0 \leq |\mu| < \varepsilon < 1$, according to inequality (1.31).

Since $\Psi(\zeta)$ is an analytic function for $\text{Im } \zeta \geq 0$, it satisfies the Beltrami equation

$$
\frac{\partial\Psi}{\partial\bar{\zeta}} = \mu\frac{\partial\Psi}{\partial\zeta}, \qquad 0 \leq |\mu| < \varepsilon < 1
$$

everywhere on the ball $B_\lambda(\zeta_0)$.

The properties of solutions to the latter equation have been studied in detail (cf., [169], p. 213). In particular, $\Psi \in W_p^1(B_{\lambda/2}(\zeta_0))$, where $p = p(\varepsilon) \to \infty$ as $\varepsilon \to 0$. Thus also the function φ is more regular: for any $p > 2$, $\varphi \in W_p^1(B_{\lambda/2}(\zeta_0) \cap Q)$.

In order to show that $\varphi(\zeta)$ has derivatives which are Hölder continuous on the closure of $B_{\lambda/2}(\zeta_0) \cap Q$, we shall construct another extension of this function across the boundary $\{\zeta : \text{Im } \zeta = 0\}$ onto the domain $B_{\lambda/2}(\zeta_0)$.

To this end let us consider the function $\varphi^*(\zeta)$ defined in Q by the equality

$$
h(\varphi(\zeta)) = f(z_0)(\varphi^*(\zeta) - \bar{z}_0) + R(\varphi(\zeta), z_0), \tag{1.33}
$$

where $R(z, z_0)$ is given by the relation

$$
\omega^*(z) = f(z_0)(\bar{z} - \bar{z}_0) + R(z, z_0). \tag{1.34}
$$

By definition of the function $\omega^*(z)$ it follows that

$$
|z - z_0|^{-1} |R(z, z_0)| + \left|\frac{\partial R}{\partial z}(z, z_0)\right| + \left|\frac{\partial R}{\partial\bar{z}}(z, z_0)\right| \leq c|z - z_0|^\gamma. \tag{1.35}
$$

Differentiating equality (1.33) with respect to ζ and taking into account the analyticity of the functions h and φ, we get

$$\frac{\partial \varphi^*}{\partial \bar{\zeta}}(\zeta) = -\frac{1}{f(z_0)}\frac{\partial}{\partial \bar{\zeta}}R(\varphi(\zeta), z_0) = -\frac{1}{f(z_0)}\frac{\partial R}{\partial \bar{z}}(\varphi(\zeta), z_0)\frac{\partial \overline{\varphi(\zeta)}}{\partial \bar{\zeta}}.$$

Hence, by the inequality (1.35),

$$\left|\frac{\partial \varphi^*}{\partial \bar{\zeta}}(\zeta)\right| \leq c_1 |\varphi'(\zeta)|\,|\varphi(\zeta) - z_0|^\gamma, \quad z_0 = \varphi(\zeta_0).$$

By the embedding theorem (see Theorem 1, Chapter I) applied to the embedding of the space W_p^1 into $H^{1-2/p}$, the function $\varphi(\zeta)$ is Hölder continuous for $p > 2$. Because $\varphi' \in L_p$, we conclude that

$$|\zeta - \zeta_0|^{-\beta}\left|\frac{\partial \varphi^*}{\partial \bar{\zeta}}\right| \in L_p\big(B_{\lambda/2}(\zeta_0) \cap Q\big), \quad \beta = \gamma\big(1 - 2/p\big).$$

In $B_{\lambda/2}(\zeta_0)$ let us define

$$\Psi^*(\zeta) = \begin{cases} \varphi(\zeta) & \text{if } \operatorname{Im} \zeta \geq 0 \\[2mm] \overline{\varphi^*(\bar{\zeta})} & \text{if } \operatorname{Im} \zeta < 0. \end{cases}$$

Since $\dfrac{\partial \Psi^*}{\partial \bar{\zeta}} = 0$ if $\operatorname{Im} \zeta \geq 0$ and $\Psi^*(\zeta)$ is continuous across the boundary $\{\zeta : \operatorname{Im}\zeta = 0\}$, by equalities (1.32) - (1.34) we have

$$\varphi(\zeta) = \overline{\varphi^*(\bar{\zeta})}, \quad \operatorname{Im}\zeta = 0,$$

and therefore

$$|\zeta - \zeta_0|^{-\beta}\frac{\partial \Psi^*}{\partial \bar{\zeta}} \in L_p\big(B_{\lambda/2}(\zeta_0)\big), \quad \beta = \gamma\big(1 - 2/p\big).$$

It turns out that the function $\Psi^*(\zeta)$ satisfies the estimate

$$\left|\frac{\Psi^*(\zeta) - \Psi^*(\zeta_0)}{\zeta - \zeta_0}\frac{\partial \Psi^*}{\partial \zeta}(\zeta_0)\right| \leq c_3 \Lambda |\zeta - \zeta_0|^\alpha, \tag{1.36}$$

in the ball $B_{\lambda/2}(\zeta_0)$ if $\beta - 2/p > \alpha > 0$. In (1.36),

$$\Lambda = \left\|\,|\zeta - \zeta_0|^{-\beta}\frac{\partial \Psi^*}{\partial \bar{\zeta}}\right\|_{p, B_{\lambda/2}(\zeta_0)} + \|\Psi^*\|_{1, \partial B_{\lambda/2}(\zeta_0)}.$$

In proving (1.36) we use the representation

$$u(\zeta) = \frac{1}{2\pi i}\int_{\partial B}\frac{u(y)}{y - \zeta}\,dy - \frac{1}{\pi}\iint_B \frac{\partial u(y)}{\partial \bar{y}}\frac{dy_1\,dy_2}{y - \zeta}, \tag{1.37}$$

which follows from Green's formula

$$\iint_B \frac{\partial u(y)}{\partial \bar{y}}\,dy_1 dy_2 = -\frac{i}{2}\int_{\partial B}u(y)\,dy.$$

Let us take

$$u(\zeta) = \frac{\Psi^*(\zeta) - \Psi^*(\zeta_0)}{\zeta - \zeta_0}$$

in (1.37), to conclude that the first term on its right-hand side is an analytic function in the ball $B_{\lambda/2}(\zeta_0)$ and the second term is the potential operator

$$T(U|\zeta) = \frac{1}{\pi} \iint_B \frac{U(y)}{y - \zeta} \, dy_1 dy_2$$

which maps functions $U \in L_q(B)$ into functions $T(U|\zeta) \in H^\alpha(\bar{B})$, if $q > 2$ (cf., [169], p. 203).

Therefore, for proving the bound (1.36) it is sufficient to show the boundedness of the norm

$$\left\| |\zeta - \zeta_0|^{-1} \frac{\partial \Psi^*}{\partial \bar{\zeta}} \right\|_{q, B_{\lambda/2}(\zeta_0)}$$

with some $q > 2$. Let us take $q = 2/(1 - \alpha)$. We then have

$$\left\| |\zeta - \zeta_0|^{-1} \frac{\partial \Psi^*}{\partial \bar{\zeta}} \right\|_{q, B} \leq \left\| |\zeta - \zeta_0|^{-1+\beta} \right\|_{\frac{pq}{(p-q)}, B} \left\| |\zeta - \zeta_0|^{-\beta} \frac{\partial \Psi^*}{\partial \bar{\zeta}} \right\|_{p, B}.$$

The second factor on the right-hand side of the latter inequality is by construction bounded by Λ, and the first factor is bounded if $1 - 2/q = \alpha < \beta - 2/p$. The function $u(\zeta)$ is thus Hölder continuous with index α in the domain $B_{\lambda/3}(\zeta_0)$. Clearly,

$$u(\zeta_0) = \frac{\partial \Psi^*}{\partial \zeta}(\zeta_0).$$

Due to the arbitrariness of the point ζ_0, inequality (1.36) holds for all points $\zeta_1, \zeta_2 \in B_{\lambda/3}(\zeta_0)$:

$$\left| \frac{\Psi^*(\zeta_1) - \Psi^*(\zeta_2)}{\zeta_1 - \zeta_2} - \frac{\partial \Psi^*}{\partial \zeta}(\zeta_1) \right| \leq c_4 |\zeta_1 - \zeta_2|^\alpha.$$

Interchanging ζ_1 and ζ_2, and applying the mean-value theorem

$$\Psi^*(\zeta_1) - \Psi^*(\zeta_2) = \frac{\partial \Psi^*}{\partial \zeta}(\zeta_*) (\zeta_1 - \zeta_2),$$

we conclude

$$\left| \frac{\partial \Psi^*}{\partial \zeta}(\zeta_1) - \frac{\partial \Psi^*}{\partial \zeta}(\zeta_2) \right| \leq 2c_4 |\zeta_1 - \zeta_2|^\alpha.$$

In particular, if $\operatorname{Im} \zeta = 0$, then

$$\left| \frac{\partial \varphi}{\partial \zeta}(\zeta_1) - \frac{\partial \varphi}{\partial \zeta}(\zeta_2) \right| \leq 2c_4 |\zeta_1 - \zeta_2|^\alpha,$$

the latter inequality completing the proof.   □

2. The two-phase Stefan problem. Stability of the stationary solution

2.1. Problem statement. Main result

We shall confine ourselves, as in Section 1, to the case of two space variables and to strip-like domains

$$\Omega_T = \Omega \times (0,T), \quad \Omega = \{x \; : \; |x_1| < \infty, \; |x_2| < 1\},$$

$$\Omega_T^- = \{(x,t) \in \Omega_T \; : \; x_2 < R(x_1,t)\},$$

$$\Omega_T^+ = \{(x,t) \in \Omega_T \; : \; x_2 > R(x_1,t)\},$$

where $x_2 = R(x_1,t)$ is the equation describing the free surface Γ_T. The problem data will be assumed periodic with period 1 with respect to x_1.

In each of the domains $\Omega_T^\pm$ to be determined are a solution ϑ of the heat equation

$$D_t \vartheta = \Delta \vartheta \tag{2.1}$$

and the surface $\Gamma_T = \{(x,t) \in \Omega_T \; : \; x_2 = R(x_1,t)\}$ that satisfy the following conditions.

The temperature $\vartheta(x,t)$ is continuous across the surface $\Gamma_{,}T$ and vanishes there:

$$\vartheta(x,t) = 0 \quad \text{for} \quad (x,t) \in \Gamma_T. \tag{2.2}$$

On Γ_T, the temperature $\vartheta(x,t)$ satisfies the Stefan condition formulated as

$$L D_t \vartheta^+ |D\vartheta^+|^{-1} = |D\vartheta^+| - |D\vartheta^-|, \tag{2.3}$$

where $D\vartheta^\pm(x_0,t_0) = \lim D\vartheta(x,t)$, $D_t\vartheta^\pm(x_0,t_0) = \lim D_t\vartheta(x,t)$ as the point (x,t) tends to (x_0,t_0) on the surface Γ_T from the domain $\Omega_T^\pm$; $L = \text{const.} > 0$.

On the given surfaces $\mathcal{F}_T^\pm = \mathcal{F}^\pm \times (0,T)$, $\mathcal{F}^\pm = \{x \; : \; |x_1| < \infty, \; x_2 = \pm 1\}$, the temperature $\vartheta(x,t)$ is constant,

$$\vartheta(x,t) = A^\pm \quad \text{for} \quad (x,t) \in \mathcal{F}_T^\pm, \tag{2.4}$$

$$A^+ = \text{const.} > 0, \quad A^- = \text{const.} < 0.$$

For simplicity, we take the constants $L, A^+, |A^-|$ as unity.

At the initial time, the free surface Γ_T coincides with the line $S = \{x \; : \; |x_1| < \infty, \; x_2 = 0\}$, and the temperature ϑ is prescribed,

$$\Gamma(0) = S, \quad \vartheta(x,0) = \vartheta_0(x) \quad \text{for} \quad x \in \Omega. \tag{2.5}$$

In contrast to the one-phase case, in the two-phase Stefan problem there is no proof of the existence of its classical solution on an arbitrary time interval. Neither of the constructions presented in Section 1 of this chapter (the Duvaut and Legendre transformations) applies to the two-phase Stefan problem. Nevertheless, by introducing von Mises variables the existence of the classical solution of the two-phase problem on the infinite time interval $(0,\infty)$ can immediately be proved, provided that the initial function is close to the appropriate stationary solution

$$\vartheta_\infty(x) = x_2 . \tag{2.6}$$

This yields the stability of the simplest stationary solution in the class of spatial perturbations. We remark that the one-phase problem does not admit stationary solutions, but an analogous result is true for the self-similar solution [155].

Theorem 3. *Let*

$$\vartheta_0(x) = \vartheta_\infty(x) + \delta V(x),$$

where $\vartheta_\infty(x)$ is defined by equality (2.6), $V(x)$ is a periodic function with period 1 in x_1, belonging to the space $H^{9+\alpha}$ on the closure of each of the domains $\Omega^\pm(0)$, and δ is taken sufficiently small. Further, let $V(x)$ be identically zero in the vicinity of the lines $\mathcal{F}^\pm$ and S.

Then there exists a number $\delta_* > 0$, dependent only on the value

$$M_0 = \max\{|V|_{\Omega^-(0)}^{(9+\alpha)}, |V|_{\Omega^+(0)}^{(9+\alpha)}\},$$

such that, for $|\delta| < \delta_*$, a classical solution

$$\vartheta \in C^{2,1}(\overline{\Omega_T^\pm}), \quad R \in C^{2,1}(\mathbf{R}_T^1), \quad T < \infty$$

of the problem (2.1)–(2.5) exists on the infinite time interval $(0, \infty)$. This solution is periodic with period 1 in x_1 and satisfies the estimates

$$|\vartheta - \vartheta_\infty|_{\Omega^\pm(t)}^{(2)} \leq M_1 e^{-\beta t}, \tag{2.7}$$

$$|R(t)|_{\mathbf{R}^1}^{(1)} + |D_t R(t)|_{\mathbf{R}^1}^{(0)} \leq M_1 e^{-\beta t}, \tag{2.8}$$

with the constants M_1 and β dependent only on the value M_0.

Remark. The condition $V(x) \equiv 0$ imposed in the vicinity of the lines $\mathcal{F}^\pm, S$ was assumed only in order to simplify the exposition by automatically fulfilling the compatibility conditions on these lines.

2.2. Formulation of the equivalent boundary value problem

Let the parameter δ be small enough to ensure that

$$D_2 \vartheta_0(x) \geq 1/2 \quad \text{for} \quad x \in \Omega,$$

for $|\delta| < \delta_1(M_0)$. Of course, in this case $(\pm 1)\vartheta_0(x) > 0$ at $x \in \Omega^\pm(0)$.

Assuming that the inequality $D_2\vartheta > 0$ holds in each of the domains $\Omega^{\pm T}$ (with $T > 0$ arbitrary but fixed), we introduce new independent variables

$$\tau = t, \quad y_1 = x_1, \quad y_2 = \vartheta(x, t), \quad \text{for} \quad x \in \Omega_T^\pm.$$

Under such a change of variables, the domains $\mathcal{F}_T^\pm = \mathcal{F}^\pm \times (0, T)$ are transformed onto themselves, the free boundary Γ_T is mapped onto the surface $S_T = S \times (0, T)$ and the domains $\Omega_T^{\pm}$ onto $G_T^{\pm} = G^{\pm} \times (0,T)$, $G^{\pm} = \Omega^{\pm}(0)$. The new unknown function $u(y,\tau) = x_2$ in each of the domains $G_T^{\pm}$ satisfies the equation

$$D_\tau u - D_1^2 u + D_2\left\{\frac{1}{p_2}(1 + p_1^2)\right\} = 0 \tag{2.9}$$

and is prescribed on the surfaces $\mathcal{F}_T^{\pm}$,

$$u(y,\tau) = \pm 1 \qquad \text{for} \quad (y,\tau) \in \mathcal{F}_T^{\pm}. \tag{2.10}$$

The function $u(y,\tau)$ is continuous on the image S_T of the free boundary Γ_T,

$$u^+(y,\tau) = u^-(y,\tau) = u(y,\tau) \qquad \text{for} \quad (y,\tau) \in S_T, \tag{2.11}$$

and, for $(y,\tau) \in S_T$,

$$D_\tau u + \{1 + (p_1)^2\}\left(\frac{1}{p_2^+} - \frac{1}{p_2^-}\right) = 0. \tag{2.12}$$

In (2.9)–(2.12), $p_i = D_i u$, $u^{\pm}$ ($p_i^{\pm}$, respectively) are the limit values of the function $u(y,\tau)$ (functions $p_i(y,\tau)$) as $(y,\tau) \in G_T^{\pm}$ tends to the point $(y_0, \tau_0) \in S_T$.
Finally, at the initial time,

$$u(y,0) = u_0(y) \qquad \text{for} \quad y \in \Omega, \tag{2.13}$$

where the function $u_0(y)$ is defined by the identity $y_2 = \vartheta_0(y_1, u_0(y_1, y_2))$. This is always possible due to the derivative $D_2\vartheta_0(x)$ being strongly positive.

It follows easily that the function $u_0(y)$ has the same regularity as $\vartheta_0(y)$ in each of the domains $G^{\pm}$. Moreover,

$$u_0(y) = y_2 + \delta U(\delta, y), $$

where $U(\delta, y) \equiv 0$ in a vicinity of the lines $\mathcal{F}^{\pm}, S$, and its norm in $H^{9+\alpha}(\overline{G^{\pm}})$ is uniformly bounded by a constant dependent only on M_0, if $|\delta| < \delta_1$.

It is also obvious that the function $u_0(y)$ is periodic with period 1 in y_1. Everywhere in the sequel we shall consider only such functions without making any special mention. Moreover, it will be more convenient to consider the problem in the parts of $G_T^{\pm}$ common with the domain $\{y : 0 < y_1 < 1, |y_2| < 1\} \times (0,T)$ rather than in the whole $G_T^{\pm}$; we shall preserve below the same notation for the resulting subdomains $G_T^{\pm}$ (similarly for $\mathcal{F}_T^{\pm}, S_T$).

We recall that the function $u_\infty = y_2$ corresponds to the stationary solution $\vartheta_\infty(x)$.

2.3. Construction of approximate solutions

As in Chapter II, we shall construct the solution of problem (2.9)–(2.13) as the limit, as $\varepsilon \to 0$, of the solutions $u_\varepsilon(y,\tau)$ of the initial-boundary value problem (2.9)–(2.11), (2.13). The approximate solutions satisfy on the boundary S_T the condition that follows from (2.12) by applying the ε-regularization (see Chapter II):

$$D_\tau u - \varepsilon D_1^2 u = X(p^+, p^-),$$

$$X(p^+, p^-) = -\{1 + (p_1^+)^2\}\left(\frac{1}{p_2^+} - \frac{1}{p_2^-}\right). \tag{2.14}$$

For any fixed $\varepsilon > 0$, the above problem actually represents a one-parameter family of initial-boundary value problems, continuously dependent on the parameter δ, $|\delta| < \delta_1$. For $\delta = 0$, this problem admits the unique solution $u_\infty(y)$. One may expect that the problem also remains solvable for sufficiently small values of δ.

Lemma 2. *Let the hypotheses of Theorem 1 be satisfied. Then, for each $T > 0$ and $\varepsilon > 0$ there exists $\delta_2 = \delta_2(\varepsilon, T)$ such that the initial-boundary value problem (2.9)–(2.11), (2.13), (2.14) has a unique solution $u_\varepsilon \in H^{9+\alpha,(9+\alpha)/2}\left(\overline{G_T^\pm}\right)$, provided that $|\delta| < \delta_2$. For this solution,*

$$\left|\log|D_2 u_\varepsilon|\right|_{G_T^\pm}^{(0)} < \infty. \tag{2.15}$$

Proof. The assertion follows from the implicit function theorem, if the linear initial-boundary value problem

$$\begin{align}
D_\tau H - \Delta H &= f, & (y, \tau) &\in G_T^\pm; \tag{2.16}\\
H^+ = H^- &= H, & (y, \tau) &\in S_T; \tag{2.17}\\
D_\tau H - \varepsilon D_1^2 H - D_2 H^- + D_2 H^+ &= \varphi, & (y, \tau) &\in S_T; \tag{2.18}\\
H(y, \tau) &= 0, & (y, \tau) &\in \mathcal{F}_T^\pm; \tag{2.19}\\
H(y, 0) &= 0, & y &\in \Omega; \tag{2.20}
\end{align}$$

obtained from the original nonlinear problem by linearization around its exact solution $u_\infty(y)$, is uniquely solvable for all

$$f \in H^{\alpha,\alpha/2}\left(\overline{G_T^\pm}\right), \quad \varphi \in H^{\alpha,\alpha/2}(\bar{S}_T)$$

such that

$$f(y, 0) = 0, \; y \in \mathcal{F}^\pm; \; f^\pm(y, 0) = \varphi(y, 0) \qquad \text{for} \quad y \in S. \tag{2.21}$$

The relations (2.21) represent first-order compatibility conditions for the initial-boundary value problem (2.16)–(2.20). For the latter problem, the existence of its solution is equivalent to the availability of the following a priori estimate:

$$|H|_{G_T^+}^{(2+\alpha)} + |H|_{G_T^-}^{(2+\alpha)} \le c\left(|f|_{G_T^+}^{(\alpha)} + |f|_{G_T^-}^{(\alpha)} + |\varphi|_{S_T}^{(\alpha)}\right). \tag{2.22}$$

This can proved in a standard way, in particular by using a parametric continuation method.

To derive the first estimate (for the maximum of the absolute value of the solution), we multiply equation (2.16) by H^{2k-1} and for each section $\{\tau = \text{const.}\}$, integrate on the domain $G^\pm$. Then we take the sum of the integrals and perform integration by parts in the terms containing the highest-order derivatives, to get

$$\frac{1}{2k}\frac{d}{d\tau}\left(\|H(\tau)\|_{2,\Omega}^2 + \|H(\tau)\|_{2,S}^2\right) + \varepsilon(2k - 1)\|H^{k-1}D_1 H\|_{2,S}^2$$

$$+(2k-1)\|H^{k-1}DH\|_{2,\Omega}^2 = \|fH^{2k-1}\|_{1,\Omega} + \|\varphi H^{2k-1}\|_{1,S} \equiv \mathcal{J}_0. \qquad (2.23)$$

The boundary integrals over S result from the integration by parts using condition (2.18), while those over $\{y_1 = 0\}$, $\{y_1 = 1\}$ have mutually cancelled each other due to the periodicity of the solution with respect to y_1.

Let us denote

$$F = \max\{|f|_{\Omega_T}^{(0)}, |\varphi|_{S_T}^{(0)}\}.$$

Since $a^\gamma + b^\gamma \le 2^{1-\gamma}(a+b)^\gamma$ for all positive a, b and $\gamma < 1$, we have

$$\mathcal{J}_0 \le F\left\{\|H(\tau)\|_{2k,\Omega}^{2k-1} + \|H(\tau)\|_{2k,S}^{2k-1}\right\} \le 2F\left\{\|H(\tau)\|_{2k,\Omega}^{2k} + \|H(\tau)\|_{2k,S}^{2k}\right\}^{(2k-1)/2k}.$$

Thus (2.23), in a standard way, implies the bound

$$\max_{\tau \in (0,T)}\left\{\|H(\tau)\|_{2k,\Omega}^{2k} + \|H(\tau)\|_{2k,S}^{2k}\right\}^{1/2k} \le 2FT.$$

Hence, as $k \to \infty$,

$$|H|_{\Omega_T}^{(0)} \le 2FT. \qquad (2.24)$$

By treating the boundary condition (2.18) as a parabolic equation on S_T with the right-hand side $\varphi_0 = \varphi + D_2 H^- - D_2 H^+$ and the initial condition (2.20), we achieve (see Theorem 6, Section I.2) the bound

$$|H|_{S_T}^{(2+\alpha)} \le c\{|\varphi|_{S_T}^{(\alpha)} + |H|_{G_T^-}^{(1+\alpha)} + |H|_{G_T^+}^{(1+\alpha)}\}. \qquad (2.25)$$

From the equation (2.16) considered in each of the domains $G_T^\pm$, boundary conditions (2.19) and initial condition (2.20), we can obtain the following estimate analogous to (2.25):

$$|H|_{G_T^\pm}^{(2+\alpha)} \le c\{|f|_{G_T^\pm}^{(\alpha)} + |H|_{S_T}^{(2+\alpha)}\}.$$

By adding the latter inequalities, with (2.25) taken into account, we get

$$|H|_{G_T^+}^{(2+\alpha)} + |H|_{G_T^-}^{(2+\alpha)} \le c\left\{|\varphi|_{S_T}^{(\alpha)} + |f|_{G_T^+}^{(\alpha)} + |f|_{G_T^-}^{(\alpha)} + |H|_{G_T^+}^{(1+\alpha)} + |H|_{G_T^-}^{(1+\alpha)}\right\}. \qquad (2.26)$$

The desired estimate (2.22) follows by (2.24), (2.26) and the inequality (see Lemma 4, Section I.2)

$$|H|_{G_T}^{(1+\alpha)} \le \mu|H|_{G_T}^{(2+\alpha)} + c\mu^{-(1+\alpha)}|H|_{G_T}^{(0)}$$

where $\mu > 0$ is arbitrarily small.

The lemma then follows by the results concerning the dependence of the differential properties of the solutions of linear uniformly parabolic equations on the differential properties of their coefficients and right-hand sides (see Theorem 6, Section I.2). In Chapter II, an analogous scheme has been implemented in the construction of approximate solutions to the Stefan problem on a small time interval. $\qquad\square$

Corollary. *Let $N > M_0$ be arbitrary fixed. Then there is a constant $\delta_3 = \delta_3(\varepsilon, T, N) > 0$ such that, for $|\delta| < \delta_3$,*

$$\left|\log|D_2 u_\varepsilon|\right|_{G_T^\pm}^{(0)} < N, \qquad (2.27)$$

$$\sum_{|\bar{k}|=1}^{2} \left|\hat{D}^{\bar{k}}(Du_\varepsilon)\right|_{G_T^\pm}^{(0)} < N. \quad \square \qquad (2.28)$$

2.4. A lower bound for the constant δ_3

In this section we shall prove that the constant δ_3 indeed remains bounded from below as $T \to \infty$ and $\varepsilon \to 0$. The outline of the proof is in principle the same as in Chapter II, where the general Stefan problem was studied on a small time interval.

The deviation $H = u_\varepsilon(y, \tau) - u_\infty(y)$ of the solution $u(y, \tau)$ of the initial-boundary value problem (2.9)–(2.11), (2.13), (2.14) from the solution $u_\infty(y) = y_2$ of the corresponding stationary problem satisfies the following initial-boundary value problem

$$D_\tau H = D_1^2 H + D_2 \left\{\frac{P_2 - P_1^2}{1 + P_2}\right\}, \quad (y, \tau) \in G_T^\pm; \quad (2.29)$$

$$H(y, \tau) = 0, \quad (y, \tau) \in \mathcal{F}_T^\pm; \quad (2.30)$$

$$H^+(y, \tau) = H^-(y, \tau) = H(y, \tau), \quad (y, \tau) \in S_T; \quad (2.31)$$

$$(1 + P_1^2)\left(\frac{1}{1 + P_2^-} - \frac{1}{1 + P_2^+}\right) = D_\tau H - \varepsilon D_1^2 H, \quad (y, \tau) \in S_T; \quad (2.32)$$

$$H(y, 0) = \delta U(\delta, y), \quad y \in \Omega. \qquad (2.33)$$

In the above problem, $P_i = D_i H$, $i = 1, 2$, and the following relations were used in deriving (2.29):

$$D_2\left\{\frac{1 + p_1^2}{p_2}\right\} = D_2\left\{\frac{1 + P_1^2}{1 + P_2}\right\} = D_2\left\{\frac{1 + P_1^2}{1 + P_2} - 1\right\} = -D_2\left\{\frac{P_2 - P_1^2}{1 + P_2}\right\},$$

where $p_i = D_i u$, $i = 1, 2$.

We multiply equation (2.29) by H, then integrate over each of the domains G^+ and G^- at $\tau > 0$ fixed, and finally add the results. After elementary calculations we obtain the equality

$$\frac{1}{2}\frac{d}{d\tau}\{\|H(\tau)\|_{2,\Omega}^2 + \|H(\tau)\|_{2,S}^2\} + \varepsilon\|D_1 H(\tau)\|_{2,S}^2 + \left\|(1 + P_2)^{-1}|DH|^2\right\|_{1,\Omega} = 0. \quad (2.34)$$

Due to (2.27) this implies

$$\{\|DH\|_{2,\Omega_T}, \max_{\tau \in (0,T)} \|H(\tau)\|_{2,\Omega}\} \leq \delta M, \qquad (2.35)$$

where, from now on, M will represent constants which depend only on M_0, N and are independent of ε and T.

The boundary integrals over S have been obtained by integration by parts using condition (2.32) and the equality

$$\frac{P_2 - P_1^2}{1 + P_2} = 1 - \frac{1 + P_1^2}{1 + P_2}.$$

We apply the differential operator $D_0^{\bar{k}}$, $|\bar{k}| = 4$ to equation (2.29) and boundary conditions (2.30)–(2.32), and multiply the results by $D_0^{\bar{k}} H$. Then, for fixed $\tau > 0$ integrate these expressions over the domains G^+, G^- and the boundary S, respectively. Finally taking the sum of the relations obtained yields

$$\mathcal{J}_1 \equiv \frac{1}{2}\frac{d}{d\tau}\{\|D_0^{\bar{k}}H(\tau)\|_{2,\Omega}^2 + \|D_0^{\bar{k}}H(\tau)\|_{2,S}^2\} + \varepsilon\|D_0^{\bar{k}}P_1(\tau)\|_{2,S}^2$$
$$+ \iint_\Omega \{|D_0^{\bar{k}}P_1|^2 - \frac{2P_1}{1+P_2}D_0^{\bar{k}}P_1 D_0^{\bar{k}}P_2 + \frac{1+P_1^2}{(1+P_2)^2}|D_0^{\bar{k}}P_2|^2\}\,dy$$
$$= \iint_\Omega \Psi D_0^{\bar{k}}P_2\,dy \equiv \mathcal{J}_2, \tag{2.36}$$

where

$$\Psi = \Psi_1 DPD_0^r P + \Psi_2 D_0^{\bar{r}_1}PD_0^{r_2}P + \Psi_3 DPD_0^{\bar{r}_1}PDP + \Psi_4(P, DP),$$

$$P = (P_1, P_2),\ \Psi_i = \Psi_i(P),\ i = 1,2,3,\ |\bar{r}| = 3,\ |\bar{r}_1| = |\bar{r}_2| = 2$$

and the functions Ψ_i are bounded by a constant M independent of ε and T for all $\delta, |\delta| < \delta_2$.

It follows from inequalities (2.27) and (2.28) that the quadratic form

$$X^2 - 2P_1(1+P_2)^{-1}XY + (1+P_1^2)(1+P_2)^{-2}Y^2 = K(X,Y)$$

is strongly positive definite:

$$K(X,Y) \geq M^{-1}(X^2 + Y^2).$$

We integrate equality (2.36) in time, discard the positive terms

$$\varepsilon\|D_0^{\bar{k}}P_1\|_{2,S_T}^2 + \frac{1}{2}\{\|D_0^{\bar{k}}H(T)\|_{2,\Omega}^2 + \|D_0^{\bar{k}}H(T)\|_{2,S}^2\}$$

on the left-hand side of the expression obtained, and estimate from below the corresponding quadratic form $K(D_0^{\bar{k}}P_1, D_0^{\bar{k}}P_2)$ to conclude

$$\int_0^T \mathcal{J}_1(\tau)\,d\tau \geq M^{-1}\|D_0^{\bar{k}}P\|_{2,\Omega_T}^2. \tag{2.37}$$

Notice that the parameter ε does not enter the latter and the following inequalities. The main term in the expression $\mathcal{J}_2$ is

$$\mathcal{J}_{02} = \int_G \Psi_1 DPD_0^{\bar{r}}PD_0^{\bar{k}}P\,dy,\quad |\bar{r}| = 3,\quad |\bar{k}| = 4,$$

with the maximum of the first two coefficients in the integrand bounded by the constant N (see inequality (2.28)). Hence, by Hölder's inequality, the bound (see Lemma 2,

Section I.2)

$$\left\|\hat{D}^{\bar{r}}P\right\|_{2,\Omega} \leq \mu\left\|\hat{D}^{\bar{k}}P\right\|_{2,\Omega} + c\mu^{-3}\|P\|_{2,\Omega}$$

holds, with a uniform constant c and an arbitrary positive number μ. Finally using inequality (2.35), we estimate the time integral of $\mathcal{J}_{02}$ from above, to obtain

$$\int_0^T \mathcal{J}_{02}(\tau)\,d\tau \leq M\{\mu\left\|\hat{D}^{\bar{k}}P\right\|_{2,\Omega}^2 + \mu^{-3}\delta^2\}.$$

The remaining terms of $\mathcal{J}_2$ can be estimated similarly. Taking the sum of all the estimates, we obtain for $|\bar{k}| = 4$:

$$\left\|D_0^{\bar{k}}P\right\|_{2,\Omega_T}^2 \leq M\{\mu\left\|\hat{D}^{\bar{k}}P\right\|_{2,\Omega_T}^2 + c(\mu)\delta^2\}. \tag{2.38}$$

As in Chapter II, using equation (2.29) we can estimate the "non-tangential" derivatives $\hat{D}^k P$ by the "tangential" derivatives $D_0^{\bar{k}}P$:

$$\left\|\hat{D}^{\bar{k}}P\right\|_{2,\Omega_T}^2 \leq M\{\left\|D_0^{\bar{k}}P\right\|_{2,\Omega_T}^2 + \delta^2\}.$$

Substituting into (2.38) and taking μ sufficiently small, we obtain

$$\left\|\hat{D}^{\bar{k}}P\right\|_{2,\Omega_T}^2 \leq M\delta^2. \tag{2.39}$$

We once more recall that the constant M in (2.39) is independent of ε and T.

The bound (2.39) together with the embedding theorem (cf., [73], p. 424 and [137], p. 78) applied to the embedding of $W_2^2(G_\infty^\pm)$ into $H^\gamma(\overline{G_\infty^\pm})$,

$$\left|\hat{D}^{\bar{r}}P\right|_{G_\infty^\pm}^{(\gamma)} \leq c\sum_{|\bar{m}|=1}^4 \left\|\hat{D}^{\bar{m}}P\right\|_{2,G_\infty^\pm}, \quad |\bar{r}| = 2,$$

with the uniform constant c, yield

$$\left|\hat{D}^{\bar{r}}P\right|_{G_\infty^\pm}^{(\gamma)} \leq M\delta, \quad |\bar{r}| \leq 2. \tag{2.40}$$

By choosing δ sufficiently small ($|\delta| < \delta_4(M_0, N)$), we easily fulfil the inequalities defining the interval $(-\delta_3, \delta_3)$. Hence, every interval $(-\delta_3, \delta_3)$ contains an interval $(-\delta_4, \delta_4)$, where δ_4 is independent of ε and T.

This implies that the solution of the initial-boundary value problem (2.9)–(2.11), (2.13), (2.14) exists on the infinite time interval.

Indeed, suppose this does not hold. Then a certain $T_* > 0$ would exist such that the value

$$\left|\log|D_2 u_\varepsilon|\right|_{G_T^\pm}^{(0)} + |u_\varepsilon|_{G_T^\pm}^{(9+\alpha)} = \mathcal{J}_3(T)$$

is unbounded as $T \to T_*$. However, the first term in $\mathcal{J}_3(T)$ is bounded by the constant N for all $T < T_*$ (see inequality (2.27)) and the second term is bounded by a constant dependent only on M_0, N, ε and T_*, because the same constant N bounds the norm $|u_\varepsilon|_{G_T^\pm}^{(2+\alpha)}$ (see inequality (2.28)), and equation (2.29) is uniformly parabolic.

It follows from the proofs given in this section that the estimate (2.40) (and, thus also the bounds (2.27), (2.28)) remains true for the domains $G_\infty^\pm = G^\pm \times (0, \infty)$.   □

2.5. Proof of the main result

The estimates (2.27) and (2.28) make it possible to select by diagonalization subsequences $\{u_{\varepsilon_k}(y, \tau)\}$ strongly convergent in the space $C^{2,1}(G_T^\pm)$ for any $T < \infty$ to a function $u \in C^{2,1}(G_\infty^\pm)$. Because the component $\varepsilon D_1^2 u_\varepsilon(y_1, 0, \tau)$ in the boundary condition (2.14) converges to 0 as $\varepsilon \to 0$, the limit function $u(y, \tau)$ is obviously the desired solution of the initial-boundary value problem (2.9)–(2.13).

To prove the bounds (2.7) and (2.8), let us multiply the equation for $H = u(y, \tau) - u_\infty(y)$ by H and integrate the result at fixed $\tau > 0$ over each of the domains $G^\pm$. Analogously to the derivation of (2.34) we obtain the same equality corresponding to $\varepsilon = 0$.

As usual in asymptotic analysis of the solutions of parabolic equations, the term

$$\mathcal{J}_4 = \iint_\Omega (1 + P_2)^{-1} |DH|^2 \, dy,$$

where $P_i = D_i H$, $i = 1, 2$, can be estimated from below by the norm $\|DH(\tau)\|_{2,\Omega}^2$:

$$\mathcal{J}_4 \geq M^{-1} \|DH(\tau)\|_{2,\Omega}^2,$$

and the expression $z(\tau) = \|H(\tau)\|_{2,\Omega}^2 + \|H(\tau)\|_{2,S}^2$ can be estimated from above by the same norm:

$$z(\tau) \leq M \|DH(\tau)\|_{2,\Omega}^2.$$

The latter follows from the representations

$$H^2(y_1, y_2, \tau) = 2 \int_{-1}^{y_2} H(y_1, s, \tau) D_2 H(y_1, s, \tau) \, ds,$$

$$H_2(y_1, 0, \tau) = H^2(y_1, y_2, \tau) - 2 \int_0^{y_2} H(y_1, s, \tau) D_2 H(y_1, s, \tau) \, ds,$$

applying the Hölder inequality, and integrating in y_1 over $0 < y_1 < 1$ and in y_2 over $|y_2| < 1$.

Taking the sum of the above estimates, we get

$$\frac{dz}{dt} + Mz \leq 0.$$

Hence, by Gronwall's inequality,

$$\|H(\tau)\|_{2,\Omega}^2 + \|H(\tau)\|_{2,S}^2 \leq M e^{-\beta \tau}. \tag{2.41}$$

From the representation

$$\left| D^{\bar{k}} H(y, 2) \right|^2 = \left| D^{\bar{k}} H(y, 0) \right|^2 + 2 \int_0^\tau \hat{D}^{\bar{k}} H \hat{D}^{\bar{k}} P \, dt$$

with $|\bar{k}| = 4$, and the bound (2.39), it follows that

$$\max_{\tau \in (0, \infty)} \left\| D^{\bar{k}} H(\tau) \right\|_{2, \Omega} \leq M, \quad |\bar{k}| = 4. \tag{2.42}$$

Let us now use inequalities (2.35), (2.41), (2.42), the inequality (2.1) of Section I.2,

$$\left\| D^{\bar{r}} H(\tau) \right\|_{2, G^\pm} \leq c \left\| D^{\bar{k}} H(\tau) \right\|_{2, G^\pm}^{1/4} \left\| H(\tau) \right\|_{2, G^\pm}^{3/4},$$

where $|\bar{k}| = 4$, $|\bar{r}| = 3$, as well as the bound (2.40), the embedding of the space $W_2^3(G)$ into $C^1(\bar{G})$ (see Theorem 1, Section I.2):

$$|H(\tau)|_{G^\pm}^{(1)} \leq c \sum_{|\bar{r}|=3} \left\| D^{\bar{r}} H(\tau) \right\|_{2, G^\pm}$$

and the inequality (cf. [138], p. 162)

$$|H(\tau)|_{G^\pm}^{(2)} \leq c \left(|H(\tau)|_{G^\pm}^{(2+\gamma)} \right)^{1/(1+\gamma)} \left(|H(\tau)|_{G^\pm}^{(1)} \right)^{\gamma/(1+\gamma)},$$

to conclude with the estimate

$$|H(\tau)|_{G^\pm}^{(2)} \leq M e^{-\beta \tau}. \tag{2.43}$$

The solution $\vartheta(x, t)$ of the Stefan problem can be determined from the identity

$$u(x_1, \vartheta(x, t), t) = x_2.$$

This is always possible due to estimate (2.27). The equation $x_2 = R(x_1, t)$ of the free boundary is explicitly characterized by $u(y, \tau)$: $R = u(x_1, 0, t)$.

By transformation of variables and the estimates (2.40), (2.27) and (2.43), it follows that $\vartheta(x, t)$ and $R(x_1, t)$ are regular as stated in Theorem 3, and the estimates (2.7), (2.8) hold for them.   □

Chapter IV

Lagrange variables in the multidimensional one-phase Stefan problem

---

In this chapter we propose a new approach to the analysis of the multidimensional one-phase Stefan problem for the homogeneous heat equation. In its simplest version, this problem admits the following formulation.

Denote by $\Omega(t)$ a time-dependent domain in $\mathrm{I\!R}^n$ with boundary $\Gamma(t)$. To be determined are the surface $\Gamma(t)$ and a function $\vartheta(x, t)$ representing the temperature which satisfy the system

$$D_t\vartheta = \Delta\vartheta \qquad \text{for} \quad x \in \Omega(t), \quad t \in (0, T), \tag{0.1}$$

$$\vartheta = 0, \quad V_\nu = -\sum_{i=1}^{n} \nu_i D_i\vartheta \qquad \text{for} \quad x \in \Gamma(t), \quad t \in (0, T), \tag{0.2}$$

$$\vartheta(x, 0) = \vartheta_0(x) \qquad \text{for} \quad x \in \Omega(0). \tag{0.3}$$

In (0.2), V_ν is the normal velocity of the surface $\Gamma(t)$ towards the normal vector $\nu = (\nu_1, \ldots, \nu_n)$. We assume that the domain $G = \Omega(0)$ is given, $\vartheta_0(x) > 0$ if $x \in G$, and $\vartheta_0 = 0$ if $x \in \Gamma(0) = S = \partial G$.

The proposed method of analysis of the Stefan problem is based on its analogy with the unsteady movement of an ideal liquid with a free boundary [173]. The analysis of this problem is especially convenient in the Lagrange coordinates, where the domain is a priori known.

In Lagrange variables, the Stefan problem is transformed into an initial-boundary value problem for a nonlinear system of n equations on a fixed domain. Such a system does not fit any of the types studied so far; in a certain sense it can be treated as a degenerate parabolic system. As indirect evidence of this degeneracy, just one boundary condition is imposed for n unknown functions on the lateral boundary of the cylinder $G_T = G \times (0, T)$.

1. Formulation of the problem in Lagrange variables

To formulate the Stefan problem in Lagrange variables, let us consider (0.1) as the continuity equation for a compressible liquid with density ρ and velocity $v = (v_1, \ldots, v_n)$:

$$D_t\rho + \sum_{i=1}^{n} D_i(\rho v_i) = 0. \tag{1.1}$$

We shall consider (0.2) as the kinematic condition on the surface $\Gamma(t)$ (the zero level set of the function ϑ):

$$D_t\vartheta + \sum_{i=1}^{n} v_i D_i\vartheta = 0 \qquad \text{for} \quad \vartheta = 0 . \tag{1.2}$$

Equation (1.1) will coincide with (0.1), provided that

$$\rho = \vartheta + \psi, \quad v_i = -(1/\rho)\, D_i\vartheta, \tag{1.3}$$

where ψ is a function independent of t. Due to the equalities

$$V_\nu = D_t\vartheta|D\vartheta|^{-1}, \quad \sum_{i=1}^{n} v_i D_i\vartheta = -|D\vartheta| \qquad \text{for} \quad \vartheta = 0,$$

the second condition of (0.2) reduces to

$$D_t\vartheta - |D\vartheta|^2 = 0 \qquad \text{for} \quad \vartheta = 0.$$

By (1.3), this equality is equivalent to (1.2) if $\psi = 1$.

We shall now write the equation of the trajectory of the "liquid particle" with the coordinates x and velocity vector v at time t as

$$\frac{dx}{dt} = v = -\frac{1}{1+\vartheta} D\vartheta. \tag{1.4}$$

This equality follows from (1.3). Treating (1.4) as a system of ordinary differential equations with respect to the components of the vector x, we shall complement this system by the initial conditions

$$x = \xi \qquad \text{for} \quad t = 0. \tag{1.5}$$

By the *Lagrange coordinates* we shall mean the initial coordinates ξ of the "liquid particle".

Our next objective is to reformulate the relations (0.1)–(0.3) in the new independent variables (ξ, t). Set

$$\vartheta(x(\xi, t), t) = u(\xi, t)$$

and note that $D\vartheta = (M^*)^{-1}\langle Du\rangle$, where $M^* = \nabla_\xi x$ is the matrix obtained by transposing the Jacobian M with the elements $M_{ij} = D_j x_i$, $i, j = 1, \ldots, n$. Inserting the last equality in (1.4), we obtain

$$M^* \langle D_t x \rangle \ = \ D \Big\{ \log \frac{1}{1+u} \Big\}. \tag{1.6}$$

It remains to express u in terms of M and ξ. To this end we apply the Cauchy identity

$$D_t(\hat{\rho} \cdot \mathcal{J}) \ = \ 0, \qquad \hat{\rho}(\xi, t) = \rho(x, t), \tag{1.7}$$

which is well-known in continuum mechanics, where the notation $\mathcal{J} = \det M$ is used. Equation (1.7) is a consequence of the continuity equation (1.1). Prescribing the initial conditions (0.3), (1.5) and using the equality $\hat{\rho} = u + 1$, we come to the relations

$$(u+1)\mathcal{J} = \hat{\rho}_0(\xi) = 1 + \vartheta_0(\xi). \tag{1.8}$$

The substitution of (1.8) into (1.6) leads to the sought system of equations for the functions $x_i(\xi, t)$, $i = 1, \dots, n$:

$$M^* \langle D_t x \rangle \ = \ D \Big\{ \log \frac{\det M}{1 + \vartheta_0(\xi)} \Big\}. \tag{1.9}$$

Equalities (1.5) impose initial conditions on the system (1.9). The appropriate boundary condition can be established from the obvious identity

$$D_t u = D_t \vartheta + \sum_{i=1}^{n} v_i D_i \vartheta.$$

By (1.2), this identity implies that the equality $D_t u = 0$ holds on the free boundary whose equation in Lagrange coordinates is $u = 0$. The zero level surface of the function $u(\xi, t)$ is thus time-independent. By (0.3), (1.5) and the definition of $\Gamma(t)$, the equation of the free surface may be given the form $\vartheta_0(\xi) = 0$. Therefore, with $\Gamma(t)$ treated as the image of $\Gamma(0)$ under the transformation $\xi \to x$, condition (0.2) will be automatically satisfied. Recall now condition (1.8) and take into account the equalities $u = \vartheta_0 = 0$, to conclude that

$$\mathcal{J} \ = \ 1 \qquad \text{for} \quad \xi \in \Gamma(0). \tag{1.10}$$

We now give the complete formulation of the Stefan problem in Lagrange variables. In the cylinder $G_T = G \times (0, T)$ a solution $x = (x_1, \dots, x_n)$ of the system (1.9) is to be determined that fulfils condition (1.5) on the base of the cylinder and condition (1.10) on its lateral surface $S_T = S \times (0, T)$, $S = \Gamma(0)$.

In the sequel we shall refer to the above problem as to *Problem (A)*.

2. Linearization

Let $x(\xi, t)$ and $X(\xi, t)$ be smooth vector functions, with x not necessarily satisfying system (1.9). Set

$$\tilde{x} = x + \varepsilon X, \quad \tilde{M}^* = \nabla_\xi \tilde{x}, \quad M^* = \nabla_\xi x,$$

where ε is a formally introduced small parameter.

We shall compute the terms in $\tilde{\mathcal{J}} = \det \tilde{M}$ and $\tilde{M}^*\langle D_t \tilde{x}\rangle$ which are linear in ε. We have

$$\tilde{M}^*\langle D_t \tilde{x}\rangle \;=\; M^*\langle D_t x\rangle + \varepsilon\{M^*\langle D_t X\rangle + N^*\langle D_t x\rangle\} + \dots\,, \qquad (2.1)$$

where $N^* = \nabla_\xi X$. We shall assume that $\mathcal{J} = \det M \neq 0$ and define the function Y by

$$Y = M^{-1}\langle X\rangle.$$

Applying the obvious identity

$$N^*\langle D_t x\rangle \;=\; D\{x D_t x\} - D_t\{M^*\langle x\rangle\}$$

and the definition of Y, we get

$$M^*\langle D_t X\rangle + N^*\langle D_t x\rangle \;=\; M^* M\langle D_t Y\rangle +$$

$$+ D\{Y M^*\langle D_t x\rangle\} + (M^* D_t M - D_t M^* M)\langle Y\rangle. \qquad (2.2)$$

Further, as $\varepsilon \to 0$,

$$\det \tilde{M} \;=\; \mathcal{J}\,\det(I + \varepsilon M^{-1} N) \;=\; \mathcal{J} + \varepsilon \mathcal{J}\,\mathrm{tr}(M^{-1} N) + o(\varepsilon)\,.$$

In terms of Y, this expression can be written as

$$\det \tilde{M} \;=\; \mathcal{J}\Big\{1 + \varepsilon\Big(\sum_{i=1}^{n}(D_i Y_i + b_i Y_i)\Big) + o(\varepsilon)\Big\}, \qquad (2.3)$$

where $b = (b_1, \dots, b_n)$,

$$b_i \;=\; \sum_{j,k=1}^{n} \frac{\partial \xi_i}{\partial x_k}\frac{\partial}{\partial \xi_j}\Big(\frac{\partial x_k}{\partial \xi_i}\Big). \qquad (2.4)$$

Indeed,

$$\mathrm{tr}(M^{-1}N) \;=\; \sum_{i,k=1}^{n} \frac{\partial \xi_i}{\partial x_k}\frac{\partial X_k}{\partial \xi_i} \;=\; \sum_{i,k=1}^{n} \frac{\partial \xi_i}{\partial x_k}\frac{\partial}{\partial \xi_i}\Big(\sum_{j=1}^{n} Y_j \frac{\partial x_k}{\partial \xi_i}\Big)$$

$$= \sum_{i,j,k=1}^{n} \frac{\partial \xi_i}{\partial x_k}\frac{\partial x_k}{\partial \xi_j}\frac{\partial Y_j}{\partial \xi_i} + \sum_{i,j,k=1}^{n} \frac{\partial \xi_i}{\partial x_k}\frac{\partial}{\partial \xi_i}\Big(\frac{\partial x_k}{\partial \xi_j}\Big)Y_j = \sum_{i=1}^{n} D_i Y_i + \sum_{j=1}^{n} b_j Y_j.$$

We transform the right-hand side of (2.4), using the formula (2.9) of [173]:

$$\sum_{i,k=1}^{n} \frac{\partial \xi_i}{\partial x_k}\frac{\partial}{\partial \xi_i}\Big(\frac{\partial x_k}{\partial \xi_j}\Big) \;=\; \sum_{k=1}^{n} \frac{\partial}{\partial x_k}\Big(\frac{\partial x_k}{\partial \xi_j}\Big) \;=\; \frac{1}{\mathcal{J}}D_j \mathcal{J}.$$

Hence, $b = D\{\log \mathcal{J}\}$.

From (2.1)–(2.3), the linearized system (1.9), expressed in terms of $Y = (Y_1, \ldots, Y_n)$, can be written in the form

$$A\langle D_t Y\rangle \;=\; \nabla\{\operatorname{div} Y + aY\} + B\langle Y\rangle + f \;, \tag{2.5}$$

where

$$A = M^*M, \quad B = MD_tM^* - M^*D_tM,$$

$$a = D\{\log \mathcal{J}\} - M^*\langle D_t x\rangle, \quad \operatorname{div} Y = \sum_{i=1}^{n} D_i Y_i, \quad \nabla U = DU.$$

By (2.3), the linearization of the boundary condition (1.10) leads to the relation

$$\operatorname{div} Y + bY \;=\; \varphi \qquad \text{for} \quad \xi \in S. \tag{2.6}$$

Equalities (2.5), (2.6) are to be complemented by the initial condition

$$Y(\xi, 0) \;=\; Y_0(\xi), \quad \xi \in G. \tag{2.7}$$

The complete form of the linearized problem (in the sequel referred to as *Problem (B)*) is the following: determine a solution Y of the system (2.5) in the cylinder G_T which fulfils condition (2.6) on the lateral surface S_T of G_T and condition (2.7) on its base G.

In (2.5)–(2.7), the functions f, φ and Y_0, defined on G_T, S_T and G, respectively, are assumed to be given.

It is worth noting that Problem (B) can be reduced to a formulation with one unknown scalar function. Let $Y = (Y_1, \ldots, Y_n)$; set

$$U \;=\; \operatorname{div} Y + aY \tag{2.8}$$

and consider the following system of ordinary differential equations for Y, parametrized by ξ :

$$A\langle D_t Y\rangle - B\langle Y\rangle \;=\; F, \qquad F \;=\; \nabla U + f \;. \tag{2.9}$$

Let $\Psi(\xi, t)$ be the fundamental matrix of the homogeneous system $D_t Y = A^{-1}B\langle Y\rangle$. By the method of variation of constants, we get the following representation of the solution to the Cauchy problem (2.9), (2.7) :

$$Y(\xi, t) \;=\; \int_0^t \Psi(\xi, t)\Psi^{-1}(\xi, \tau)A^{-1}(\xi, \tau)\langle F(\xi, \tau)\rangle \, d\tau$$

$$+ \; \Psi(\xi, t)\Psi^{-1}(\xi, 0)\langle Y_0\rangle \;\equiv\; \int_0^t Q_0(\xi, t, \tau)\langle \nabla U(\xi, \tau)\rangle \, d\tau + g(\xi, t) \;. \tag{2.10}$$

Further, using equation (2.9) and representation (2.10) we compute the derivative $D_t u = \operatorname{div}(D_t Y) + aD_t Y + D_t aY$ to obtain

$$D_t U \;=\; \operatorname{div}\left(A^{-1}\langle \nabla U\rangle\right) + \nabla U A^{-1}\langle a\rangle \;+\; \operatorname{div}\left(\int_0^t Q(\xi, t, \tau)\langle \nabla U(\xi, \tau)\rangle \, d\tau\right)$$

$$+ \, h(\xi, t) \, + \, \int_0^t \omega(\xi, t, \tau | a) \nabla U(\xi, \tau) \, d\tau \, , \tag{2.11}$$

where

$$Q(\xi, t, \tau) \; = \; A^{-1}(\xi, t) B(\xi, t) Q_0(\xi, t, \tau),$$

$$\omega(\xi, t, \tau | a) \; = \; Q_0^*(\xi, t, \tau) \langle D_t a(\xi, t) \rangle - A^{-1}(\xi, t) B(\xi, t) \langle a(\xi, t) \rangle,$$

$$h(\xi, t) \; = \; f A^{-1} \langle a \rangle + \mathrm{div} \, (A^{-1} B \langle g \rangle) + (D_t a - A^{-1} B \langle a \rangle) \, g.$$

Combining (2.6) and (2.8) leads to the equality $U \, = \, (a - b) Y + \psi$, which by differentiation in time gives the boundary condition for the potential U on the lateral surface S_T :

$$D_t U \; = \; \nabla U c + \int_0^t \omega(\xi, t, \tau | a - b) \, \nabla U(\xi, \tau) \, d\tau \, + \, \psi \, , \tag{2.12}$$

where $c = a - b$, $\psi = g(D_t c - A^{-1} B \langle c \rangle) + f + A^{-1} \langle c \rangle + D_t \psi$.

Moreover, the function U satisfies the initial condition

$$U(\xi, 0) \; = \; U_0(\xi) \; \equiv \; \mathrm{div} \, Y_0(\xi) + a(\xi, 0) Y_0(\xi) \, . \tag{2.13}$$

We are now able to formulate the linearized Problem (A) in terms of the function U : *determine a solution of the equation (2.11) in the cylinder G_T, satisfying condition (2.12) on the lateral surface S_T of the cylinder G_T and condition (2.13) on its base.*

3. Correctness of the linear model

In this section we shall prove a theorem on the existence and uniqueness of the generalized solution of Problem (B), with the linearization applied to the exact solution of Problem (A).

Lemma 1. *Let $x(\xi, t)$ be the exact solution of Problem (A), with the third derivatives in ξ and t continuous in the domain G_T. Then*

$$B \; \equiv \; M D_t M^* - M^* D_t M \; = \; 0 \qquad for \quad (\xi, t) \in G_T, \tag{3.1}$$

$$a - b \; = \; -q \nu, \quad q \, \geq \, a_0 = \mathrm{const.} > 0 \qquad for \quad (\xi, t) \in S_T, \tag{3.2}$$

$$y A \langle y \rangle \; \geq \; a_0 |y|^2 \qquad for \quad (\xi, t) \in G_T, \tag{3.3}$$

where ν is the unit outward normal to S.

Proof. The first relation follows from the equalities

$$B_{jk} \; = \; \sum_{i=1}^n \{ D_j (D_t x_i) D_k x_i - D_j x_i D_k (D_t x_i) \} \; =$$

$$= \sum_{i=1}^{n} \{D_j(D_t x_i D_k x_i) - D_k(D_j x_i D_t x_i)\} + \sum_{i=1}^{n} \{D_t x_i(D_j D_k x_i - D_k D_j x_i)\}$$

$$= D_j D_k \{\log(\mathcal{J}/\rho_0)\} - D_k D_j \{\log(\mathcal{J}/\rho_0)\} = 0 ,$$

which we have obtained by using equation (1.3), and the equality of the second derivatives of the scalar functions x_i and $(\log \mathcal{J} - \log \rho_0)$.

To derive (3.2), let us note that the function $w = -\log(1 + u)$ satisfies the homogeneous parabolic equation

$$D_t w = \text{div} \{M^{-1} M^{*-1} \langle \nabla w \rangle\} - M^{-1} \langle \text{div} (M^{-1})^* \rangle \nabla w \qquad (3.4)$$

in G, it vanishes on the lateral surface S_T, is strictly negative on the base G of the
cylinder G_T and

$$a - b = -M^* \langle D_t x \rangle = -\nabla w \qquad \text{for} \quad \xi \in S.$$

S_T is the minimal value level surface of the function w, while ∇w is directed along the
outward normal to the surface S and vanishes nowhere due to the maximum principle.
Equation (3.4) results from the relations

$$D_t \hat{\rho} = D_t \rho + \sum_{i=1}^{n} v_i D_i \rho = -\rho \sum_{i=1}^{n} D_i v_i = -\rho \text{div}_x v ,$$

$$D_t w = D_t \{-\log \hat{\rho}\} = \text{div}_x v, \quad v = M^{*-1} \langle \nabla_\xi w \rangle$$

by applying the operator div_x expressed in Lagrange variables to the latter equality.

Finally, inequality (3.3) follows from the definition of the matrix A, the identity
$w = \log \mathcal{J} - \log(\vartheta_0 + 1)$, and the boundedness of $|w|$ and $|\log(\vartheta_0 + 1)|$. Let us define

$$(X, Y) = \int_G X(\xi, t) Y(\xi, t) d\xi , \quad (X, Y)_S = \int_S X(\xi, t) Y(\xi, t) d\sigma_\xi ,$$

where X, Y are either vector functions (then $XY = \sum_{i=1}^{n} X_i Y_i$) or scalar functions.

Denote by $L_2(G_T)$ $(L_2(S_T))$ the space of either vector or scalar functions with modulus
square integrable in G_T (on S_T), without specifying whether a given function is vector-
or scalar-valued.   □

Under the hypotheses of Lemma 1, Problem (B) can be reformulated as follows:
determine a vector function $Y(\xi, t)$ which in G_T fulfils the system of equations

$$A \langle D_t Y \rangle = \nabla(\text{div} Y + aY) + f , \qquad (3.5)$$

the condition

$$\text{div} Y + aY = -q(Y\nu) + \varphi \qquad \text{on} \quad S , \qquad (3.6)$$

and the initial condition

$$Y(\xi, 0) \; = \; 0 \qquad \text{for} \quad \xi \in G \; . \tag{3.7}$$

We shall refer to the problem (3.5)–(3.7) as *Problem (C)*.

By taking the scalar product of equation (3.5) with an arbitrary smooth vector function Ψ and integrating the result by parts over the domain G, we arrive at the identity

$$(A\langle D_t Y\rangle, \; \Psi) + (\text{div } Y + aY, \; \text{div } \Psi) \; + \; \big(q(Y\nu), \; (\Psi\nu)\big)_S \; = \; (f, \Psi) + \big(\psi, (\Psi\nu)\big)_S, \tag{3.8}$$

which leads to the definition of a generalized solution.

Definition. By the *generalized solution of Problem (C)* we shall mean the vector function $Y \in L_2(G_T)$ such that

$$Y\nu \in L_2(S_T), \qquad |D_t Y|, \; \text{div } Y \in L_2(G_T) \; ,$$

and Y fulfils the identity (3.8) with an arbitrary smooth function Ψ and initial condition (3.7).

Theorem 1. *Let the conditions (3.2), (3.3) be satisfied and let*

$$K_0 \; = \; \max\{|a|_{G_T}^{(0)}, \; |D_t A|_{G_T}^{(0)}\} \; < \; \infty \; .$$

Then the generalized solution of Problem (C) is unique.

Proof. The difference of two possible solutions $Y_{(1)}$ and $Y_{(2)}$ of Problem (C) fulfils the identity (3.8) with $f = 0$, $\varphi = 0$. Since there always exists a sequence of smooth functions Ψ_n strongly convergent in the space $L_2(G_T)$ to the function Y, such that div Ψ_n converges strongly to div Y in the space $L_2(G_T)$ and $\Psi_n \nu$ converges strongly to $Y\nu$ in $L_2(S_T)$, we can assume $\Psi = Y$ in (3.8) :

$$(A\langle D_t Y\rangle, Y) + (\text{div } Y + aY, \text{div } Y) + \big(q(Y\nu), (Y\nu)\big)_S \; = \; 0 \; .$$

Hence, by applying the Hölder and Young inequalities, we obtain

$$\frac{1}{2}\|\text{div } Y\|_{2,G}^2 - \frac{1}{2}\frac{d}{dt}(Y, A\langle Y\rangle) + a_0\|Y\nu\|_{2,S}^2 \; \leq \; \frac{1}{2}K_0\|Y\|_{2,G}^2 \; . \tag{3.9}$$

Integrating (3.9) in time, using (3.2), (3.3) and the initial condition (3.7) gives us the inequality

$$\|Y(t)\|_{2,G}^2 \; \leq \; K_0 a_0^{-1} \int_0^t \|Y(\tau)\|_{2,G}^2 \, d\tau,$$

which implies the assertion of the theorem. □

Theorem 2. *Let hypotheses (3.2), (3.3) be satisfied, the boundary S be in the class C^2 and let*

$$K_1 \; = \; \max\{|A, D_t A|_{G_T}^{(0)}, |a, D_t a|_{G_T}^{(0)}, |q, D_t q|_{S_T}^{(0)}\} \; < \; \infty \; , \tag{3.10}$$

$$K_2 \; = \; \max\{\|f, D_t f\|_{2,G_T}, \|\varphi, D_t \varphi\|_{2,S_T}\} \; < \; \infty \; . \tag{3.11}$$

Then there exists a generalized solution Y of Problem (C) such that

$$\max_{t\in(0,T)}\left\{\|D_tY\|_{2,G},\|D_tU\|_{2,G},\|U(t)\|^{(1)}_{2,G},\|Y\nu\|_{2,S}\right\} < \infty, \qquad (3.12)$$

where $U = \operatorname{div} Y + aY$.

Proof. Let $\Psi_n(\xi)$ be sufficiently regular vector functions which form an orthonormal basis in the space $L_2(G)$. We construct the generalized solution of Problem (C) as the limit, as $N \to \infty$, of the Galerkin approximations

$$Y^N = \sum_{k=1}^{N} l_k^N(t)\Psi_k(\xi)$$

with the functions $l_k^N(t)$ satisfying the Cauchy problem

$$\sum_{j=1}^{N} a_{kj}\frac{dl_j^N}{dt} = \sum_{j=1}^{N} b_{kj}l_j^N + \varphi_k, \quad l_k^N(0) = 0, \quad k = 1,\dots,N. \qquad (3.13)$$

Equations (3.13) represent the conditions ensuring the orthogonality of the expression $\{A\langle D_tY^N\rangle - \nabla(\operatorname{div} Y^N + aY^N) - f\}$ to the first N basis functions $\Psi_1,\dots,\Psi_N$ with the boundary condition (3.6) taken into account.

By these conditions, the functions $a_{kj}, b_{kj}, \varphi_k$ can be represented by

$$a_{kj} = (\Psi_k, A\langle\Psi_j\rangle), \quad -b_{kj} = (\operatorname{div}\Psi_j + a\Psi_j, \operatorname{div}\Psi_k) + \big(q(\Psi_j\nu),(\Psi_k\nu)\big)_S,$$

$$\psi_k = (f,\Psi_k) + \big(\varphi,\Psi_k\nu\big)_S.$$

The existence of solutions to the Cauchy problem (3.13) follows by the standard theory of ordinary differential equations. Indeed, by condition (3.3) the matrix $\{a_{kj}\}$ admits the bounded inverse:

$$\sum_{k,j=1}^{N} a_{kj}y_ky_j = \Big(\sum_{k=1}^{N} y_k\Psi_k, A\langle\sum_{j=1}^{N} y_j\Psi_j\rangle\Big) \geq a_0\Big\|\sum_{k=1}^{N} y_k\Psi_k\Big\|^2_{2,G} = a_0|y|^2,$$

hence the system (3.13) can be solved with respect to the derivatives in time.

Conditions (3.10) and (3.11) indicate that the derivatives of the functions a_{kj}, b_{kj} are bounded and the functions φ_k belong to the space $W_2^1(0,T)$. Consequently, there exists a unique solution of problem (3.13) in the space $W_2^2(0,T)$.

It follows easily that the functions Y^N fulfil the identity (3.8) for an arbitrary function

$$\Psi = \sum_{k=1}^{N} c_k(t)\,\Psi_k(\xi).$$

To justify the passage to the limit as $N \to \infty$ in that identity, we obtain uniform estimates on the Galerkin approximations Y^N, independent of N :

$$\left\{\max_{t\in(0,T)}\|Y^N(t)\|_{2,G},\|Y^N\nu\|_{2,S_T},\|\operatorname{div} Y^N\|_{2,G_T}\right\} \leq C_1, \qquad (3.14)$$

$$\left\{ \max_{t\in(0,T)} \|D_t Y^N(t)\|_{2,G}, \|D_t Y^N \nu\|_{2,S_T}, \|\operatorname{div} D_t Y^N\|_{2,G_T} \right\} \le C_2, \qquad (3.15)$$

$$C_i = C_i(K_1, K_2), \quad i = 1, 2 \ .$$

The estimates (3.14), (3.15) can be derived in a standard way. Since the proof of (3.15) is more difficult, we present it in detail. Let us differentiate (3.13) in time, multiply by $D_t l_k^N$ and take the sum in k from 1 to N. We have

$$\frac{1}{2}\frac{d}{dt}\left(D_t Y^N, A\langle D_t Y^N\rangle\right) + \|\operatorname{div} D_t Y^N\|_{2,G}^2 + \left(q(D_t Y^N \nu), (D_t - Y^N \nu)\right)_S$$

$$= -\frac{1}{2}\left(D_t Y^N, D_t A\langle D_t Y^N\rangle\right) + \left(D_t f, D_t Y^N\right) - \left(a D_t Y^N + D_t a Y^N, \operatorname{div} D_t Y^N\right)$$

$$-\left(D_t q(Y^N \nu), (D_t Y^N \nu)\right)_S + \left(D_t \varphi, (D_t Y^N \nu)\right)_S. \qquad (3.16)$$

We integrate (3.16) in time over $(0, t)$, bound the right-hand side from above using the Hölder and Young inequalities, and the left-hand side with the use of (3.2), (3.3), to deduce the inequality

$$\|D_t Y^N(t)\|_{2,G}^2 + \|\operatorname{div} D_\tau Y^N\|_{2,G_t}^2 + \|D_\tau Y^N \nu\|_{2,S_t}^2$$

$$\le c(K_1)\{\|D_\tau Y^N\|_{2,G_t}^2 + K_2^2\},$$

which, by Gronwall's lemma, yields (3.15).

(3.14) and (3.15) imply in a standard way the existence of the generalized solution of Problem (C) for which these estimates also hold.

Coming back to the identity (3.8) we see that

$$(A\langle D_t Y\rangle - f, \Psi) + (U, \operatorname{div} \Psi) = 0 ,$$

for arbitrary finite functions Ψ, hence equivalently

$$A\langle D_t Y\rangle - f = \nabla U. \qquad (3.17)$$

The last relation, combined with (3.15) and (3.11), implies that

$$\max_{t\in(0,T)} \|\nabla U\|_{2,G} < C.$$

Thus we can speak of the trace of the function U on the boundary S for almost all $t \in (0, T)$.

We finally take an arbitrary function Ψ, not necessarily finite, to infer from (3.8) by (3.17) that the identity

$$\left(U - q(Y\nu) - \varphi, \Psi\nu)\right)_S = 0$$

holds. Hence, the boundary condition (3.6) is satisfied almost everywhere on S. $\square$

Classical solution of the one-dimensional Stefan problem for the homogeneous heat equation

---

1. The one-phase Stefan problem. Existence of the solution

At the initial time let the liquid phase occupy the domain $\Omega^+(0) = \{x : 0 < x < x_0\}$ and let the solid phase occupy the domain $\Omega^-(0) = \{x : x_0 < x < \infty\}$. We assume there that the temperature in the solid phase is equal to the melting value, i.e. $\vartheta_0 = 0$ everywhere in $\Omega^-(0)$ and the specific internal energy $U_0 = -1$.

Then in $\Omega_T^+ = \{(x,t) : 0 < x < R(t), \ 0 < t < T\}$ the temperature $\vartheta(x,t)$ satisfies the heat equation

$$\frac{\partial}{\partial t}\Phi(\vartheta) = \frac{\partial^2 \vartheta}{\partial x^2} \qquad \text{for} \quad (x,t) \in \Omega_T^+ \tag{1.1}$$

and two conditions on the free boundary:

$$\vartheta = 0, \quad \frac{dR}{dt} = -\frac{\partial \vartheta}{\partial x} \qquad \text{for} \quad x = R(t), \ t \in (0,T). \tag{1.2}$$

In addition, on the fixed boundary $x = 0$,

$$\vartheta = \vartheta^0(t) \quad \text{or} \quad \frac{\partial \vartheta}{\partial x} + b(t)\vartheta = \vartheta^1(t) \qquad \text{for} \quad t \in (0,T), \tag{1.3}$$

and, at the initial time,

$$R(0) = x_0, \qquad \vartheta(x,0) = \vartheta_0(x) \qquad \text{for} \quad x \in \Omega^+(0). \tag{1.4}$$

According to the results of Chapter II, for sufficiently regular data, the problem just formulated admits a unique classical solution on a small time interval. We now investigate the conditions which permit the possibility of extending the solution onto arbitrary time intervals. By Theorem 4, Chapter II, it is enough to show that $\vartheta(x,t)$ is positive in Ω_T^+ and the norm

$$J_0(T) = |\vartheta|_{\Omega_T^+}^{(1+\gamma)}$$

is bounded for some $\gamma \in (0,1)$.

Indeed, if $J_0(T) < \infty$, then on the top $\{t = T\}$ of the domain Ω_T^+ the function $\tilde{\vartheta}_0(x) = \vartheta(x,T)$ belongs to the space $H^{2+\gamma}(\overline{\Omega(T)})$ and satisfies all the hypotheses of Theorem 4, Chapter II. The solution $\vartheta(x,t)$ can thus be extended onto an interval $(T, T+\delta)$ for some positive δ.

Let

$$M_2 = \max\left\{|\vartheta^0|^{(r/2)}_{[0,\infty)}, |\vartheta_0|^{(r)}_{\Omega^+(0)}, |\vartheta^1, b|^{(1+\gamma)/2}_{[0,\infty)}\right\}, \quad r = 2 + \gamma.$$

Lemma 1. *Let the function $\vartheta^0(t)$ be non-negative, and let $b(t)$ and $\vartheta^1(t)$ be such that the solution of problem (1.1)–(1.4) which fulfils the second of the conditions (1.3) is non-negative in Ω^+_T. Then, under the hypotheses of Theorem 4, Chapter II, the solution $\vartheta(x,t)$ of problem (1.1)–(1.4) satisfies the estimates*

$$0 \leq \vartheta(x,t) \leq N_1(M_2)\exp\left(N_2(M_2)T\right), \tag{1.5}$$

$$-N_3(M_2)\left(|\vartheta|^{(0)}_{\Omega_T} + K\right) \leq \frac{\partial\vartheta}{\partial x}(R(t),t) \leq 0, \tag{1.6}$$

where $K = \max\limits_{x\in\Omega^+(0)}\left|(x - x_0)^{-1}\vartheta_0(x)\right|.$

Proof. For the first of conditions (1.3), the estimate (1.5) follows from the maximum principle. Considering the second of conditions (1.3) at $x = 0$, the function

$$u(x,t) = \vartheta(x,t)\exp\left\{2M_2 x - (4M_2^2 + 1)t\right\}$$

satisfies a homogeneous parabolic equation in Ω^+_T for which the maximum principle holds. On the boundary $x = R(t)$ and at the initial time this function is bounded by constants dependent only on M_2. On the boundary $x = 0$ the function $u(x,t)$ fulfils the second of conditions (1.3) with $\tilde{b}(t) = b(t) - 2M_2 < 0$ and $\tilde{\vartheta} = \vartheta(t)\exp\{-(4M_2^2+1)t\}$. Therefore, at the possible points of positive maximum or negative minimum, $u(0,t)$ is bounded by known quantities.

Non-positiveness of the derivative $\frac{\partial\vartheta}{\partial x}(R(t),t)$ follows from the positiveness of the solution $\vartheta(x,t)$ in the domain Ω^+_T and its vanishing on the free boundary. To estimate the derivative $\frac{\partial\vartheta}{\partial x}(R(t),t)$ from below, consider the function

$$u(x,t) = \vartheta(x,t) + N(x - R(t)), \quad N = \text{const.} > 0.$$

Since the derivative dR/dt is non-negative by the Stefan condition, the expression

$$Lu \equiv a(\vartheta)\frac{\partial u}{\partial t} - \frac{\partial^2 u}{\partial x^2}, \quad a(s) = \frac{d\Phi}{ds}(s),$$

is non-positive in Ω^+_T. The function $u(x,t)$ therefore cannot attain its positive maximum in the interior of Ω^+_T. Because, further, $R(t) \geq R(0)$, by choosing appropriately large N, dependent only on K and $|\vartheta|^{(0)}_{\Omega^+_T}$, we can easily succeed in making $u(x,t)$ non-positive on the boundary $x = 0$ and at the initial time. The function $u(x,t)$ is thus non-positive everywhere in Ω^+_T, and it vanishes on the boundary $x = R(t)$. Hence, the derivative $\frac{\partial u}{\partial x}$ is non-negative on the free boundary, this being equivalent to the left part of (1.6). $\quad\square$

Remark. If the functions $b(t)$ and $\vartheta^1(t)$ in the second of conditions (1.3) are non-positive, then the solution $\vartheta(x,t)$ of problem (1.1)–(1.4), satisfying the second of conditions (1.3), is non-negative in the domain Ω^+_T.

Lemma 2. *Let the hypotheses of Lemma 1 hold. Then the solution $\vartheta(x,t)$ of problem (1.1)–(1.4) satisfies the bound*

$$J_0(T) = |\vartheta|_{\Omega_T^+}^{(1+\gamma)} \leq N_4, \tag{1.7}$$

where the constant N_4 depends on M_2 and T, and is finite if the constants M_2 and T are finite.

Proof. Throughout this proof, without any further specification all constants N will depend on M_2 and T.

First of all let us note that the free boundary $x = R(t)$ is monotone increasing in time, and its distance from the fixed boundary $x = 0$ is at least x_0. This makes it possible to apply Theorem 9 of Section I.3 on the regularity of generalized solutions of the Stefan problem in the case of one space variable, to claim that

$$|\vartheta|_{\Omega_T^+}^{(\beta)} \leq N_5 \tag{1.8}$$

for some $\beta \in (0,1)$. Also we can use local estimates for linear parabolic equations (see Theorem 6, Section I.2).

In particular, we can assume that the derivative $\frac{\partial\vartheta}{\partial x}$ on the boundary $x = 0$ has modulus bounded by the same constant N_5. In the domain Ω_T^+, the derivative $\frac{\partial\vartheta}{\partial x}$ satisfies a homogeneous parabolic equation for which the maximum principle holds. This derivative is bounded on the boundaries $x = 0$, $x = R(t)$ and at the initial time. The derivative $\frac{\partial\vartheta}{\partial x}$ has bounded modulus on the domain Ω_T^+.

In the new variables

$$\tau = t, \quad y = \frac{x}{R(t)}, \tag{1.9}$$

the domain $Q_T = Q \times (0,T)$, $Q = \{y : 0 < y < 1\}$, corresponds to Ω_T^+. The bounded function $u(y,\tau) = \vartheta(yR(\tau),\tau)$ is a solution of the initial-boundary value problem

$$\tilde{a}\frac{\partial u}{\partial \tau} - \frac{1}{R^2(\tau)}\frac{\partial^2 u}{\partial x^2} = f \quad \text{for} \quad (y,\tau) \in Q_T, \tag{1.10}$$

$$u = \vartheta^0 \quad \text{or} \quad \frac{1}{R}\frac{\partial u}{\partial y} + bu = \vartheta^1 \quad \text{for} \quad y = 0, \ \tau \in (0,T), \tag{1.11}$$

$$u(1,\tau) = 0 \quad \text{for} \quad \tau \in (0,T),$$

$$u(y,0) = \vartheta_0(yx_0) \quad \text{for} \quad y \in Q, \tag{1.12}$$

where

$$f = -\frac{y}{R^2(\tau)}\frac{\partial u}{\partial y}(1,\tau)\frac{\partial u}{\partial y}(y,\tau)\tilde{a} .$$

The function $\tilde{a} = a\big(u(y,\tau)\big)$ belongs to the space $H^{\beta,\beta/2}(\bar{Q}_T)$ (by (1.8)), and $R(\tau)$ is an element of $H^\alpha[0,T]$ for arbitrary $\alpha \in (0,1)$. Hence, by Theorems 3 and 4, Section I.2,

$$|u|_{Q_T}^{(2+\beta)} \leq N_6\big(|f|_{Q_T}^{(\beta)} + 1\big). \tag{1.13}$$

Because of the boundedness of $|\tilde{a}|_{Q_T}^{(\beta)}$, $|\log R|_{[0,T]}^{(0)}$ and the derivatives $\frac{dR}{d\tau}$, $\frac{\partial u}{\partial y}$, the $H^{\beta,\beta/2}(\bar{Q}_T)$-norm of f can be estimated in the standard way as

$$|f|_{Q_T}^{(\beta)} \leq N_7\left(|u|_{Q_T}^{(1+\beta)} + 1\right).$$

Applying the interpolation inequality (2.3), Section I.2,

$$|u|_{Q_T}^{(1+\beta)} \leq \mu|u|_{Q_T}^{(2+\beta)} + c\mu^{-(1+\beta)}|u|_{Q_T}^{(0)}$$

with uniform constant c and arbitrarily small positive number μ to the last relation, we see that

$$|f|_{Q_T}^{(\beta)} \leq N_8\left(\mu|u|_{Q_T}^{(2+\beta)} + \mu^{-2}\right).$$

Recalling the bound (1.13) and using the latter inequality, for $\mu N_6 N_8 < 1$ we conclude that

$$|u|_{Q_T}^{(2+\beta)} \leq N_9.$$

The above inequality combined with the straight-forward relation

$$|\vartheta|_{Q_T^+}^{(1+\gamma)} \leq N_{10}|u|_{Q_T}^{(2+\beta)}$$

completes the proof.   □

Hence we have also proved the following theorem:

Theorem 1. *Let the hypotheses of Theorem 4, Chapter II, be satisfied, and let the functions $\vartheta^0, \vartheta^1, b$ be such that the solution $\vartheta(x,t)$ of problem (1.1)–(1.4) is non-negative in the domain Ω_∞^+. Then the constructed solution exists for all values of $t \in (0,\infty)$, and it belongs to the space $H^{r,r/2}$, $r = 2 + \gamma$, in an arbitrary closed bounded subdomain of Ω_∞^+.*   □

The postulated regularity of the functions $\vartheta^0, \vartheta^1, b$ is necessary for the construction of the solution; in turn, the regularity of the solution $\vartheta(x,t)$ and $R(t)$ at the points of Ω_∞^+ that do not belong to the lines $\{t = 0\}$ and $\{x = 0\}$ is independent of the regularity of the functions $\vartheta^0, \vartheta^1, b$ and ϑ_0. Indeed, let $\Phi(s)$ be infinitely differentiable and $G_{\delta,T}^n$ be a sequence of the domains $\{(x,t) : 0 < \delta - 1/n < x < R(t), \delta - 1/n < t < T\}$ such that $G_{\delta,T}^n \supset G_{\delta,T}^{n+1}$. Because $R \in H^{(3+\gamma)/2}[\delta - 1/n, T]$, the solution $\vartheta(x,t)$ belongs to the space $H^{3+\gamma,(3+\gamma)/2}\left(\overline{G_{\delta,T}^{n+1}}\right)$ (by local estimates for solutions of linear parabolic equations). By the Stefan condition (1.2), the function $R(t)$ belongs to the space $H^{(4+\gamma)/2}[\delta - 1/(n + 1), T]$ on the smaller interval $[\delta - 1/(n + 1), T]$. By successive repetitions of the above process, we conclude the infinite differentiability of the solution $\vartheta(x,t)$ in the domain $\{\delta < x \leq R(t), \delta < t \leq T\}$. While the infinite differentiability of $\vartheta(x,t)$ in the interior of Ω_∞^+ followed from the properties of parabolic equations, the same property of the free boundary $x = R(t)$ at $t > 0$ is a consequence of the Stefan condition (1.2) which "lifts" the regularity of the solution in case of one space variable.

The one-dimensional one-phase Stefan problem offers advantageous features: the free boundary $x = R(t)$ is always monotone there, and the boundedness of the derivative $\frac{dR}{dt}$ is ensured by the boundedness of the solution $\vartheta(x, t)$ and the positiveness of

$$J_1(x) \; = \; K\,(x - x_0) - \vartheta_0(x) \qquad (1.14)$$

for some constant $K > 0$. It turns out that these conditions are sufficient for constructing the classical solution of the Stefan problem (1.1)–(1.4) on an infinite time interval.

Theorem 2. *Let $\Phi \in C^2[0, \infty)$, $\Phi'(s) \geq a_0 > 0$ and let the non-negative functions $\vartheta^0(t)$, $\vartheta_0(x)$ be such that*

$$\vartheta^0(t) \; \leq \; Kx_0, \quad \vartheta_0(x) \; \leq \; K\,(x_0 - x)\,. \qquad (1.15)$$

Then, for all times in $(0, \infty)$ there exists a unique classical solution $\{R(t), \vartheta(x, t)\}$, with the derivative dR/dt Hölder continuous for all $t > 0$, and $\vartheta(x, t)$ belonging to the space $H^{r, r/2}\left(\overline{G^{\infty}_{\delta, T}}\right)$, $r > 2$, in each bounded domain $G^{\infty}_{\delta, T} = \{(x, t) \in \Omega^+_T : x > \delta, \ t > \delta\}$.

If $\Phi(s)$ is an infinitely differentiable function, then $R(t)$ is also infinitely differentiable for all $t > 0$.

Proof. As already mentioned, it is enough to prove that the solution $\vartheta(x, t)$ belongs to the space $H^{r, r/2}$ for every domain $\overline{G^{\infty}_{\delta, T}}$.

The functions $\vartheta(x, t)$ and $R(t)$ can be constructed in a natural way as follows: $\vartheta^0(t)$ and $\vartheta_0(x)$ are approximated by sufficiently regular functions $\vartheta^0_\varepsilon(t)$ and $\vartheta^\varepsilon_0(x)$ which satisfy the hypotheses of Theorem 1. Then fundamental subsequences convergent to $\vartheta(x, t)$ and $R(t)$, respectively, are selected from the sequences $\{\vartheta_\varepsilon\}, \{R_\varepsilon\}$, where $\{\vartheta_\varepsilon, R_\varepsilon\}$ is the solution of problem (1.1)–(1.4), corresponding to $\vartheta^0_\varepsilon(t)$ and $\vartheta^\varepsilon_0(x)$ as data.

We now formulate (without proof) the following convergence result.

Lemma 3. *Let the positive functions $\vartheta^0(t)$ and $\vartheta_0(x)$ satisfy condition (1.15). Then there exist sequences $\{\vartheta^0_\varepsilon(t)\}$ and $\{\vartheta^\varepsilon_0(x)\}$, pointwise convergent to the functions $\vartheta^0(t)$ and $\vartheta_0(x)$, respectively, for which the hypotheses of Theorem 1 and inequality (1.15) hold.* $\square$

We need estimates uniform with respect to ε for the solution. The first of these estimates follows by Lemma 1:

$$0 \leq \vartheta_\varepsilon \leq Kx_0, \ (x, t) \in \Omega^\varepsilon_\infty; \quad 0 \leq \frac{dR_\varepsilon}{dt} \leq K, \ 0 < t < \infty, \qquad (1.16)$$

where $\Omega^\varepsilon_\infty = \{(x, t) : 0 < x < R_\varepsilon(t), \ 0 < t < \infty\}$.

We further show that the solution $\vartheta_\varepsilon(x, t)$ is Hölder continuous in the domain $G^{\infty}_{\delta, T}$ for some index $\beta > 0$ and Hölder constant uniformly bounded for all $\varepsilon > 0$ (although dependent on δ).

The function $u(x, t) = \zeta(t)\vartheta(x, t)$, where $\zeta \in C^2[0, \infty)$, $\zeta = 1$ for $t > \delta$ and $\zeta = 0$ for $t < \delta/2$, satisfies the equation

$$a(\vartheta_\varepsilon)\frac{\partial u}{\partial t} - \frac{\partial^2 u}{\partial x^2} \; = \; f(x, t)$$

with bounded right-hand side $f = \frac{d\zeta}{dt} a(\vartheta_\varepsilon)\vartheta_\varepsilon$ in $\Omega_\infty^\varepsilon$. On $x = R_\varepsilon(t)$ we have

$$u = 0, \quad \frac{\partial u}{\partial x} = -\zeta^{-1}\frac{dR}{dt}, \quad \frac{\partial u}{\partial t} = \zeta^{-1}\left|\frac{\partial u}{\partial x}\right|^2 \qquad \text{for} \quad x = R_\varepsilon(t).$$

Multiply the equation for $u(x,t)$ by the derivative $\frac{\partial u}{\partial t}$ and then integrate over $x \in [\delta_1, R_\varepsilon(t)]$:

$$\int_{\delta_1}^{R_\varepsilon(t)} a\left|\frac{\partial u}{\partial t}\right|^2 dx + \frac{1}{2}\frac{d}{dt}\int_{\delta_1}^{R_\varepsilon(t)}\left|\frac{\partial u}{\partial x}\right|^2 dx + \frac{1}{2\zeta}\left|\frac{\partial u}{\partial x}\right|^3\bigg|_{x=R_\varepsilon(t)}$$

$$= \int_0^{R_\varepsilon(t)} f\frac{\partial u}{\partial t} dx - \left(\frac{\partial u}{\partial x}\frac{\partial u}{\partial t}\right)\bigg|_{x=\delta}.$$

Applying Hölder's inequality to the right-hand side of this identity, from the boundedness of $\left(\frac{\partial u}{\partial x}\frac{\partial u}{\partial t}\right)\big|_{x=\delta}$ (following from local bounds on the solutions of parabolic equations), we deduce the inequality

$$\int_0^T \int_\delta^{R_\varepsilon(t)}\left|\frac{\partial u}{\partial t}\right|^2 dxdt + \max_{t\in(\delta,T)}\int_\delta^{R_\varepsilon(t)}\left|\frac{\partial u}{\partial x}(x,t)\right|^2 dx \leq M_1(K,T,\delta),$$

which, as in the study of the regularity of the generalized solution of the one-dimensional Stefan problem (see Section I.3), implies that $u(x,t)$ is in the space $H^{\beta,\beta/2}(\overline{G_{\delta,T}^\infty})$ for some $\beta > 0$. Because $u = \vartheta_\varepsilon$ in $G_{\delta,T}^\infty$, we have

$$\left|\vartheta_\varepsilon\right|_{G_{\delta,T}^\infty}^{(\beta)} \leq M_3(K,T,\delta). \tag{1.17}$$

Each function $\vartheta_\varepsilon(x,t)$ is defined on its own domain $\Omega_\infty^\varepsilon$. Hence the selection of a fundamental subsequence and passing to the limit can be performed more easily in the equivalent problem formulated in terms of the variables (1.9), in which the solution $u_\varepsilon(y,\tau) = \vartheta_\varepsilon(yR_\varepsilon(\tau),\tau)$ is defined on a fixed domain, the same for all ε.

Due to (1.16) and (1.17), in the domain $Q_{\delta,T} = \{(y,\tau) : \delta < y < 1, \delta < t < T\}$, the bounded function u_ε satisfies equation (1.10) with the coefficients in the main part belonging to the space $H^{\beta,\beta/2}(\bar{Q}_{\delta,T})$. By (1.16), the coefficient of the lowest derivative is bounded. Then one can apply the standard local estimates for linear parabolic equations in $W_p^2(Q_{\delta,T})$ (see Theorem 7, Section I.2), to conclude

$$\|u_\varepsilon\|_{p,Q_{\delta,T}}^{(2)} \leq M_4(K,\delta,p)$$

for arbitrary $p > 2$.

In view of the embedding of $W_p^{2,1}(Q_{\delta,T})$ into $H^{1+\beta,(1+\beta)/2}(\bar{Q}_{\delta,T})$, for sufficiently large p we get the bound (see Lemma 5, Section I.2)

$$\left|u_\varepsilon\right|_{Q_{\delta,T}}^{(1+\beta)} \leq c\|u_\varepsilon\|_{p,Q_{\delta,T}}^{(2)} \leq cM_4.$$

Therefore, the coefficients and the right-hand side of equation (1.10) are Hölder continuous and, by local estimates for the solutions of linear parabolic equations in Hölder spaces (see Theorem 6, Section I.2), it follows that

$$|u_\varepsilon|_{Q_{\lambda,T}}^{(2+\beta)} \le M_5(K,T,\delta,\lambda) \qquad \text{for} \quad \lambda > \delta. \tag{1.18}$$

Let $\lambda = 1/n$. By the application of diagonalisation to the sequence $\{u_\varepsilon\}$, we can select a subsequence $\{u_{\varepsilon_n}\}$ fundamental in the space $H^{r,r/2}(\bar{Q}_{1/m,m})$, $r = 2+\gamma$, $\gamma < \beta$, for every fixed $m > 0$. Clearly, the functions

$$R(t) = \left(x_0^2 - 2 \int_0^t \frac{\partial u}{\partial y}(1,\tau)\, d\tau \right)^{1/2},$$

where

$$u(y,\tau) = \lim_{n\to\infty} u_{\varepsilon_n}(y,\tau),$$

and $\vartheta(x,t) = u\big(x/R(t),t\big)$, are the desired solution of problem (1.1)–(1.4). Here,

$$R \in H^{(3+\gamma)/2}[1/\delta,T], \quad \vartheta \in H^{r,r/2}(\overline{G_{\delta,T}^\infty}), \quad \delta > 0.$$

This solution is unique, because for each finite $T > 0$ the function

$$U(x,t) = \begin{cases} \Phi\big(\vartheta(x,t)\big) & \text{in } \Omega_T^+, \\ -1 & \text{otherwise (for } x > 0,\ 0 < t < T), \end{cases}$$

is a bounded generalized solution of the Stefan problem with given data $\vartheta^0(t)$ at $x = 0$ and $U_0(x)$ at $t = 0$, where

$$U_0(x) = \begin{cases} \Phi\big(\vartheta_0(x)\big) & \text{for } x \in (0,x_0), \\ -1 & \text{for } x > x_0. \end{cases}$$

Finally, we shall consider the Stefan problem (1.1)–(1.4) with the heat flux prescribed on the boundary $x = 0$:

$$\frac{\partial \vartheta}{\partial x}(0,t) = \vartheta^1(t). \tag{1.19}$$

As we noted in Chapter III, if $\vartheta^1 \equiv 0$, then Lagrange variables in which the domain Ω_T^+ is transformed onto a given fixed domain can be introduced, and the Stefan problem becomes equivalent to an initial-boundary value problem for a system of nonlinear equations. It turns out that in case of one space variable this system reduces to a nonlinear heat equation, and the condition $\vartheta^1 \equiv 0$ can be removed (the general situation $\vartheta^1 \neq 0$ can be considered, instead).

Moreover, in the case of one space variable so-called "mass" Lagrange variables

$$\tau = t, \quad y = \int_x^{R(t)} \big\{ \Phi\big(\vartheta(s,t)\big) + 1 \big\}\, ds \tag{1.20}$$

can be introduced, with the simplest resulting differential equation.

Indeed, the domain Ω_T^+ is transformed by (1.20) onto $Q_T = \{(y, \tau) \; : \; 0 < y < Y(\tau), \; 0 < \tau < T\}$, where

$$Y(\tau) = \int_0^{R(\tau)} \{\Phi(\vartheta(x, \tau)) + 1\} \; dx.$$

From $\Phi(\vartheta(R, \tau)) = \Phi(+0) = 0$ it follows that

$$\frac{dY}{d\tau} = \frac{dR}{d\tau} + \int_0^{R(\tau)} \frac{\partial \Phi}{\partial \tau} \; dx = \frac{dR}{d\tau} + \int_0^{R(\tau)} \frac{\partial^2 \vartheta}{\partial x^2} \; dx = -\frac{\partial \vartheta}{\partial x}(0, \tau).$$

Thus, $Y(\tau)$ is given by

$$Y(\tau) = Y_0 - \int_0^\tau \vartheta^1(s) \; ds, \qquad Y_0 = \int_0^{x_0} \{\Phi(\vartheta_0(x)) + 1\} \; dx.$$

The function $v(y, \tau) \equiv \vartheta(x, t)$ satisfies the nonlinear heat equation

$$b(v)\frac{\partial v}{\partial \tau} = \frac{\partial^2 v}{\partial y^2} \quad \text{in} \quad Q_T, \qquad \text{where} \qquad b(v) = \frac{d\Phi}{dv}(v)\{1 + \Phi(v)\}^{-2},$$

vanishes on the boundary $\{y = 0\}$ and

$$\frac{\partial v}{\partial y}(Y(\tau), \tau) = -\vartheta^1(\tau).$$

To determine the initial value of the function $v(y, \tau)$ ($v(y, 0) = v_0(y)$), let us note that all the quantities in (1.20) are known at the initial time. Hence the transformation

$$y = Y_0(x) \equiv \int_x^{x_0} \{\Phi(\vartheta_0(s)) + 1\} \; ds, \qquad x = Y_0^{-1}(y) \tag{1.21}$$

is well-defined, and this, in turn, determines the function

$$v_0(y) = \vartheta_0(Y_0^{-1}(y)). \tag{1.22}$$

Altogether, the Stefan problem (1.1)–(1.4), (1.19) in Lagrange variables is equivalent to an initial-boundary value problem for the nonlinear heat equation on a given domain. Since the latter class of problems has been extensively studied, this greatly simplifies the analysis of the one-dimensional one-phase Stefan problem with heat flux prescribed on the fixed boundary.

For some special functions $\Phi(s)$, the solution of the Stefan problem can be determined in an explicit analytic form. Let

$$\Phi(s) = s/(1 - s), \qquad \vartheta^1(t) \equiv 0.$$

In the cylinder $Q_T = \{(y, \tau) \; : \; 0 < y < Y_0, \; 0 < \tau < T\}$, the function $v(y, \tau)$ satisfies the heat equation

$$\frac{\partial v}{\partial \tau} = \frac{\partial^2 v}{\partial y^2}. \tag{1.23}$$

It vanishes on the boundary $\{y = 0\}$ and its derivative with respect to y vanishes on the boundary $\{y = Y_0\}$.

Let $\tilde{v}(y, \tau)$ be the function periodic in y with period $4Y_0$, defined by the even extension of the function $v(y, \tau)$ onto the domain $\{(y, \tau) \; : \; Y_0 < y < 2Y_0, \; 0 < \tau < T\}$ and further by its odd extension onto $\{(y, \tau) \; : \; -2Y_0 < y < 0, \; 0 < \tau < T\}$. This function satisfies the heat equation (1.23) in the strip $\{(y, \tau) \; : \; |y| < \infty, \; 0 < \tau < T\}$. It coincides at the initial time with the given function $\tilde{v}_0(y)$, where $\tilde{v}_0(y)$, a $4Y_0$-periodic function of y, is the analogous extension of $v_0(y)$.

Hence

$$v(y, \tau) = (4\pi t)^{-1/2} \int_{-\infty}^{\infty} \tilde{v}_0(s) \exp\left\{ -\frac{(y - s)^2}{4t} \right\} ds.$$

From the relation

$$\frac{dR}{dt}(t) = \frac{\partial v}{\partial y}(0, t),$$

it follows that

$$\frac{dR}{dt}(t) = (4\pi t)^{-1/2} \int_{-\infty}^{\infty} \frac{d\tilde{v}_0}{ds}(s) \exp\left(-\frac{s^2}{4t} \right) ds. \qquad (1.24)$$

The representation (1.24), together with (1.21) and (1.22), avoids the necessity of solving the nonlinear Stefan problem (1.1)–(1.4).

2. Asymptotic behaviour of the solution of the one-phase Stefan problem

In this section we study the behaviour, as $t \to \infty$, of the solution of problem (1.1)–(1.4) subject to the first condition (1.3). The case of heat flux prescribed on the boundary $\{x = 0\}$ has been discussed in Friedman's monograph [92]. Besides, as already mentioned, the problem we consider in Lagrange variables is equivalent to an initial-boundary value problem for the nonlinear heat equation on a given domain, for which the methods of analysis are well-known.

An outline of the study of problem (1.1)–(1.4) consists of an analysis of its simplest self-similar solutions and an application of the comparison theorem to the generalized solutions of the Stefan problem (see Theorem 11, Section I.3).

First we prove the existence of self-similar solutions of the Stefan problem.

Theorem 3. *Let $\Phi \in C^2[0, \beta]$, $\Phi'(s) > 0$ for $s \in [0, \beta]$. Then the Stefan problem (1.1)– (1.4) with given data $\vartheta^0(t) = \beta = \text{const} > 0$, $x_0 = 0$, $U_0(x) = -1$ for $x \in (0, \infty)$ has the unique solution*

$$R_*(t) = D_*(\beta) t^{1/2}, \quad \vartheta_*(x, t) = v(xt^{-1/2}, \beta).$$

Here $D_*(\beta)$ depends continuously dependent on β, and $\lim D_*(\beta) = 0$ as $\beta \to 0$.

Proof. It follows easily that the function $v(\xi, \beta)$, with $\xi = xt^{-1/2}$, and the parameter $D_* = D_*(\beta)$ can be determined from the conditions

$$\frac{d^2v}{d\xi^2} + \frac{\xi}{2}a(v)\frac{dv}{d\xi} = 0, \quad a = \Phi'(v), \quad \xi \in (0, D_*), \tag{2.1}$$

$$v(0, \beta) = \beta, \qquad v(D_*, \beta) = 0, \tag{2.2}$$

$$\frac{dv}{d\xi}(D_*, \beta) = -\frac{1}{2}D_*. \tag{2.3}$$

We will show that for each $D > 0$ there exists at least one function $V(\xi)$, satisfying equation (2.1) and boundary conditions (2.2). Further, substituting the value of $dV/d\xi$, computed at the point $\xi = D$, into the left-hand side of (2.3), we obtain an equation whose solution D_* determines the solution of the problem (2.1)–(2.3).

To define $V(\xi)$, let us consider the linear boundary value problem

$$\frac{d^2\tilde{V}}{d\xi^2} + \frac{\xi}{2}a\big(g(\xi)\big)\frac{d\tilde{V}}{d\xi} = 0, \quad \tilde{V}(0) = \beta, \quad \tilde{V}(D) = 0,$$

where the coefficient a has as argument an arbitrary non-negative function $g(\xi)$, which is continuous on the interval $(0, D)$ and bounded there by the constant β. The solution of this problem admits the representation

$$\tilde{V}(\xi) = \beta \left\{ \int_0^D \exp\left(-\int_0^\tau \frac{s}{2}a\big(g(s)\big)\,ds\right)d\tau \right\}^{-1} \int_\xi^D \exp\left(-\int_0^\tau \frac{s}{2}a\big(g(s)\big)\,ds\right)d\tau. \tag{2.4}$$

The right-hand side of the above expression defines a continuous operator $\Psi(g)$ in the set $\mathcal{M}$ of functions g with the properties specified earlier. Moreover, due to the uniform boundedness of the derivatives of $\tilde{V}(\xi)$,

$$\left|\frac{d\tilde{V}}{d\xi}(\xi)\right| \le \beta \left\{ \int_0^D \exp\left(-\frac{a_0}{4}s^2\right)ds \right\}^{-1}, \quad \text{with} \quad a_0 = \min_{s \in (0,\beta)}\{a(s), a^{-1}(s)\},$$

the operator $\Psi(g)$ is completely continuous on the set $\mathcal{M}$. Hence, by Schauder's theorem, there exists at least one fixed point V of the operator $\Psi : V \to \Psi(V)$, where $V(\xi)$ satisfies equation (2.1) and conditions (2.2).

Since $V(\xi)$ admits a representation analogous to (2.4), and thus

$$-\beta \exp\left(-\frac{a_0}{4}D^2\right)\left\{\int_0^D \exp\left(-\frac{\tau^2}{4a_0}\right)d\tau\right\}^{-1} \le \frac{dV}{d\xi}(D)$$

$$\le -\beta \exp\left(-\frac{D^2}{4a_0}\right)\left\{\int_0^D \exp\left(-\frac{a_0\tau^2}{4}\right)d\tau\right\}^{-1},$$

the equation

$$\frac{dV}{d\xi}(D) = -\frac{1}{2}D$$

has at least one positive solution D_*.

The function $U_*(x,t)$, which coincides with $\Phi(\vartheta_*(x,t))$ for $0 < x < R_*(t)$ and which is equal to -1 for $x > R_*(t)$, is the unique bounded generalized solution of the Stefan problem with data as specified in Theorem 3. This implies the uniqueness of the self-similar solution we have constructed.

The continuity of $D_*(\beta)$ with respect to β follows from the continuous dependence of solutions to ordinary differential equations on a parameter.

The last assertion of the theorem follows from the identity

$$\frac{1}{2}D_*^2(\beta) + \int_0^{D_*(\beta)} \xi\Phi\big(v(\xi,\beta)\big)\,d\xi \;=\; \beta, \tag{2.5}$$

which results from multiplying equation (2.1) by ξ and then integrating it over $\xi \in (0, D_*)$ using (2.2) and (2.3). □

We shall restrict the study of the asymptotic behaviour of the solution of the Stefan problem to the situation where

$$\lim_{t\to\infty} \vartheta^0(t) = \beta, \qquad 0 \le \beta < \infty, \tag{2.6}$$

with the case $\beta = 0$ considered separately.

Theorem 4. *Let the hypotheses of Theorem 2 be satisfied and the condition (2.6) hold with $\beta > 0$. Then the equality*

$$\lim_{t\to\infty} t^{-1/2}R(t) = D_*(\beta),$$

is fulfilled by the solution $R(t)$ of problem (1.1)–(1.4), where $D_*(\beta)$ denotes the solution of problem (2.1)–(2.3).

Proof. The existence of the solution $\{R(t), \vartheta(x,t)\}$ of problem (1.1)–(1.4) for all $t \in (0,\infty)$ follows from Theorem 2. To estimate $R(t)$ from below, let us fix an arbitrary small $\varepsilon > 0$ and assume t_0 large enough to ensure that

$$0 < \beta - \varepsilon \le \vartheta^0(t) \le \beta + \varepsilon \qquad \text{for} \quad t \in (t_0, \infty). \tag{2.7}$$

The function $\vartheta_*(x,t) = v\big(x(t-t_0)^{-1/2}, \beta - \varepsilon\big)$ defines a generalized solution $U_*(x,t)$ which exceeds neither $U^0(t) = \Phi(\vartheta^0(t))$ on the boundary $\{x = 0\}$ nor $U_0(x)$, coincides with $\Phi(\vartheta(x,t_0))$ for $0 < x < R(t_0)$ and equals -1 for $x > R(t_0)$. The function $U(x,t)$, coinciding with $\Phi(\vartheta(x,t))$ for $0 < x < R(t)$ and equal to -1 for $x > R(t)$, for $t > t_0$ is a generalized solution of the Stefan problem, and coincides with the functions $u^0(t)$ and $U_0(x)$ at $x = 0$ and $t = t_0$, respectively. Hence, by Theorem 11, Section I.3, it follows that

$$R(t) \ge (t - t_0)^{1/2}D_*(\beta - \varepsilon), \quad U(x,t) \ge U_*(x,t) \qquad \text{for} \quad t > t_0. \tag{2.8}$$

To estimate $R(t)$ from above, we shall use the identity

$$B(t) = B(t_0) + \int_{t_0}^{t} \vartheta^0(\tau)\, d\tau, \tag{2.9}$$

where $B(t) = \int_{0}^{R(t)} xU(x,t)\, dx + \frac{1}{2} R^2(t).$

We obtain (2.9) from multiplying (1.1) by x and then integrating by parts. We shall estimate $B(t)$ from below by making use of inequalities (2.8):

$$\int_{0}^{R(t)} xU(x,t)\, dx \geq \int_{0}^{R_*(t)} xU_*(x,t)\, dx = (t - t_0) \int_{0}^{D_*(\beta-\varepsilon)} \xi\Phi\big(v(\xi,\beta-\varepsilon)\big)\, d\xi.$$

Recalling identity (2.5) and inequality (2.7), we get

$$\frac{1}{2} R^2(t) + (t - t_0)\left(\beta - \varepsilon - \frac{1}{2} D_*^2(\beta - \varepsilon)\right)$$

$$\leq\; B(t_0) + \int_{t_0}^{t} \vartheta^0(\tau)\, d\tau \;\leq\; B(t_0) + (t - t_0)(\beta + \varepsilon).$$

Passing to the limit as $t \to \infty$ gives us

$$\overline{\lim_{t\to\infty}}\, \big(t^{-1} R^2(t)\big) \;\leq\; D_*^2(\beta - \varepsilon) + 4\varepsilon.$$

At the same time, (2.8) implies that

$$D_*^2(\beta - \varepsilon) \leq \varliminf_{t\to\infty}\big(t^{-1} R^2(t)\big).$$

The last two inequalities, the continuity of function $D_*(\beta)$ and the assumed arbitrary choice of ε complete the proof. $\square$

Let us denote

$$z(t, t_0) = \int_{t_0}^{t} \vartheta^0(\tau)\, d\tau.$$

Theorem 5. *Let the hypotheses of Theorem 2 be satisfied, and let the equality* (2.6) *hold with* $\beta = 0$. *If* $z(\infty, 0) < \infty$, *then*

$$\lim_{t\to\infty} R(t) = \{2B(0) + 2z(\infty, 0)\}^{1/2},$$

where $B(t)$ *is defined by* (2.9). *If* $z(\infty, 0) = \infty$, *then*

$$\lim_{t\to\infty} \big\{ R(t)\big(z(t,0)\big)^{-1/2} \big\} = \sqrt{2}.$$

Proof. Define

$$M(t) = \max_{x\in\Omega^+(t)} U(x,t), \quad M_\infty = \overline{\lim_{t\to\infty}} M(t).$$

We estimate the second term of (2.9) from below by zero. Then, dividing both sides of the obtained inequality by $z(t,0)$ and passing to the limit, we conclude that

$$\varlimsup_{t\to\infty}\left\{\frac{R^2(t)}{z(t,0)}\right\} \le 2\left\{1 + B(0)\left(\lim_{t\to\infty} z(t,0)\right)^{-1}\right\}. \tag{2.10}$$

On the other hand, the same term on the left-hand side of (2.9) can be estimated from above by the expression $(1/2)M(t)R^2(t)$, thus giving rise to the inequality

$$\varliminf_{t\to\infty}\left\{\frac{R^2(t)}{z(t,0)}\right\} \ge \frac{2}{1 + M_\infty}\left\{1 + B(0)\left(\lim_{t\to\infty} z(t,0)\right)^{-1}\right\}. \tag{2.11}$$

The assertion of the theorem is a consequence of (2.10) and (2.11), provided that $M_\infty = 0$. To prove that indeed $M_\infty = 0$, we shall use the following obvious result.

Lemma 4. *The solution of the initial-boundary value problem*

$$a(u)\frac{\partial u}{\partial t} = \frac{\partial^2 u}{\partial x^2} \quad \text{for } (x,t) \in \mathbb{R}_\infty^+, \quad u|_{t=0} = u_0, \quad u|_{x=0} = u^0, \tag{2.12}$$

where $u_0, u^0 = \text{const.}$, admits the representation $u(x,t) = v(\xi)$, $\xi = xt^{-1/2}$, with $v(\xi)$ given by

$$v(\xi) = u_0 + (u^0 - u_0) \cdot \frac{\displaystyle\int_\xi^\infty \exp\left(-\int_0^\tau \frac{s}{2}a(v(s))\, ds\right) d\tau}{\displaystyle\int_0^\infty \exp\left(-\int_0^\tau \frac{s}{2}a(v(s))\, ds\right) d\tau}.$$

In particular, $\min(u_0, u^0) \le u(x,t) \le \max(u_0, u^0)$, and, if $u_0 u^0 > 0$, then $\Phi(u(x,t))$ is the generalized solution of the Stefan problem.

Let $u(x,t)$ be the solution of the problem (2.12) with given data $u^0 = \varepsilon$, $u_0 = |\vartheta|_{\Omega_\infty^+}^{(0)}$, where $\varepsilon > 0$ is arbitrary but fixed. We shall assume the time t_0 large enough to ensure that

$$0 \le \vartheta^0(t) < \varepsilon \quad \text{for } t \in (t_0, \infty).$$

By construction, the function $\Phi(u(x, t - t_0))$ is a generalized solution of the Stefan problem such that $\Phi(u(x, t - t_0)) > U(x,t)$ on the boundary $\{x = 0\}$ and for $t = t_0$. By the comparison theorem applied to generalized solutions of the Stefan problem,

$$0 \le U(x,t) \le \Phi(u(x, t - t_0)) \quad \text{for } (x,t) \in \Omega_\infty^+ \setminus \Omega_{t_0}^+. \tag{2.13}$$

Let us consider the domain

$$Q^\delta = \{(x,t) : 0 < x < \delta(t - t_0)^{1/2}, \, t_0 < t < \infty\}.$$

By construction, the function $u(x,t)$ assumes values close to $u^0 = \varepsilon$ in the domain Q^δ, as long as δ remains sufficiently small, i.e., there exists $\delta = \delta(\varepsilon)$ such that

$$\varepsilon \le u(x,t) \le 2\varepsilon \quad \text{for } (x,t) \in Q^\delta. \tag{2.14}$$

It remains to show that, for some $t_* > t_0$, the whole domain $\Omega_\infty^+ \setminus \Omega_{t_*}^+$ is contained in Q^δ. By the continuity of the function Φ and due to the arbitrary choice of ε, the latter property will imply that $M_\infty = 0$.

We now take $\beta < \varepsilon$ small enough to ensure that $D_*(\beta) < \delta/2$. In turn, t_0 is to be chosen sufficiently large to ensure that

$$\vartheta^0(t) < \beta, \qquad \text{for} \quad t \in (t_0, \infty).$$

By comparing the generalized solutions of the Stefan problem it follows that, for $t \geq t_0$, the solution $\{R(t), \vartheta(x,t)\}$ of problem (1.1)–(1.4) can be estimated from above by the solution $\{\tilde{R}(t), \tilde{\vartheta}(x,t)\}$ of the problem (1.1)–(1.4) with $\tilde{\vartheta}(0,t) = \beta$, $\tilde{\vartheta}(x,t_0) \geq \vartheta(x,t_0)$:

$$R(t) \leq \tilde{R}(t) \qquad \text{for} \quad t \in (t_0, \infty). \tag{2.15}$$

In turn, the assertion of Theorem 4 holds for the solution $\{\tilde{R}(t), \tilde{\vartheta}(x,t)\}$:

$$\lim_{t \to \infty} t^{-1/2} \tilde{R}(t) = D_*(\beta) < \delta/2.$$

Hence, there exists $t_* > t_0$ such that

$$\tilde{R}(t) < \delta(t - t_0)^{1/2} \qquad \text{for} \quad t \in (t_*, \infty).$$

This implies that the whole domain $\Omega_\infty^+ \setminus \Omega_{t_*}^+$ is contained in Q^δ. □

3. The two-phase Stefan problem

In this section, we shall derive conditions on the data $\vartheta^\pm(t)$, $\vartheta_0(x)$ which ensure the existence of the classical solution of the two-phase Stefan problem with temperature prescribed on the fixed boundaries $\{x = \pm 1\}$ for all positive time. The case of heat flux prescribed on the boundaries $\{x = \pm 1\}$ can be treated analogously, provided that the temperature has been shown to be positive (negative) in Ω_T^+ (in Ω_T^-) and the distance between the free boundary $\{x = R(t)\}$ and the fixed boundaries $\{x = \pm 1\}$ remains strictly positive throughout.

Recall the formulation of the problem: find a pair $\{R(t), \vartheta(x,t)\}$, where the function $R(t)$ defines the domains $\Omega_T^- = \{(x,t) : -1 < x < R(t), 0 < t < T\}$ and $\Omega_T^+ = \{(x,t) : R(t) < x < 1, 0 < t < T\}$, and $\vartheta(x,t)$ is continuous in $\Omega_T = \Omega \times (0,T)$, $\Omega = \{x : |x| < 1\}$ is positive in Ω_T^+ and negative in Ω_T^-. Moreover, $\vartheta(x,t)$ satisfies the nonlinear heat equation

$$a(\vartheta)\frac{\partial \vartheta}{\partial t} = \frac{\partial^2 \vartheta}{\partial x^2} \qquad \text{for} \quad (x,t) \in \Omega_T^\pm, \tag{3.1}$$

as well as the following two conditions on the free boundary $x = R(t)$, $t \in (0,T)$:

$$\vartheta = 0, \qquad -\frac{dR}{dt} = \frac{\partial \vartheta}{\partial x}\bigg|_{x=R(t)+0} - \frac{\partial \vartheta}{\partial x}\bigg|_{x=R(t)-0}. \tag{3.2}$$

On the fixed boundaries $\{x = \pm 1\}$,

$$\vartheta(\pm 1, t) = \vartheta^{\pm}(t) \qquad \text{for} \quad t \in (0, T), \tag{3.3}$$

and at the initial time instant

$$R(0) = x_0 \in \Omega, \quad \vartheta(x, 0) = \vartheta_0(x) \qquad \text{for} \quad x \in \Omega. \tag{3.4}$$

Lemma 5. *Let $\vartheta^-(t) < 0$, $\vartheta^+(t) > 0$ under the hypotheses of Theorem 6, Chapter II, and set*

$$M_0 = \max\{|\vartheta^-, \vartheta^+|^{(0)}_{[0,\infty)}, |\vartheta_0|^{(0)}_{\Omega}\}.$$

Then the solution of problem (3.1)–(3.4) satisfies the bounds

$$\begin{aligned} 0 < \vartheta(x, t) \leq M_0 \qquad &\text{for} \quad (x, t) \in \Omega_T^+, \\ -M_0 \leq \vartheta(x, t) < 0 \qquad &\text{for} \quad (x, t) \in \Omega_T^-, \end{aligned} \tag{3.5}$$

$$\frac{\partial \vartheta}{\partial x}(R(t) \pm 0, t) \geq 0 \qquad \text{for} \quad t \in (0, T). \tag{3.6}$$

The inequalities (3.5) follow by the maximum principle. (3.6) is a consequence of the positiveness (negativeness) of $\vartheta(x, t)$ in the domain Ω_T^+ (in Ω_T^-) and its vanishing on the boundary $x = R(t)$. ☐

Lemma 6. *Let the hypotheses of Lemma 5 be satisfied, and let*

$$M_1 = \max\{\|\vartheta^-, \vartheta^+\|^{(1)}_{2,[0,T]}, |\vartheta_0|^{(1)}_{2,\Omega}\}.$$

Then the solution $\{R(t), \vartheta(x, t)\}$ of problem (3.1)–(3.4) satisfies the bounds

$$\left\{ \left\| \frac{\partial \vartheta}{\partial t}, \frac{\partial^2 \vartheta}{\partial x^2} \right\|_{2,\Omega_T^{\pm}}, \max_{t \in (0,T)} \left\| \frac{\partial \vartheta}{\partial x}(t) \right\|_{2,\Omega} \right\} \leq N_1, \tag{3.7}$$

$$-1 + N_1^{-1} \leq R(t) \leq 1 - N_1^{-1} \qquad \text{for} \quad t \in (0, T), \tag{3.8}$$

where the constant N_1 depends on M_1, a_0, T, $\delta = \min_{t \in (0,T)} |\vartheta^{\pm}(t)|$, and N_1 is finite for finite values of M_1, T, δ^{-1}.

Proof. We split the proof of inequality (3.7) into three steps. First we shall estimate $\|\partial \vartheta / \partial x\|_{2,\Omega_T}$. Let

$$\vartheta^0(x, t) = \vartheta^-(t) + \big((1 + x)/2\big)\big(\vartheta^+(t) - \vartheta^-(t)\big).$$

Multiply equation (3.1) by the difference $(\vartheta - \vartheta_0)$, integrate over each of the domains $\Omega^{\pm}(t)$ (recall that $\Omega^{\pm}(t)$ is the intersection of the domain $\Omega_T^{\pm}$ with the line $\{t = \text{const.}\}$ and then take the sum of the integrals. After simple transformations we obtain the equality

$$\frac{d}{dt} \int_{\Omega} b(\vartheta) \, dx + \int_{\Omega} \left| \frac{\partial \vartheta}{\partial x} \right|^2 dx = - \int_{\Omega} \frac{\partial \vartheta}{\partial x} \frac{\partial \vartheta_0}{\partial x} \, dx$$

$$+\frac{d}{dt}\int_\Omega \vartheta^0\left(\int_0^\vartheta a(s)\,ds\right)dx - \int_\Omega \frac{\partial\vartheta^0}{\partial t}\left(\int_0^\vartheta a(s)\,ds\right)dx, \tag{3.9}$$

where $a(s) = \Phi'(s)$ for $s \neq 0$, $b(s) = \int_0^s \xi a(\xi)\,d\xi$.

To estimate the first term on the right-hand side of (3.9), we use Hölder's inequality:

$$\int_\Omega \frac{\partial\vartheta}{\partial x}\frac{\partial\vartheta^0}{\partial x}\,dx \leq \frac{1}{2}\int_\Omega\left|\frac{\partial\vartheta}{\partial x}\right|^2 dx + \frac{1}{2}\int_\Omega\left|\frac{\partial\vartheta^0}{\partial x}\right|^2 dx.$$

The expression under derivative on the right-hand side of (3.9) is summable in time. Thus, Gronwall's lemma is applicable to estimate the left-hand side of equality (3.9) from below. By using the inequality

$$\frac{a_0}{2}s^2 \leq b(s) \leq \frac{1}{2a_0}s^2, \qquad \|\vartheta\|_{2,\Omega} \leq 4\left\|\frac{\partial\vartheta}{\partial x}\right\|_{2,\Omega}. \tag{3.10}$$

we then conclude

$$\left\|\frac{\partial\vartheta}{\partial x}\right\|_{2,\Omega_T} \leq N_2(M_1, a_0, T). \tag{3.11}$$

Apart from (3.11), we need an estimate of $\left|\frac{\partial\vartheta}{\partial x}(\pm 1, t)\right|$. By estimating from above the right-hand side of the equality

$$\left|\frac{\partial\vartheta}{\partial x}(-1,t)\right|^2 = \left|\frac{\partial\vartheta}{\partial x}(x,t)\right|^2 + 2\int_{-1}^x \frac{\partial\vartheta}{\partial x}\frac{\partial^2\vartheta}{\partial s^2}\,ds$$

with the use of Hölder's inequality, and integrating over $x \in (-1, R(t))$, in view of equation (3.1) we get

$$\left|\frac{\partial\vartheta}{\partial x}(-1,t)\right|^2 \leq \left\|\frac{\partial\vartheta}{\partial x}(t)\right\|_{2,\Omega^-(t)}^2 + (R(t)+1)^{-1} + \frac{2}{a_0}\left\|\frac{\partial\vartheta}{\partial x}(t)\right\|_{2,\Omega^-(t)}\left\|\frac{\partial\vartheta}{\partial t}(t)\right\|_{2,\Omega^-(t)}.$$

Because of the inequality

$$\delta \leq |\vartheta^-(t)| = \int_{-1}^{R(t)}\frac{\partial\vartheta}{\partial x}(x,t)\,dx \leq (R(t)+1)^{1/2}\left\|\frac{\partial\vartheta}{\partial x}(t)\right\|_{2,\Omega^-(t)},$$

we have

$$(R(t)+1)^{-1} \leq \delta^{-2}\left\|\frac{\partial\vartheta}{\partial x}(t)\right\|_{2,\Omega^-(t)}^2. \tag{3.12}$$

Hence

$$\left|\frac{\partial\vartheta}{\partial x}(-1,t)\right|^2 \leq \delta^{-2}\left\|\frac{\partial\vartheta}{\partial x}(t)\right\|_{2,\Omega^-(t)}^4 + a_0^{-2}\left\|\frac{\partial\vartheta}{\partial x}(t)\right\|_{2,\Omega^-(t)}^2 + \left\|\frac{\partial\vartheta}{\partial t}(t)\right\|_{2,\Omega^-(t)}^2.$$

An analogous inequality also holds for the derivative $\left|\frac{\partial\vartheta}{\partial x}(1,t)\right|$. Finally we get the bound

$$\left|\frac{\partial\vartheta}{\partial x}(-1,t)\right|^2 + \left|\frac{\partial\vartheta}{\partial x}(1,t)\right|^2 \le \left\|\frac{\partial\vartheta}{\partial t}\right\|^2_{2,\Omega^-(t)} + \left\|\frac{\partial\vartheta}{\partial t}\right\|^2_{2,\Omega^+(t)} + \lambda\left\|\frac{\partial\vartheta}{\partial x}\right\|^4_{2,\Omega}, \qquad (3.13)$$

with $\lambda = \min(\delta^{-2}, a_0^{-2})$.

Let us now multiply equation (3.1) by $\partial\vartheta/\partial t$, integrate over each of the domains $\Omega^\pm(t)$ and take the sum of the resulting integrals. After integrating by parts in the terms

$$\int_{\Omega^\pm(t)} \frac{\partial\vartheta}{\partial t}\frac{\partial^2\vartheta}{\partial x^2}\, dx,$$

making use of the identity

$$\frac{\partial\vartheta}{\partial t}\left(R(t)\pm 0, t\right) + \frac{dR}{dt}(t)\frac{\partial\vartheta}{\partial x}\left(R(t)\pm 0, t\right) = 0$$

and of the Stefan condition (3.2), we come to the equality

$$\mathcal{J}_0(t) \equiv \frac{1}{2}\left|\frac{dR}{dt}\right|^2\left\{\left.\frac{\partial\vartheta}{\partial x}\right|_{x=R(t)+0} + \left.\frac{\partial\vartheta}{\partial x}\right|_{x=R(t)-0}\right\} + \frac{1}{2}\frac{d}{dt}\left\|\frac{\partial\vartheta}{\partial x}(t)\right\|^2_{2,\Omega}$$

$$+\left\|\frac{\partial\vartheta}{\partial t}\right\|^2_{2,\Omega^-(t)} + \left\|\frac{\partial\vartheta}{\partial t}\right\|^2_{2,\Omega^+(t)} = \frac{\partial\vartheta}{\partial x}(1,t)\frac{d\vartheta^+}{dt}(t) - \frac{\partial\vartheta}{\partial x}(-1,t)\frac{d\vartheta^-}{dt}(t) \equiv \mathcal{J}_1(t).$$

By (3.6), $\mathcal{J}_0(t)$ can be estimated from below:

$$\mathcal{J}_0(t) \ge \frac{1}{2}\frac{d}{dt}\left\|\frac{\partial\vartheta}{\partial x}(t)\right\|^2_{2,\Omega} + \left\|\frac{\partial\vartheta}{\partial t}\right\|^2_{2,\Omega^-(t)} + \left\|\frac{\partial\vartheta}{\partial t}\right\|^2_{2,\Omega^+(t)}.$$

Using (3.13), we now estimate $\mathcal{J}_1(t)$ from above:

$$\mathcal{J}_1(t) \le \frac{1}{2}\left\{\left|\frac{d\vartheta^+}{dt}(t)\right|^2 + \left|\frac{d\vartheta^-}{dt}(t)\right|^2\right\} + \frac{1}{2}\left\{\left|\frac{\partial\vartheta}{\partial x}(1,t)\right|^2 + \left|\frac{\partial\vartheta}{\partial x}(-1,t)\right|^2\right\}$$

$$\le \max\left\{\left|\frac{d\vartheta^+}{dt}(t)\right|^2, \left|\frac{d\vartheta^-}{dt}(t)\right|^2\right\} + \frac{1}{2\lambda}\left\|\frac{\partial\vartheta}{\partial x}(t)\right\|^4_{2,\Omega} + \frac{1}{2}\left\|\frac{\partial\vartheta}{\partial t}(t)\right\|^2_{2,\Omega}.$$

Therefore,

$$\left\|\frac{\partial\vartheta}{\partial t}(t)\right\|^2_{2,\Omega} + \frac{d}{dt}\left\|\frac{\partial\vartheta}{\partial x}(t)\right\|^2_{2,\Omega} \le 2\max\left\{\left|\frac{d\vartheta^+}{dt}(t)\right|^2, \left|\frac{d\vartheta^-}{dt}(t)\right|^2\right\} + \frac{1}{\lambda}\left\|\frac{\partial\vartheta}{\partial x}(t)\right\|^4_{2,\Omega}$$

and the estimate (3.7) follows from (3.11) by Gronwall's lemma (see Lemma 1, Section I.2). Recalling (3.12) and the analogous estimate for $\left(1 - R(t)\right)^{-1}$, we eventually conclude (3.8). $\qquad\square$

Theorem 6. *Let the hypotheses of Theorem 6, Chapter II, be satisfied and let $\vartheta^+(t) >$*
0, $\vartheta^-(t) < 0$ for $t \in [0, \infty)$. Then, for all $T \in (0, \infty)$, there exists a classical solution
of the Stefan problem (3.1)–(3.4) such that

$$R \in H^{(3+\gamma)/2}[0, T], \qquad \vartheta \in H^{2+\gamma,(2+\gamma)/2}\left(\overline{\Omega_T^{\pm}}\right).$$

Proof. By Theorem 6 of Chapter II, it is sufficient to show that $|\vartheta|_{\Omega_T^+}^{(1+\beta)} + |\vartheta|_{\Omega_T^-}^{(1+\beta)}$ is
bounded for some $\beta \in (0, 1)$. From (3.8), each of the domains $\Omega^\pm(t)$ is non-degenerate.
(3.5), in turn, yields the desired sign of $\vartheta(x, t)$ in each of the domains $\Omega_T^\pm$.

First, observe that, as has also been shown for the generalized solution of the one-
dimensional Stefan problem, (3.7) implies that the solution $\vartheta(x, t)$ is Hölder continuous:

$$|\vartheta|_{\Omega_T}^{(\beta)} \le N_3(M_1, T, a_0, \delta). \tag{3.14}$$

Moreover, (3.7) implies an estimate for the norm of $R(t)$ in $W_4^1[0, T]$:

$$\|R\|_{4,[0,T]}^{(1)} \le N_4(M_1, T, a_0, \delta). \tag{3.15}$$

Indeed, the distance $2\delta_0$ between the free boundary $\{x = R(t)\}$ and the boundaries
$\{x = \pm 1\}$ is positive by (3.8). Hence the regularity of the solution in each of the domains
$\{(x, t) : -1 < x < -1 + \delta_0, \ 0 < t < T\}$ and $\{(x, t) : 1 - \delta_0 < x < 1, \ 0 < t < T\}$
is determined by the differentiability of functions $\Phi(s), \vartheta^\pm(t), \vartheta_0(x)$ (see Theorem 6,
Chapter I, on local estimates of the solutions of linear parabolic equations), because the
solution on these domains belongs to the space $H^{r,r/2}$. Correspondingly, the absolute
value of $\partial\vartheta/\partial x$ is bounded on the boundaries $\{x = \pm 1\}$. From this property, using the
representation

$$\left|\frac{\partial\vartheta}{\partial x}\right|_{x=R(t)\pm 0}^2 = \left|\frac{\partial\vartheta}{\partial x}\right|_{x=\pm 1}^2 + 2\int_{\Omega^\pm(t)} \frac{\partial\vartheta}{\partial x}\frac{\partial^2\vartheta}{\partial x^2}\,dx,$$

the Stefan condition (3.2) and (3.7), we conclude that (3.15) holds.

The further considerations do not differ from those for the one-phase Stefan problem.
In each of the domains $\Omega_T^\pm$, the transformation

$$\tau = t, \qquad y = (2x - R \mp 1)/(1 \mp R)$$

maps $\Omega_T^\pm$ into the rectangle Ω_T. The corresponding function $u(y, \tau) = \vartheta(x, t)$ in Ω_T
satisfies a parabolic equation with the coefficients of the highest-order derivatives in the
Hölder space $H^{\beta,\beta/2}(\overline{\Omega}_T)$ (see estimates (3.14) and (3.15)). Since the solution $u(y, \tau)$
coincides with a function from $H^{r,r/2}$ on the boundaries $y = \pm 1$ and at the initial time
instant, it belongs to the Hölder space $H^{q,q/2}(\overline{\Omega}_T)$ for some $q > 2$, if the coefficient of
$\partial u/\partial y$ in the equation for $u(y, \tau)$ is Hölder continuous (see Theorem 3, Section I.2).
But this coefficient belongs to $L_4(\Omega_T)$ (see estimate (3.15)). This, according to Theorem
5, Section I.2, means that $u \in W_4^{2,1}(\Omega_T)$ and

$$\|u\|_{4,\Omega_T}^{(2)} \le N_5(M_2, T, a_0, \delta), \tag{3.16}$$

where $M_2 = \max\{|\vartheta^-, \vartheta^+|_{[0,T]}^{(r/2)}, |\vartheta_0|_{\Omega^\pm(0)}^{(r)}\}$.

The space $W_4^{2,1}(\Omega_T)$ is embedded into $H^{1+\beta,(1+\beta)/2}(\bar{\Omega}_T)$ for some $\beta > 0$ (see Lemma 5, Section I.2). Thus, the derivative of the function $u(y,\tau)$, as well as the derivative of $\vartheta(x,t)$ satisfy the Hölder condition

$$|\vartheta|_{\Omega_T^\pm}^{(1+\beta)} \leq N_6(M_2, T, a_0, \delta).$$

This completes the proof of the theorem.   □

As in case of the one-phase Stefan problem, the regularity requirements on the data in the two-phase Stefan problem can be relaxed. If the functions $(\pm 1)\vartheta^\pm(t)$ are strongly positive, then the non-degeneracy of each of the domains $\Omega^\pm(\iota)$ is an underlying condition for the construction of the solution. If this is provided, then, as in Section 1 of the present chapter, it can be shown that the regularity of the solution $\vartheta(x,t)$ in each closed subdomain of $\bar{\Omega}_T^\pm$, which intersects the boundary $\{x = R(t)\}$ and does not contain any points of the lines $\{x = \pm 1\}$ and $\{t = 0\}$, depends only on the regularity of Φ. The hypotheses on the problem data which guarantee the non-degeneracy of domains $\Omega^\pm(t)$ have been formulated in Lemma 6. Therefore, the following result holds.

Theorem 7. *Let $\Phi \in C^2(-\infty, 0] \cap C^2[0, \infty)$, $\Phi'(s) \geq a_0 > 0$, let $\vartheta^\pm(t), \vartheta_0(x)$ be such that $\vartheta^\pm \in W_{2,\mathrm{loc}}^1[0, \infty)$, $\vartheta_0 \in W_2^1(\Omega)$, $(\pm 1)\vartheta^\pm(t) > 0$ for $t > 0$, $(\pm 1)\vartheta_0(x) \geq 0$ in $\Omega^\pm(0)$, $|x_0| < 1$, and let the compatibility conditions of zero-order $\vartheta^\pm(0) = \vartheta_0(\pm 1)$ hold at the points $x = \pm 1$. Then there exists a unique classical solution $\{R(t), \vartheta(x,t)\}$ of problem (3.1) - (3.4) for all $t \in (0, \infty)$. For all $t \in (0, \infty)$, the derivative $\frac{dR}{dt}(t)$ is Hölder continuous, and $R(t), \vartheta(x,t)$ are Hölder continuous on each bounded subset of $[0, \infty)$ and $\bar{\Omega}_\infty$, respectively. Moreover, the function $\vartheta(x,t)$ belongs to the space $H^{q,q/2}(\bar{G})$ with some $q > 2$ in each closed bounded domain $\bar{G} \subset \bar{\Omega}_\infty^\pm$ which does not intersect with the lines $\{x = \pm 1\}$ and $\{t = 0\}$.*

If $\Phi \in C^\infty(-\infty, 0] \cap C^\infty[0, \infty)$, then $R(t)$ is infinitely differentiable for $t > 0$.

In contrast to Theorem 2, Section I.2, it has been assumed in this statement that the derivatives of the problem data exist, are square integrable and that zero-order compatibility conditions hold. However, the regularity of $\vartheta_0(x)$ and $\vartheta^\pm(t)$ assumed in Lemma 6 does not yet imply such compatibility conditions. They result as a consequence of a specific construction of the solution, where the data $\vartheta^\pm(t), \vartheta_0(x)$ are approximated by smooth functions $\vartheta_\varepsilon^\pm(t), \vartheta_0^\varepsilon(x)$. It can be easily shown that if $\vartheta^\pm(0) \neq \vartheta_0(\pm 1)$ and $\vartheta_\varepsilon^\pm(0) = \vartheta_0^\varepsilon(\pm 1)$, then the norms of the functions $\vartheta_\varepsilon^\pm, \vartheta_0^\varepsilon$ are not uniformly bounded in W_2^1 and, hence, the estimate (3.8) needed for the approximate solutions $\{R_\varepsilon(t), \vartheta_\varepsilon(x,t)\}$ will not be uniform with respect to ε. In turn, the construction of the approximate solutions $\vartheta_\varepsilon(x,t)$ requires that the compatibility conditions hold.

Let us now consider the two-phase Stefan problem with heat flux prescribed on the boundary of Ω. To be determined are $R(t)$ and $\vartheta(x,t)$, the latter continuous in Ω_T and satisfying equation (3.1) in each of the domains $\Omega_T^\pm$, conditions (3.2) on the boundary $\{x = R(t)\}$, initial condition (3.4) and

$$\frac{\partial \vartheta}{\partial x}(\pm 1, t) = \vartheta_0^\pm(t) \qquad \text{for} \quad t \in (0, T). \tag{3.17}$$

As mentioned above, in this problem it is enough to specify requirements for the functions $\vartheta^{\pm}(t), \vartheta_0(x)$ which will ensure that the solution $\vartheta(x,t)$ is strictly positive in Ω_T^+, strictly negative in Ω_T^- and the distance between the free boundary $\{x = R(t)\}$ and the boundary of Ω remains positive throughout.

Lemma 7. *Assume that the hypotheses of Theorem 6, Chapter II, are satisfied with*

$$0 \leq \vartheta_0^{\pm}(t) \leq M_0, \quad |\vartheta_0(x)| \leq M_0, \quad (\pm 1)\vartheta_0(x) \geq 0 \quad in \quad \Omega^{\pm}(0).$$

Then the solution $\vartheta(x,t)$ of the problem (3.1), (3.2), (3.4), (3.17) satisfies the bounds

$$(\pm 1)\vartheta(x,t) > 0 \quad in \quad \Omega_T^{\pm}, \quad |\vartheta|_{\Omega_T}^{(0)} \leq N_1(M_0). \tag{3.18}$$

Suppose further that, for $t \in (0,\infty)$,

$$-1 < R^-(t) = A(t) - 2M(t), \quad A(t) + 2M(t) = R^+(t) < 1, \tag{3.19}$$

where

$$A(t) = x_0 + \int_{-1}^{1} \tilde{U}_0(x) \, dx + \int_{0}^{1} \left\{ \vartheta_0^+(\tau) - \vartheta_0^-(\tau) \right\} d\tau,$$

$$M(t) = \max_{x \in \Omega} |\tilde{U}(x,t)|, \quad \tilde{U}_0 = \tilde{U}(x,0),$$

$$\tilde{U} = \begin{cases} U(x,t) \equiv \Phi\big(\vartheta(x,t)\big) & for \ x \in \Omega^+(t), \\ U(x,t) + 1 & for \ x \in \Omega^-(t). \end{cases}$$

Then

$$R^-(t) \leq R(t) \leq R^+(t). \tag{3.20}$$

Proof. The function $\vartheta(x,t)$ is positive in Ω_T^+ by the maximum principle. To estimate $\vartheta(x,t)$ from above, observe that the function $v(x,t) = (x+2)\vartheta(x,t)$ satisfies in Ω_T^+ a parabolic equation whose solution cannot attain its maximum in the interior of Ω_T^+. Since

$$\frac{\partial v}{\partial x}(1,t) + v(1,t) = \vartheta_0^+(t),$$

we have

$$\frac{\partial v}{\partial x}(1,t) \geq 0 \quad and \quad v(1,t) \leq \vartheta_0^+(t) \leq M_0$$

at the points where $v(x,t)$ can attain its positive maximum on the boundary $\{x = 1\}$.

In an analogous way we can estimate the function $\vartheta(x,t)$ in Ω_T^-.

To prove the inequalities (3.20), we integrate equation (3.2) in x over each of the domains $\Omega^{\pm}(t)$ and then in time. After taking the sum of the resulting integrals, we get

$$\int_{\Omega^-(t)} \tilde{U}(x,t) \, dx + \int_{\Omega^+(t)} \tilde{U}(x,t) \, dx + R(t) = A(t).$$

The inequalities

$$\int_{\Omega^-(t)} \tilde{U}(x,t)\, dx \le 0, \qquad \int_{\Omega^+(t)} \tilde{U}(x,t)\, dx \le 2M(t),$$

imply that $R(t) \ge A(t) - 2M(t) = R^-(t) > -1$. Analogously, it follows from the inequalities

$$\int_{\Omega^-(t)} \tilde{U}(x,t) \ge -2M(t), \qquad \int_{\Omega^+(t)} \tilde{U}(x,t)\, dx \ge 0$$

that $R(t) \le A(t) + 2M(t) = R^+(t) < 1$. □

Hence, under the hypotheses of Theorem 6, Chapter II and those of Lemma 7 of the present chapter, the solution $\{R(t), \vartheta(x,t)\}$ of problem (3.1), (3.2), (3.4), (3.17) exists for all positive t.

Assume now that the functions $\vartheta_0^\pm(t)$, $\vartheta_0(x)$ satisfy only the hypotheses of Lemma 7. There are always sequences of functions $\vartheta_{0\varepsilon}^\pm(t)$ and $\vartheta_0^\varepsilon(x)$, pointwise convergent to $\vartheta_0^\pm(t)$ and $\vartheta_0(x)$, respectively, such that they satisfy the hypotheses of Theorem 6, Chapter II and Lemma 7 of the present chapter. Let $\{R_\varepsilon(t), \vartheta_\varepsilon(x,t)\}$ be the solution of problem (3.1), (3.2), (3.4), (3.17) with given functions $\vartheta_{0\varepsilon}^\pm(t)$ and $\vartheta_0^\varepsilon(x)$. As in Section 1 of this chapter, we can show that there exist subsequences of $\{R_\varepsilon(t), \vartheta_\varepsilon(x,t)\}$, fundamental in each space $H^{r,r/2}(\bar{G})$, where G is an arbitrary closed bounded domain in $\bar{\Omega}_\infty^\pm$ which does not contain points of the lines $\{x = \pm 1\}$ and $\{t = 0\}$. Clearly, the limit pair $\{R(t), \vartheta(x,t)\}$ is a bounded solution of the Stefan problem (3.1), (3.2), (3.4), (3.17). In this problem, although equation (3.1) and the boundary conditions (3.2) are fulfilled in the classical sense, the conditions (3.4) and (3.17) hold only in the sense of an appropriate integral identity. We have proved the following.

Theorem 8. *Let* $\Phi \in C^2(-\infty, 0] \cap C^2[0, \infty)$, $\Phi'(s) \ge a_0 > 0$, $0 \le \vartheta^\pm(t) \le M_0$ *for* $t \in (0, \infty)$. *Let* $(\pm 1)\vartheta_0(x) \ge 0$ *in* $\Omega^\pm(0)$, $|\vartheta^0(x)| \le M_0$ *for* $x \in \Omega$, *and let* $-1 < A(t) - 2M(t)$, $A(t) + 2M(t) < 1$ *for* $t \in (0, \infty)$, *where* $A(t)$ *and* $M(t)$ *are defined by (3.19). Then there exists a unique bounded classical solution* $\{R(t), \vartheta(x,t)\}$ *of problem (3.1), (3.2), (3.4), (3.17) for all positive* $t > 0$, *such that* dR/dt *is Hölder continuous for all* t. $\vartheta(x,t)$ *belongs to the space* $H^{r,r/2}(\bar{G})$ *with some* $r > 2$ *on each bounded closed domain* $\bar{G} \subset \bar{\Omega}_\infty^\pm$ *that does not intersect the boundary of* Ω_∞. □

We shall restrict the study of the asymptotic behaviour of the solution of the two-phase Stefan problem to the case where the temperature is prescribed on the boundary of Ω so that

$$\lim_{t \to \infty} \vartheta^+(t) = \beta^+ > 0, \qquad \lim_{t \to \infty} \vartheta^-(t) = \beta^- < 0.$$

It is then quite natural to assume that the solution $\{R(t), \vartheta(x,t)\}$ of problem (3.1)–(3.4) converges as $t \to \infty$ to the solution $\{R_\infty, \vartheta_\infty(x)\}$ of the stationary Stefan problem, explicitly given by

$$R_\infty = \frac{\beta^+ + \beta^-}{\beta^- - \beta^+}, \qquad \vartheta_\infty(x) = \frac{\beta^+ - \beta^-}{2}(x - R_\infty). \tag{3.21}$$

We begin the study of the asymptotic behaviour with the simplest case where $\vartheta^\pm = $ const. General results follow from these considerations and comparison theorems for the generalized solutions of the Stefan problem.

Lemma 8. *Let $\vartheta^\pm(t) = $ const. in the hypotheses of Theorem 7. Then*

$$\|\vartheta(t) - \vartheta_\infty\|_{2,\Omega} + |R(t) - R_\infty| \leq c_1 \exp(-c_2 t), \tag{3.22}$$

with positive constants c_1, c_2 dependent only on a_0, M_0 and $\vartheta^\pm$.

Proof. Denote

$$u(x, t) = \vartheta(x, t) - \vartheta_\infty(x).$$

Multiply the equation for $\vartheta(x, t)$ by $u(x, t)$, then integrate over each of the domains $\Omega^\pm(t)$ and take the sum of the resulting integrals, to get after simple transformations

$$\frac{d}{dt}\left\{\int_\Omega b \, dx + \alpha^2 |R(t) - R_\infty|^2\right\} + \left\|\frac{\partial u}{\partial x}(t)\right\|_{2,\Omega}^2 = 0, \tag{3.23}$$

where

$$b(u, x) = \int_0^u sa(s + \vartheta_\infty(x)) \, ds, \qquad \alpha^2 = \frac{\beta^+ - \beta^-}{2} > 0.$$

If $a_0 \leq a(s) \leq a_0^{-1}$, then

$$\frac{a_0}{2} u^2 \leq b(u, x) \leq \frac{1}{2a_0} u^2.$$

The lemma then follows from the inequalities

$$\int_\Omega u^2(x, t) \, dx \leq 16 \left\|\frac{\partial u}{\partial x}(t)\right\|_{2,\Omega}^2,$$

$$4\alpha^4 |R(t) - R_\infty|^2 = |u(R(t), t)| = \left|\int_{-1}^{R(t)} \frac{\partial u}{\partial x}(x, t) \, dx\right| \leq 4 \left\|\frac{\partial u}{\partial x}(t)\right\|_{2,\Omega}^2$$

by appying Gronwall's lemma. Indeed, let us take

$$z(t) = \int_\Omega b(u(x, t), x) \, dx + \alpha^2 |R(t) - R_\infty|^2.$$

Then

$$z(t) \leq c_3 \left\|\frac{\partial u}{\partial x}(t)\right\|_{2,\Omega}^2,$$

and, by (3.23), we claim that

$$\frac{\partial z}{\partial t} + c_3^{-1} z \leq 0, \qquad z(0) = c_4.$$

Hence, $z(t) \leq c_4 \exp(-c_3 t)$.   □

Theorem 9. *Let the hypotheses of Theorem 7 be satisfied with*

$$\lim_{t \to \infty} \vartheta^+(t) = \beta^+ > 0, \quad \lim_{t \to \infty} \vartheta^-(t) = \beta^- < 0.$$

Then

$$\lim_{t \to \infty} \left| R(t) - (\beta^+ + \beta^-)/(\beta^- - \beta^+) \right| = 0.$$

Proof. Let us take an arbitrary fixed small $\varepsilon > 0$ and $t_0 > 0$ such that

$$\beta^- - \varepsilon \leq \vartheta^-(t) \leq \beta^- + \varepsilon < 0,$$
$$0 < \beta^+ - \varepsilon \leq \vartheta^+(t) \leq \beta^+ + \varepsilon, \qquad \text{for} \quad t \in (t_0, \infty).$$

There always exists a function $\tilde{\vartheta}_0(x)$ such that $\tilde{\vartheta}_0 \in W_2^1(\Omega)$, $\tilde{\vartheta}_0(x) \geq \vartheta(x, t_0)$ for $x \in \Omega$; $(\pm 1)\tilde{\vartheta}_0(x) > 0$ for $x \in \Omega^\pm(t_0)$; $\tilde{\vartheta}_0(\pm 1) = \beta^\pm + \varepsilon$.

The solution $\{\tilde{R}(t), \tilde{\vartheta}(x, t)\}$ of the Stefan problem (3.1)–(3.4), with the data $\{\beta^\pm + \varepsilon, \tilde{\vartheta}_0(x)\}$, exists on the interval (t_0, ∞) and, by the comparison theorem, it bounds from above the solution $\{R(t), \vartheta(x, t)\}$:

$$\tilde{\vartheta}(x, t) \geq \vartheta(x, t), \quad \tilde{R}(t) > R(t), \qquad \text{for} \quad t \in (t_0, \infty)$$

Passing to the limit as $t \to \infty$ and taking into account that

$$\lim_{t \to \infty} \tilde{R}(t) = R_\infty + 2\varepsilon/(\beta^- - \beta^+),$$

we obtain the inequality

$$\varliminf_{t \to \infty} R(t) \geq R_\infty + 2\varepsilon/(\beta^- - \beta^+).$$

Similarly we obtain the bound

$$\varlimsup_{t \to \infty} R(t) \leq R_\infty - 2\varepsilon/(\beta^- - \beta^+).$$

Since ε is arbitrary, we have

$$\varliminf_{t \to \infty} R(t) = \varlimsup_{t \to \infty} R(t) = R_\infty.$$ □

4. Special cases: one-phase initial state, violation of compatibility conditions, unbounded domains

In the preceding sections, we have considered neither the situation of just one phase at the initial time instant nor incompatible input data of the problem, i.e., $\vartheta_0(x_0 - 0) \neq 0$ in the one-phase Stefan problem or

$$\vartheta_0(\pm 1) \neq \vartheta^\pm(0), \quad \vartheta_0(x_0 \pm 0) \neq 0$$

in the two-phase problem.

There are two possible approaches to the resulting problems. The first of them consists of a direct treatment of the "singular" problems: properties of the solutions to the heat equation are studied there in "triangular" domains $\Omega_T^\pm$ whose intersection $\Omega^+(t)$ with the lines $\{t = \text{const.}\}$ reduces to a single point at the initial time instant, or those properties are explored for the formulation of the problem without compatibility conditions of the boundary and initial data. The second approach, in our opinion more natural, exploits an approximation of the original problem by a regularized Stefan problem, an analysis of the differential properties of the solutions to the latter and an appropriate limit passage. Such an approach does not require any exact knowledge of the behaviour of solutions at points where either the compatibility conditions are violated or one of the phases degenerates. In other words, the construction of solutions is separated there from the local regularity analysis of these solutions, the latter becoming now an independent problem. The above treatment is possible due to the specifics of parabolic equations: if any "weak" estimate is known for the whole domain of the solution, then the differential properties of the solution in any subdomain depend on the corresponding properties of the problem data in a slightly larger subdomain only. They do not depend on the character of the input data anywhere far away from this subdomain.

Let us first consider the one-phase Stefan problem (1.1)–(1.4) with the temperature prescribed on the boundary $\{x = 0\}$, where either $x_0 = R(0) = 0$ or condition (1.14) has been violated.

Theorem 10. *Let* $\Phi \in C^2[0, \infty)$, $\Phi'(s) = a(s) \geq a_0 > 0$. *If* $x_0 > 0$, *then let* $\vartheta^0(t)$ *and* $\vartheta_0(x)$ *be non-negative functions, bounded by finite constant* K. *If* $x_0 = 0$, *then assume there exists a strongly positive function* $\vartheta^+(t)$ *in the space* $W_2^1[0, T]$, *where* $T > 0$ *is arbitrarily small, such that*

$$\vartheta^+(0) = 0, \quad 0 < \vartheta^+(t) \leq \vartheta^0(t) \leq K \quad \text{for} \quad t \in (0, T). \tag{4.1}$$

Under the above hypotheses, problem (1.1)–(1.4) with fixed temperature $\vartheta^0(t)$ *on the boundary* $\{x = 0\}$ *admits a unique bounded classical solution* $\{R(t), \vartheta(x, t)\}$, *such that* $\frac{dR}{dt}(t)$ *is Hölder continuous for all* $t > 0$. $\vartheta(x, t)$ *belongs to the space* $H^{r, r/2}(\bar{G})$, $r > 2$, *in any closed bounded subdomain* $\bar{G}$ *of* $\bar{\Omega}_\infty^+$ *which contains neither the point* $x = 0$ *nor* $t = 0$.

Proof. Let us consider the sequence $\{R_\varepsilon(t), \vartheta_\varepsilon(x,t)\}$ of solutions to the regular Stefan problems (1.1)–(1.4), where

$$R_\varepsilon(0) = x_0^\varepsilon = x_0 \qquad \text{if} \quad x_0 > 0,$$

$$\vartheta_0^\varepsilon \in C^1[0, x_0^\varepsilon], \qquad \vartheta_0^\varepsilon(x_0^\varepsilon) = 0,$$

and the non-negative functions ϑ_0^ε, uniformly bounded by constant K, converge uniformly to the function $\vartheta_0(x)$.

If $x_0 = 0$, then $x_0^\varepsilon = \varepsilon$ and $\vartheta_0^\varepsilon(x) = 0$ for $0 < x < \varepsilon$. In the first case, the functions $R_\varepsilon(t)$ are bounded from below by x_0, because $R_\varepsilon(t)$ are monotone increasing and coincide with x_0 at the initial time. In the second case, where $x_0^\varepsilon = \varepsilon$, the lower barrier $R_\varepsilon^+(t)$ of the functions $R_\varepsilon(t)$ can be constructed as the solution of the Stefan problem (1.1)–(1.4), with $\vartheta_\varepsilon^+(x,t)$ equal to zero at the initial time and coinciding with the function $\vartheta^+(t)$ on the boundary $\{x = 0\}$. Let

$$U_\varepsilon(x,t) = \begin{cases} \Phi[\vartheta_\varepsilon(x,t)] & \text{for } 0 < x < R_\varepsilon(t) \\ -1 & \text{for } x > R_\varepsilon(t) \end{cases}$$

and let

$$U_\varepsilon^+(x,t) = \begin{cases} \Phi[\vartheta_\varepsilon^+(x,t)] & \text{for } 0 < x < R^+(t) \\ -1 & \text{for } x > R_\varepsilon^+(t). \end{cases}$$

By construction, the functions $U_\varepsilon, U_\varepsilon^+$ are generalized solutions of the Stefan problem. U_ε^+ does not exceed U_ε on the boundary $\{x = 0\}$ and at the initial time instant. Therefore, by the comparison theorem applied to the generalized solutions of the Stefan problem,

$$R_\varepsilon(0) = R_\varepsilon^+(0); \qquad R_\varepsilon(t) \geq R_\varepsilon^+(t) \qquad \text{for} \quad t \in (0,T). \qquad (4.2)$$

The functions $\vartheta_\varepsilon^+(x,t)$ are uniformly in ε Hölder continuous with respect to the space variable:

$$|\vartheta_\varepsilon^+(x',t) - \vartheta_\varepsilon^+(x'',t)| \leq N_1\left(\|\vartheta^+\|_{2,[0,T]}^{(1)}, a_0, K\right)|x' - x''|^\beta. \qquad (4.3)$$

This can be shown analogously to the method in Section 3 of the present chapter: the equation for ϑ_ε^+ is to be multiplied by the derivative $\partial\vartheta_\varepsilon^+/\partial t$, the equality obtained is then integrated in x, and, after some manipulation, the L_2-norm of the highest derivatives is estimated, giving the bound (4.3).

Hence

$$|\vartheta^+(t)| = |\vartheta_\varepsilon^+(0,t) - \vartheta_\varepsilon^+(R_\varepsilon^+(t),t)| \leq N_1|R_\varepsilon^+(t)|^\beta,$$

and eventually we obtain

$$R_\varepsilon(t) \geq \max\left\{x_0^\varepsilon, \left|\frac{\vartheta^+(t)}{N_1}\right|^{1/\beta}\right\} > 0 \qquad \text{for} \quad t \in (0,T). \qquad (4.4)$$

The upper estimate of the functions $\{R_\varepsilon(t)\}$ follows from the same comparison theorem for the generalized solutions of the Stefan problem as above. Indeed, the functions $\vartheta_\varepsilon(x,t)$ are uniformly bounded by the constant K (by the maximum principle). Thus on the line $\{x = x_0^\varepsilon\}$ and at the initial time, the functions $U_\varepsilon(x,t)$ do not exceed the values of $U_\varepsilon^*(x,t)$, where

$$U_\varepsilon^* = \begin{cases} \Phi(\vartheta_\varepsilon^*) & \text{for } x_0^\varepsilon < x < R_\varepsilon^*(t), \\ -1 & \text{for } x > R_\varepsilon^*(t) \end{cases}$$

and $\{R_\varepsilon^*(t),\ \vartheta_\varepsilon^*(x,t)\}$ is the self-similar solution of the one-phase Stefan problem, given by

$$\vartheta_\varepsilon^*(x_0^\varepsilon, t) = K,\ R_\varepsilon^*(0) = x_0^\varepsilon,\ R_\varepsilon^*(t) = x_0^\varepsilon + D_*(K)t^{1/2}.$$

Therefore

$$R_\varepsilon(t) = x_0^\varepsilon + D_*(K)t^{1/2}. \tag{4.5}$$

Due to (4.4) and (4.5), we can use the local parabolic estimates:

$$\vartheta_\varepsilon(\delta, t) = \vartheta_\varepsilon^0(t) \in H^{r/2}[\delta, T], \qquad \delta > 0, \quad r > 2;$$

$$|\vartheta_\varepsilon^0|_{[\delta,T]}^{(r/2)} \le N_2(\delta, K, a_0).$$

Further, as in our proof of Theorem 2, Section 1 of the present Chapter, it follows that

$$\int_\delta^T \int_\delta^{R_\varepsilon(t)} \left\{ \left| \frac{\partial \vartheta_\varepsilon}{\partial t} \right|^2 + \left| \frac{\partial^2 \vartheta_\varepsilon}{\partial x^2} \right|^2 \right\} dx\,dt + \max_{t \in (\delta, T)} \int_\delta^{R_\varepsilon(t)} \left| \frac{\partial \vartheta_\varepsilon}{\partial x}(x, t) \right|^2 dx \le N_3(\delta, K, a_0).$$

By the proof of Theorem 6, Section 3 of the present chapter, the last estimate implies the Hölder continuity of the functions $\vartheta_\varepsilon(x, t)$ and the summability of the fourth power of the derivative dR_ε/dt:

$$\left\{ |\vartheta_\varepsilon|_{G_{\delta,T}^\infty}^{(\beta)},\ \|R_\varepsilon\|_{4,[\delta,T]}^{(1)} \right\} \le N_4(\delta, K, a_0),$$

where

$$G_{\delta,T}^\infty = \{(x,t)\ :\ \delta < x < R_\varepsilon(t),\ \delta < t < T\}.$$

In the variables (1.9), the functions $v_\varepsilon(y, \tau) = \vartheta_\varepsilon(yR_\varepsilon(\tau), \tau)$ are the solutions of a uniformly (for $\tau \ge \delta$) parabolic equation with coefficients that satisfy all the hypotheses of Theorem 7, Section I.2. Hence, $v_\varepsilon \in W_4^{2,1}(Q_{\delta_1,T})$, in a slightly smaller domain $Q_{\delta_1,T} = \{(y,\tau)\ :\ \delta_1 < y < 1,\ \delta_1 < \tau < T\}$. The space $W_4^{2,1}(Q_{\delta_1,T})$ is embedded into the space $H^{\gamma,\gamma/2}(\bar{Q}_{\delta,T})$ for some $\gamma > 1$, thus the coefficients of the equation for $v_\varepsilon(y, \tau)$ are Hölder continuous. Therefore, by Theorem 6, Section I.2,

$$|v_\varepsilon|_{Q_{\delta_2,T}}^{(r)} \le N_5(\delta_2, \delta, K, a_0), \qquad r > 2, \quad \delta_2 > \delta_1 > \delta.$$

This implies by standard diagonalization that the sequence $\{v_\varepsilon(y, \tau)\}$ contains a fundamental subsequence which in every space $H^{r,r/2}(\bar{Q}_{\lambda,T})$ is convergent to the function $v(y, \tau)$. The pair

$$R(t) = \left(x_0^2 - 2 \int_0^t \frac{\partial v}{\partial y}(1, \tau)\,d\tau \right)^{1/2}, \qquad \vartheta(x,t) = v\left(\frac{x}{R(t)}, t \right)$$

is then the desired solution of the Stefan problem under consideration. $\square$

As already mentioned, even if equation (1.1) and the free boundary conditions (1.2) are satisfied at $t > 0$ in the classical sense, the boundary condition at $x = 0$ and the initial condition (for $x_0 > 0$) hold only in sense of the appropriate integral identity.

The above scheme applied to studying the differential properties of solutions remains in principle applicable also to the two-phase Stefan problem, provided that the following estimate holds:

$$-1 < R^-(t) \le R_\varepsilon(t) \le R^+(t) < 1. \tag{4.6}$$

Theorem 11. *Let* $\Phi \in C^2(-\infty, 0] \cap C^2[0, \infty)$ *and* $\Phi'(s) = a(s) \ge a_0 > 0$, $\vartheta^\pm \in W^1_{2,\text{loc}}(0, \infty)$. *Let* $(\pm 1)\vartheta^\pm(t) > 0$ *for* $t \in (0, \infty)$, *and let the bounded function* $\vartheta_0(x)$ *be non-positive for* $x \in \Omega^-(0)$ *and non-negative for* $x \in \Omega^+(0)$.

Then there exists a unique bounded classical solution $\{R(t), \vartheta(x, t)\}$ *of the problem* (3.1)–(3.4) *for all positive t. The derivative* $\frac{dR}{dt}(t)$ *is Hölder continuous for positive t, and the function* $\vartheta(x, t)$ *belongs to* $H^{r, r/2}(\bar{G}), r > 2$, *in each closed bounded subdomain* $\bar{G}$ *of* $\overline{\Omega^\pm_\infty}$ *which does not intersect the boundary of* $\bar{\Omega}_\infty$.

Proof. Suppose that $\{R_\varepsilon(t), \vartheta_\varepsilon(x, t)\}$ is the solution of the problem (3.1)–(3.4) with data $\{\vartheta^-(t), \vartheta^+(t), \vartheta^\varepsilon_0(x)\}$, where the functions $\vartheta^\varepsilon_0(x)$ belong to $W^1_2(\Omega)$, are non-positive for $x \in [-1, x^\varepsilon_0]$, non-negative for $x \in [x^\varepsilon_0, 1]$ and satisfy compatibility conditions of zero-th order at the points $x = \pm 1$. The points x^ε_0 coincide with x_0 as long as $|x_0| < 1$, and $x^\varepsilon_0 = \mp 1 \pm \varepsilon$, when $x_0 = \mp 1$.

The existence of the regularized solutions $\{R_\varepsilon(t), v_\varepsilon(x, t)\}$ follows from Theorem 7, Section 3 of this chapter.

Let us define

$$u(x, t) = \zeta(t)\vartheta_\varepsilon(x, t),$$

where $\zeta \in C^1[0, \infty)$, $\zeta = 0$ if $t \in (0, \delta/2)$, $\zeta = 1$ for $t \in (\delta, \infty)$.

In each of the domains $\Omega^\pm_T = \{(x, t) \in \Omega_T : (\pm 1)\vartheta_\varepsilon(x, t) > 0\}$, the function $u(x, t)$ satisfies a uniformly parabolic equation with bounded right-hand side. Multiplying the equation for $u(x, t)$ by the derivative $\frac{\partial u}{\partial t}$ and then integrating over each of $\Omega^\pm_T$, in the same way as before we can prove that $u(x, t)$ belongs to the space $W^{2,1}_2(\Omega^\pm_T)$ and, thus, is Hölder continuous with respect to x. At the same time, $\zeta = 1$ for $t > \delta$. Hence

$$|\vartheta_\varepsilon(x', t) - \vartheta_\varepsilon(x'', t)| \le N_6 |x' - x''|^\beta \qquad \text{for} \quad t > \delta, \tag{4.7}$$

with N_6 dependent only on the norms $\|\vartheta^\pm\|^{(1)}_{2, [0, T]}$, $|\vartheta^\varepsilon_0|^{(0)}_\Omega$ and constants a_0, δ. The last estimate and the strict positiveness of the functions $|\vartheta^\pm(t)|$ imply the inequalities (4.6).

Now, as in the proof of Theorem 6 of Section 3, we shall show that the functions $R_\varepsilon(t)$ belong to $W^1_4[\delta, T]$ with the appropriate norms uniformly bounded for all $\varepsilon > 0$. The further considerations repeat exactly the arguments used in the proof of Theorem 10.

In contrast to the one-phase Stefan problem, (4.7) holds up to the points $x = \pm 1$ in the two-phase problem, i.e., the solution $\vartheta(x, t)$ of problem (3.1)–(3.4) satisfies boundary conditions at $x = \pm 1$ in the classical sense. The initial condition $\vartheta(x,0) = \vartheta_0(x)$, as in the one-phase Stefan problem, is satisfied in the L_2-sense:

$$\lim_{t \to 0} \int_\Omega |\vartheta(x,t) - \vartheta_0(x)|^2 \, dx = 0.$$

This follows from the obvious estimate

$$\int\int_{\Omega_T} \left| \frac{\partial \vartheta}{\partial x}(x,t) \right|^2 \, dx dt < \infty. \qquad \square$$

The case we have not yet considered is related to Stefan problems in unbounded domains. Let us assume

$$\Omega^+(0) = \{(x : -\infty < x < 0\}, \qquad x_0 = 0.$$

Theorem 12. *Let $\Phi \in C^2[0, \infty)$, $\Phi'(s) \geq a_0 > 0$, and let $\vartheta_0(x)$ be a non-negative, measurable function bounded from above by a finite constant K.*

Then, for all $t > 0$, there exists a unique classical solution $\{R(t), \vartheta(x,t)\}$ of the Stefan problem (1.1), (1.2), (1.4), such that

$$R \in H^{(3+\gamma)/2}[\delta, T], \quad \vartheta \in H^{r,r/2}(\overline{\Omega_T^+} \setminus \Omega_\delta^+),$$

$$0 < \delta < T < \infty, \quad r = 2 + \gamma, \ \gamma \in (0, 1),$$

$$\Omega_T^+ = \{(x,t) : -\infty < x < R(t), \ 0 < t < T\}.$$

In proving the above theorem, it is enough to observe that the solution $\{R(t), \vartheta(x,t)\}$ we are looking for is bounded above by the self-similar solution $\{R_*(t), \vartheta_*(x,t)\}$ of the problem (1.1), (1.2), (1.4) with $\vartheta_0(x) = K = $ const. The existence of this solution can be shown as in Theorem 3, Section 2.

Theorem 13. *Let $\Phi \in C^2(-\infty, 0] \cap C^2[0, \infty)$, $\Phi'(s) \geq a_0 > 0$, and let $\vartheta_0(x)$ be a measurable function with absolute value bounded by a constant K, non-positive in the domain $\Omega^-(0) = \{x : -\infty < x < 0\}$ and non-negative in $\Omega^+(0) = \{x : 0 < x < \infty\}$, $x_0 = 0$.*

Then, for all $t > 0$, there exists a unique classical solution $\{R(t), \vartheta(x,t)\}$ of the Stefan problem (3.1), (3.2), (3.4), such that

$$R \in H^{(3+\gamma)/2}[\delta, T], \quad \vartheta \in H^{r,r/2}(\overline{\Omega_T^\pm} \setminus \Omega_\delta^\pm), \quad r = 2 + \gamma,$$

$$\gamma \in (0, 1), \quad 0 < \delta < T < \infty,$$

$$\Omega_T^+ = \{(x,t) : R(t) < x < \infty, \ 0 < t < T\},$$
$$\Omega_T^- = \{(x,t) : -\infty < x < R(t), \ 0 < t < T\}.$$

As before, it is enough for the proof to find continuous functions $R_*^{\pm}(t)$ which bound $R(t)$:

$$-\infty < R_*^-(t) < R(t) < R_*^+(t) < \infty.$$

It follows easily that the self-similar solution of the Stefan problem (1.1), (1.2), (1.4) with $\vartheta_0(x) = -K$ for $x \in \Omega^-(0)$ bounds the function $R(t)$ from above:

$$R(t) \le R_*^+(t) \qquad \big(\text{then } \vartheta_*^+(x,t) \le \vartheta(x,t) \big).$$

Analogously, the self-similar solution $\{R_*^-(t), \vartheta_*^-(x,t)\}$ of the one-phase problem (1.1), (1.2), (1.4) with $\vartheta_0(x) = K$ for $x \in \Omega^+(0)$ approximates $R(t)$ from below.
In both cases, the condition (1.2) reduces to

$$\frac{dR_*^{\pm}}{dt}(t) = \frac{\partial \vartheta_*^{\pm}}{\partial x}\big(R_*^{\pm}(t) \mp 0, t\big). \qquad\qquad \square$$

5. The two-phase multi-front Stefan problem

Let $\Omega = \{x : |x| < 1\}$. We shall consider the following Stefan problem:

$$\frac{\partial U}{\partial t} = \frac{\partial^2 \vartheta}{\partial x^2}, \quad \vartheta = \chi(U) \qquad \text{for} \quad (x,t) \in \Omega_T, \tag{5.1}$$

$$\vartheta(\pm 1, t) = \vartheta^{\pm}(t) \qquad \text{for} \quad t \in (0,T), \tag{5.2}$$

$$U(x,0) = U_0(x) \qquad \text{for} \quad x \in \Omega, \tag{5.3}$$

where a finite number of connected components of the liquid and solid phases exist at the initial time instant, i.e., the domain Ω is the union of a finite number of intervals $\Omega^{(k)}(0), k = 1, \ldots, m+1$, each of them occupied either by the liquid or solid phase.

The input data $\vartheta^{\pm}(t), U_0(x)$ are such that the generalized solution of the Stefan problem (5.1)–(5.3) exists for all positive t. A sufficient condition for this is that $\vartheta^{\pm} \in W^2_{2,\mathrm{loc}}(0,\infty)$, $\vartheta_0 = \chi(U_0) \in W^1_2(\Omega)$, $\Phi \in C^2(-\infty, 0] \cap C^2[0, \infty)$ and the compatibility condition $\vartheta^{\pm}(0) = \vartheta_0(\pm 1)$ holds at the initial time instant.

Definition 1. *The generalized solution of Stefan problem (5.1) - (5.3) has (C, m)-structure at the time moment $t = \tau$, if there exist numbers $r_1(\tau), \ldots, r_m(\tau)$, $-1 = r_0 < r_1 < \ldots < r_m < r_{m+1} = 1$, such that the domains $\Omega^{(k)}(\tau) = \{x : r_{k-1} < x < r_k\}$, $k = 1, \ldots, m+1$, are alternately occupied by the liquid and solid phases.*

It is also said that the generalized solution of the Stefan problem has (C, m)-structure at time $t = \tau$.

Definition 2. *The generalized solution of the Stefan problem (5.1)–(5.3) on an interval (T_1, T_2) preserves the (C, m)-structure imposed at time $t = T_1$, if there are continuous functions $r_k(t)$, $k = 1, \ldots, m$, $-1 = r_0 < r_1(t) < \ldots < r_m(t) < r_{m+1}(t) = 1$ on*

$[T_1, T_2)$ such that, for $t \in [T_1, T_2)$, the domains $\Omega^{(k)}(t)$, $k = 1, \ldots, m+1$, are alternately occupied by the liquid and solid phases.

If $(0, T)$ is decomposed into a finite number of subintervals (T_{k-1}, T_k), $0 = T_0 < T_1 < \ldots < T_{p+1} = T$, on which the generalized solution of the Stefan problem preserves the (C, m_k)-structure, imposed at $t = T_k$, $k = 0, 1, \ldots, p$, then the generalized solution of problem (5.1)–(5.3) is referred to as the *classical solution of the two-phase multi-front Stefan problem on the time interval* $(0, T)$ (or, shortly, the *classical solution*).

Our aim here is to explore the requirements on the input data under which the generalized solution of the problem (5.1)–(5.3) will become classical for all positive t.

In contrast to the former sections, we shall not assume that the functions $\vartheta^{\pm}(t)$ have constant sign. We shall postulate only that their zeros can accumulate at infinity, at most. The entire time axis $(0, \infty)$ will then admit decomposition into at most a countable number of subintervals

$$(T_{k-1}, T_k), \quad k = 1, 2, \ldots, \quad 0 = T_0 < T_1 < \ldots < T_k < \ldots, \quad \lim_{k \to \infty} T_k = \infty$$

such that $\vartheta^{\pm}(t)$ have constant sign on each of them. This justifies the restriction of further considerations to an interval $(0, T)$, where the functions $\vartheta^{\pm}(t)$ do not change sign.

Theorem 14. *Let* $\Phi \in C^2[-\infty, 0) \cap C^2[0, \infty)$, $\Phi'(s) \geq a_0 > 0$, *and let* $\vartheta_0 \in W^1(\Omega)$, $\vartheta^{\pm} \in W_2^1(0, T)$, *and* $|\vartheta^{\pm}(t)| > 0$ *for* $t \in (0, T)$, *let* (C, m)-*structure be imposed at the initial time and the compatibility conditions* $\vartheta^{\pm}(0) = \vartheta_0(\pm 1)$ *hold at the points* $x = \pm 1$.

Then the generalized solution of the problem (5.1)–(5.3) is classical on the whole interval $(0, T)$.

Proof. We shall at first assume that $\vartheta^{\pm} \in W_2^2(0, T)$, and that, for $m \geq 2$ and all $k = 2, \ldots, m$, there exist points $y_k \in \Omega^{(k)}(0)$ for which $|\vartheta_0(y_k)| > 0$.

As proved in Chapter I, the generalized solution of the Stefan problem, with Hölder continuous temperature $\vartheta \in H^{\gamma, \gamma/2}(\bar{\Omega}_T)$ exists. Hence, there is a time interval $(0, \tau)$ such that

$$|\vartheta^{(k)}(t)| = |\vartheta(y_k, t)| > 0, \quad k = 2, \ldots, m, \quad \text{for} \quad t \in (0, \tau).$$

By the local parabolic estimates, the functions $\vartheta^{(k)}(t)$ are at least once continuously differentiable on the interval $(0, \tau)$.

Set $\vartheta^{(1)}(t) = \vartheta^-(t)$, $\vartheta^{(m+1)}(t) = \vartheta^+(t)$. In the domain $G_\tau^{(k)} = G^{(k)} \times (0, \tau)$, where $G^{(k)} = \{x : y_{k-1} < x < y_k\}$, $y_1 = -1$, $y_{m+1} = 1$, let us consider the following Stefan problem: determine a pair of functions $\{\bar{U}, \bar{\vartheta}\}$, $\bar{\vartheta} = \chi(\bar{U})$, which satisfy equation (5.1), the initial condition (5.3) and the boundary conditions

$$\bar{\vartheta}(y_j, t) = \vartheta^{(j)}(t), \quad j = k-1, k, \quad \text{for} \quad t \in (0, \tau).$$

The initial state includes two phases in each of these problems. Hence, as already shown, there exists a unique classical solution of the two-phase single-front Stefan problem in the domain $G_T^{(k)}$. Concerning the uniqueness of the generalized solution, we can conclude by the construction that the obtained $\{\tilde{U}, \tilde{\vartheta}\}$ in each of the domains $G_T^{(k)}$ coincide with the generalized solution of problem (5.1)–(5.3) that has been found earlier. Therefore, on a sufficiently small time interval, the generalized solution of problem (5.1)–(5.3) is its classical solution.   □

As for the two-phase single-front Stefan problem, the norms

$$\|\vartheta\|_{2,\Omega_\tau}^2, \qquad \max_{t\in(0,\tau)} \|\vartheta(t)\|_{2,\Omega}^{(1)} \tag{5.4}$$

are bounded on $(0,\tau)$ by a constant dependent only on

$$M = \max\left\{\|\vartheta_0\|_{2,\Omega}^{(1)}, \|\vartheta^+, \vartheta^-\|_{2,(0,T)}^{(1)}\right\}.$$

As we have proved in Section I.3, in the case of one space variable the boundedness of the norm (5.4) implies that $\vartheta(x,t)$ belongs to the Hölder space $H^{\gamma,\gamma/2}(\bar{\Omega}_\tau)$. However, as seen in the proof, this requirement was essential for the generalized solution of the Stefan problem. The hypothesis $\vartheta^\pm \in W_2^2(0,T)$ can be thus replaced by the weaker condition $\vartheta^\pm \in W_2^1(0,T)$.

Since $\vartheta(x,\tau) \in W_2^1(\Omega)$, compatibility conditions of zero-order hold at the points $x = \pm 1$ and the solution has (C,m)-structure at time $t = \tau$. By assuming $t = \tau$ as new initial time instant, the classical solution of the Stefan problem can be extended onto an interval $(\tau, \tau + \delta)$, $\delta > 0$.

Is there any obstacle preventing the extension of the classical solution onto the whole interval $(0,T)$? The only reason might be the disappearance of a connected component $\Omega^{(k)}(t)$ of the liquid or solid phase at $t = t_1$. By (5.4), all the hypotheses of the theorem are satisfied at $t = t_1$, except for the generalized solution has (C,m_1)-structure rather than (C,m)-structure, with $m_1 < m$.

Then, $t = t_1$ is taken as the new initial time, and the classical solution can be extended onto the maximal interval (t_1, t_2) where the (C,m_1)-structure of the solution, imposed at $t = t_1$, is preserved.

Because the total number of the liquid and solid components is finite, the whole interval $(0,T)$ admits a finite decomposition into subintervals (t_{k-1},t_k), $k = 1,\ldots,p+1$, $0 = t_0 < t_1 < \ldots < t_p < t_{p+1} = T$, such that the (C,m_{k-1})-structure of the solution, imposed at $t = t_{k-1}$, is preserved.

Suppose now that there is a connected component $\Omega^{(k)}(0)$ which does not include any point y_k where $|\vartheta_0(y_k)| > 0$, i.e., $\vartheta_0(x) = 0$ for all $x \in \Omega^{(k)}(0)$. Then, providing $|\vartheta_0(y_{k-1})| > 0$ and $|\vartheta_0(y_{k+1})| > 0$, one-phase problems are to be solved in the domains $G_T^{(k)}$ and $G_T^{(k+1)}$, and all the considerations can be repeated with obvious changes.

If, in particular, $\vartheta_0(x) \equiv 0$ in the domain $\Omega^{(k-1)}(0)$, then the components $\Omega^{(k-1)}(t)$ and $\Omega^{(k)}(t)$ of different phases remain in equilibrium at $t > 0$ and the free boundary remains "frozen": $r_{k-1}(t) \equiv r_k(0)$. This equilibrium lasts up to the disappearance of either of the components $\Omega^{(k-1)}(t)$ and $\Omega^{(k)}(t)$.   □

6. Filtration of a viscid compressible liquid in a vertical porous layer

By introducing the mass Lagrange coordinates, the original problem reduces to a nonlinear boundary value problem for a quasilinear parabolic equation in a fixed domain. The global-in-time existence of the classical solution is implied there by the local-in-time solvability of the problem and non-local Hölder estimates of the solution. These follow from the results of [137] by estimates on the sup-norm of the solution which have been obtained for all functions $f(p)$ such that

$$\int_1^\infty \frac{1}{f(p)} \, dp = \infty,$$

and for functions $f(p)$ of the form p^γ or e^{p-1}. By the comparison lemma and standard energy estimates, the asymptotic behaviour of the solution as $t \to \infty$ can be explored.

6.1. Problem statement. The main result

Let a viscid compressible liquid fill the domain $\Omega(t) = \{x \; : \; 0 < x < R(t)\}$, with velocity $\bar{v}$, pressure p and mass density $\rho = f(p)$ (where $f(p)$ is a given function) which satisfy the continuity equation and Darcy's law

$$m\frac{\partial \rho}{\partial t} + \frac{\partial}{\partial x}(\rho \bar{v}) = 0. \tag{6.1}$$

$$\bar{v} = -\frac{k}{\mu}\frac{\partial p}{\partial x} - \rho g. \tag{6.2}$$

In the above equations, $m = \text{const.} > 0$ is the ground porosity, $k = \text{const.} > 0$ the permeability coefficient, $\mu = \text{const.} > 0$ the viscosity of the filtrating liquid, $g = \text{const.} > 0$ the gravitational acceleration.

On the free boundary $\{x = R(t)\}$,

$$m\frac{dR}{dt} = \bar{v}, \qquad p = p_* = \text{const.} > 0. \tag{6.3}$$

The mass outflow is prescribed on the bottom $\{x = 0\}$:

$$\rho \bar{v} = \varphi(t). \tag{6.4}$$

Furthermore, the initial location of the free boundary and the initial pressure distribution are given:

$$R(0) = R_0 > 0, \quad p(x, 0) = p_0(x) \qquad \text{for} \quad x \in \Omega(0). \tag{6.5}$$

Without loss of generality, the constants m, k, μ, g, p_* can be assumed equal to 1, and we can set $f(1) = 1$.

Theorem 15. *Let $f \in C^2[0, \infty)$, $f'(p) > 0$ for $p \in [0, \infty)$, and assume either that*

$$\int_1^\infty \frac{1}{f(p)} \, dp = \infty \tag{6.6}$$

or that $f(p)$ is equal to either $e^{(p-1)}$ or p^γ, $\gamma \geq 1$.

In addition, let $p_0 \in H^{2+\alpha}\big(\overline{\Omega(0)}\big)$, $p_0 \geq 1$, $\varphi(t) \geq 0$, $\varphi \in H^{(1+\alpha)/2}[0,\infty)$, and the compatibility conditions of first-order hold at the points $x = 0$ and $x = R_0$.

Then, on the arbitrary time interval $(0,T)$, there exists a unique solution of the problem (6.1)–(6.5),

$$p \in H^{2+\alpha,(2+\alpha)/2}(\overline{\Omega}_T), \quad R \in H^{(3+\alpha)/2}[0,T],$$

where $\Omega_T = \{(x,t) : x \in \Omega(t), t \in (0,T)\}$.

Let us consider the stationary solutions that correspond to $\varphi(t) \equiv 0$. It follows from (6.1)–(6.4) that the pressure $\tilde p(x)$ satisfies the equation

$$\frac{d\tilde p}{dx} + f(\tilde p) = 0,$$

which by the second condition of (6.3) implies that

$$\int_1^{\tilde p(x)} \frac{1}{f(s)}\, ds \; = \; R_\infty - x. \tag{6.7}$$

The position R_∞ of the free boundary or the total mass M_∞ of the liquid are free parameters, related to each other by

$$M_\infty = \int_0^{R_\infty} \tilde\rho(x)\, dx = \int_0^{R_\infty} f[\tilde p(x)]\, dx. \tag{6.8}$$

We find it more convenient to consider the total mass of the liquid as given and determine R_∞ from equation (6.8). This is, because for each value of R_∞, equation (6.7) has a solution for $x \in (0, R_\infty)$. Indeed, if $f(p)$ does not satisfy condition (6.6), then equation (6.7) has no solution for

$$R_\infty > R_* = \int_1^\infty \frac{1}{f(p)}\, dp. \tag{6.9}$$

The equation (6.8) can be inverted. To this end, let us introduce the new independent variable

$$y = \int_x^{R_\infty} \tilde\rho(s)\, ds, \quad y(0) = M_\infty, \quad y(R_\infty) = 0.$$

The function $P(y) = \tilde p(x)$ is a solution of the Cauchy problem

$$-\tilde\rho(x)\frac{dP}{dy} + \tilde\rho(x) = 0, \quad P(0) = 1,$$

hence $P(y) = 1 + y$. Because

$$\frac{dy}{dx} = -\tilde\rho(x) = -f[P(y)] = -f(1+y),$$

we conclude that

$$R_\infty = x(0) = \int_0^{M_\infty} \frac{1}{f(1+s)}\,ds = \int_1^{M_\infty+1} \frac{1}{f(s)}\,ds. \tag{6.10}$$

If the constant R_*, given by (6.9), is bounded, then we take

$$\int_1^{p_*(x)} \frac{1}{f(s)}\,ds = R_* - x. \tag{6.11}$$

Theorem 16. *Let the hypotheses of Theorem 15 hold with*

$$\int_0^\infty \varphi(t)\,dt < \infty \quad and \quad \lim_{t\to\infty} \varphi(t) = 0.$$

Then, for the solution of the problem (6.1)–(6.5) with $f(p) = \exp(p-1)$,

$$\lim_{t\to\infty} R(t) = R_\infty,$$

where R_∞ is given by (6.10), with the constant

$$M_\infty = \int_0^\infty \varphi(t)\,dt + \int_0^{R_0} f[p_0(x)]\,dx.$$

When $\int_0^\infty \varphi(t)\,dt = \infty$, we shall have $\lim_{t\to\infty} R(t) = \infty$, if $f(p)$ satisfies condition (6.6). Otherwise, when $f(p)$ does not satisfy (6.6) and for initial data such that $R_0 < R_*$, $p_0(x) < p_*(x)$, where R_* and $p_*(x)$ are defined by equalities (6.9), (6.11), $\lim_{t\to\infty} R(t) = R_*$.

The local-in-time existence of the solution of problem (6.1)–(6.5) can be proved analogously to Theorem 4, Chapter II. The problem we consider differs from that in Chapter II only in the nonlinear condition prescribed on the given boundary, a difference which is not significant for the proof of short-time solvability. Therefore, for proving the global-in-time existence of the solution it is sufficient to estimate its norm in $H^{2+\alpha,(2+\alpha)/2}(\bar\Omega_T)$ by the norms $|\varphi|_{[0,T]}^{(1+\alpha)/2}$ and $|p_0|_{\Omega(0)}^{(2+\alpha)}$.

6.2. An equivalent boundary value problem in a fixed domain

Let us define the mass Lagrange variable

$$y = \int_x^{R(t)} \varrho(s,t)\,ds.$$

The function $u(y,t) = P(y,t) - y - 1$, where $P(y,t) = p(x,t)$, is a solution of the following boundary value problem:

$$\frac{f'(u+y+1)}{f^2(y+u+1)}\,\frac{\partial u}{\partial t} = \frac{\partial}{\partial y}\left[f(u+y+1)\frac{\partial u}{\partial y}\right] \qquad \text{for} \quad (y,t)\in G_T, \tag{6.12}$$

$$f(u+y+1)\frac{\partial u}{\partial y} = \varphi(t), \quad y = Y(t) \qquad \text{for} \quad t \in (0,T), \qquad (6.13)$$

$$u = 0, \qquad y = 0 \qquad \text{for} \quad t \in (0,T), \qquad (6.14)$$

$$u(y,0) = u_0(y) \quad \text{and} \quad y \in G(0). \qquad (6.15)$$

In (6.12)–(6.15),

$$G_T = \{(y,t) : y \in G(t),\ t \in (0,T)\}, \quad G(t) = \{y : 0 < y < Y(t)\},$$

$$\frac{dY}{dt} = \varphi(t), \qquad Y(0) = y_0 = \int_0^{R_0} f[p_0(x)]\,dx,$$

with $u_0(y)$ given by

$$u_0(y) = p_0[\tilde{x}(y)] - y - 1, \quad y = \int_{\tilde{x}(y)}^{R_0} f[p_0(s)]\,ds.$$

Lemma 9. *Let $M(T) = |u|_{G_T}^{(0)}$, and let $M_0 = \max\{|u_0|_{G(0)}^{(2+\alpha)}, |\varphi|_{[0,T]}^{((1+\alpha)/2)}\}$. Then*

$$|u|_{G_T}^{(2+\alpha)} \le M_1\big(M(T), M_0, T\big), \qquad (6.16)$$

$$\max_{t\in(0,T)} |u(t)|_{G(t)}^{(1)} \le M_2\big(M(T), M_0\big). \qquad (6.17)$$

Proof. By applying the maximum principle to $P(y,t)$ (or, in the original variables, to $p(x,t)$) and due to the hypotheses of the lemma it follows that the equation (6.12) does not degenerate for

$$1 \le u(y,t) + y + 1 \le M(T) + M_0 + 1.$$

Thus, the estimates (6.16) and (6.17) are implied by results of [137] and [134], respectively. □

6.3. A comparison lemma

We shall prove the following comparison result for solutions of problem (6.1)–(6.5).

Lemma 10. *Let $\{p_i, R_i\}$ be the solution of the problem (6.1)–(6.5), corresponding to the data $\{p_{0i}(x), \varphi_i(t), R_{0i}\}$, $i = 1, 2$, with $p_i \in H^{r,r/2}(\bar{\Omega}_{iT})$, $R_i \in H^{r/2}[0,T]$, where $r = 2+\alpha$, $\Omega_{iT} = \{(x,t) : x \in \Omega_i(t),\ t \in (0,T)\}$ and $\Omega_i(t) = \{x : 0 < x < R_i(t)\}$. If*

$$p_{01}(x) \ge p_{02}(x), \quad \varphi_1(t) \ge \varphi_2(t), \quad R_{01} \ge R_{02},$$

then

$$p_1(x,t) \ge p_2(x,t), \quad R_1(t) \ge R_2(t)$$

for $x \in \Omega_2(t),\ t \in (0,T)$.

Proof. Under the hypotheses of the lemma, the result on the continuous dependence of the solution $\{p(x,t), R(t)\}$ upon input data $\{p_0(x), \varphi(t), R_0\}$ holds in the appropriate Hölder norms $|p|_{\Omega_T}^{(r)}$ and $|R|_{[0,T]}^{(r/2)}$. Hence, if we prove the assertion in the case $R_{01} > R_{02}$, $\varphi_1(t) > \varphi_2(t)$, then by passing to the limit we shall complete the proof also in the case $R_{01} \geq R_{02}$, $\varphi_1(t) \geq \varphi_2(t)$.

Now, let $R_{01} > R_{02}$. Then $R_1(t) > R_2(t)$ for sufficiently small t. Assume that there exists $t_0 > 0$ for which $R_1(t_0) = R_2(t_0)$. Because $R_1(t) > R_2(t)$ for $t < t_0$, we have

$$\frac{dR_1}{dt}(t_0) \leq \frac{dR_2}{dt}(t_0).$$

This together with condition (6.3) implies that

$$\frac{\partial p_1}{\partial x}(R_1(t_0), t_0) \geq \frac{\partial p_2}{\partial x}(R_2(t_0), t_0). \tag{6.18}$$

On the other hand, the difference $w = p_1 - p_2$ satisfies a linear, homogeneous, uniformly parabolic equation in Ω_{2T}, is non-negative at the initial time and on the line $x = R_2(t)$, and $\frac{\partial w}{\partial x} + a(t)w < 0$ for $x = 0$. By the substitution $w = u \exp(\beta_1 t + \beta_2 x)$ it is easy to conclude that $u(x,t)$ satisfies the equation

$$\frac{\partial u}{\partial t} - b_1 \frac{\partial^2 u}{\partial x^2} + b_2 \frac{\partial u}{\partial x} + b_3 u = 0 \qquad \text{for} \quad (x,t) \in \Omega_{2T}$$

and the condition

$$\frac{\partial u}{\partial x} + b_4(t)u < 0 \quad \text{for} \quad x = 0,$$

with coefficients $b_1(x,t) > 0$, $b_3(x,t) > 0$ and $b_4(t) > 0$.

Then the minimal value of the function $u(x,t)$ is equal to 0 and is achieved at the point $(R_2(t_0), t_0)$, where

$$\frac{\partial u}{\partial x}(R_2(t_0), t_0) < 0.$$

The latter inequality follows from a result in [92], p. 69, leading to a contradiction to (6.18). $\qquad\square$

6.4. The case $\displaystyle\int_1^\infty \frac{1}{f(p)}\,dp = \infty$

Assuming that (6.6) holds, we shall now construct the exact solution of problem (6.1)–(6.5), and estimate the maximum of any solution of the original problem by using the comparison lemma.

Lemma 11. *Let the hypotheses of Theorem 15 hold, and let $f(p)$ satisfy (6.6). Then*

$$1 \leq p(x,t) \leq \tilde{p}(x - ct),$$

where the function $\tilde{p}(\xi)$, bounded at $|\xi| < \infty$, is defined by the equality

$$\int_1^{\tilde{p}(\xi)} \frac{1}{f(s) + c}\, ds = R_0 - \xi, \qquad (6.19)$$

with the constant c dependent only on M_0.

Proof. Condition (6.6) implies that the equation (6.19) is invertible for all values of the argument from the interval $(-\infty, R_0)$ and the function $\tilde{p}(x - ct)$ satisfies problem (6.1)–(6.5) with input data $\{\tilde{p}(x), \tilde{\varphi}(t), R_0\}$, where $\tilde{\varphi}(t) = cf[\tilde{p}(-ct)]$. Since $f[\tilde{p}(-ct)] \geq 1$ and

$$\frac{d\tilde{p}}{dx} = -f[\tilde{p}(x)] - c < 0, \qquad \left| \frac{d\tilde{p}}{dx}(x) \right| \geq 1 + c,$$

by taking c large enough we can easily ensure that

$$\tilde{\varphi}(t) > \varphi(t), \qquad \tilde{p}(x) > p_0(x).$$

Hence the assertion follows due to the comparison lemma.   □

6.5. The case $f(p) = \exp(p - 1)$

We are going to transform the Stefan problem formulated in Lagrange variables to an equivalent boundary value problem for which the maximum principle holds.

Lemma 12. *Let the hypotheses of Theorem 15 be satisfied, and let $f(p) = \exp(p - 1)$. Then*

$$|p|_{\Omega_T}^{(0)} \leq M_3(M_0).$$

Proof. Let us apply the substitution $\tau = t$, $\xi = 1 - e^{-y}$ to the domain G_T in Lagrange variables (y, τ).

The function

$$w(\xi, \tau) = \frac{1}{1 + \xi}\, \exp(u(y, t)),$$

where $u(y, t)$ is the solution of (6.12)–(6.15), satisfies the boundary value problem

$$\frac{(1 + \xi)^{-1}}{e^2 w^2} \frac{\partial w}{\partial \tau} = (1 + \xi)\frac{\partial^2 w}{\partial \xi^2} + 2\frac{\partial w}{\partial \xi} \qquad \text{for} \quad (\xi, \tau) \in Q_T;$$

$$(1 + \xi)\frac{\partial w}{\partial \xi} + w = \frac{1}{e}\varphi(\tau), \quad \xi = \xi_0(\tau) \qquad \text{for} \quad \tau \in (0, T);$$

$$w(0, \tau) = 0 \qquad \text{for} \quad \tau \in (0, T);$$

$$w(\xi, 0) = \frac{1}{1 + \xi}\, \exp\left\{ u_0\left(\log \frac{1}{1 - \xi} \right) \right\} \qquad \text{for} \quad \xi \in (0, \xi_*).$$

Here $\xi_0(\tau) = 1 - \exp(-Y(\tau))$, $\xi_* = \xi_0(0) < 1$ and $Q_T = \{(\xi, \tau) : 0 < \xi < \xi_0(\tau), 0 < \tau < T\}$. The maximum principle holds for this problem.   □

6.6. The case $f(p) = p^\gamma$, $\gamma \geq 1$

We shall estimate the solution of problem (6.1)–(6.5) in the sup-norm by applying standard energy estimates in the original variables (x, t).

Lemma 13. *Let the hypotheses of Theorem 15 hold and let $f(p) = p^\gamma$, $\gamma \geq 1$. Then*

$$p(x, t) \leq M_4(M_0, T).$$

Proof. The function $v = p^{\gamma+1}$ satisfies the following boundary value problem:

$$\frac{\partial}{\partial t} v^{\gamma/(\gamma+1)} = \frac{\partial}{\partial x}\left\{\frac{1}{1+\gamma}\frac{\partial v}{\partial x} + v^{2\gamma/(\gamma+1)}\right\} \qquad \text{for} \quad (x, t) \in \Omega_T, \tag{6.20}$$

$$\frac{1}{1+\gamma}\frac{\partial v}{\partial x} + v^{2\gamma/(\gamma+1)} = -\varphi(t) \qquad \text{for} \quad x = 0, \ t \in (0, T), \tag{6.21}$$

$$v = 1, \quad \frac{1}{1+\gamma}\frac{\partial v}{\partial x} + 1 = -\frac{dR}{dt} \qquad \text{for} \quad x = R(t), \ t \in (0, T), \tag{6.22}$$

$$R(0) = R_0, \quad v(x, 0) = v_0(x) \qquad \text{for} \quad x \in \Omega(0). \tag{6.23}$$

Multiplying equation (6.20) by 1 and then by v, after partial integration and some simple transformations which exploit the boundary conditions (6.21) and (6.22), we obtain the bounds

$$\max_{t \in (0,T)} \|v(t)\|_{\gamma/(\gamma+1), \Omega(t)} \leq N, \tag{6.24}$$

$$\frac{d}{dt}\|v(t)\|_{2\gamma/(\gamma+1), \Omega(t)}^{2\gamma/(\gamma+1)} + \frac{1}{N}\left(\|v(t)\|_{2,\Omega(t)}^{(1)}\right)^2 \leq N\left\{1 + \left\|\frac{\partial v}{\partial x}(t)|v(t)|^{2\gamma/(\gamma+1)}\right\|_{1,\Omega(t)}\right\}. \tag{6.25}$$

Here and in the following, N is a constant which depends only on γ, T and M_0.

To estimate the right-hand side of (6.25) from above, we apply Hölder's inequality and Lemma 3 of Chapter I:

$$\left\|\frac{\partial v}{\partial x}|v|^{2\gamma/(\gamma+1)}\right\|_{1,\Omega} \leq \|v\|_{2,\Omega}^{(1)}\left\{\|v\|_{4\gamma/(\gamma+1),\Omega}\right\}^{2\gamma/(\gamma+1)},$$

$$\|v - 1\|_{4\gamma/(\gamma+1),\Omega}^{2\gamma/(\gamma+1)} \leq \beta\left\{\|v\|_{2,\Omega}^{(1)}\right\}^{6\gamma/(6\gamma+4)}\|v - 1\|_{\gamma/(\gamma+1),\Omega}^{\frac{\gamma}{\gamma+1}\frac{3\gamma+1}{3\gamma+2}}.$$

In view of the above relations and the estimate (6.24), by the obvious inequalities

$$a^\lambda \leq c(\lambda)\left(1 + |1 + a|^\lambda\right), \quad \lambda \geq 0, \ a \geq 1,$$

$$|a - 1|^\lambda \leq c(\lambda)\left(1 + a^\lambda\right), \quad \lambda \geq 0, \ a \geq 2,$$

due to estimate (6.25) and Gronwall's lemma, we conclude that

$$\max_{t \in (0,T)}\|v(t)\|_{2\gamma/(\gamma+1),\Omega(t)} + \left\|\frac{\partial v}{\partial x}\right\|_{2,\Omega_T} \leq M_5(N). \tag{6.26}$$

We now multiply (6.20) by $\partial v/\partial t$ and then integrate over $\Omega(t)$, to get

$$\frac{\gamma}{\gamma+1}\int_{\Omega(t)}v^{-\frac{1}{1+\gamma}}\left|\frac{\partial v}{\partial t}\right|^2 dx + \frac{(1+dR/dt)}{2(1+\gamma)}\left|\frac{\partial v}{\partial x}\right|^2\bigg|_{x=R(t)} + \frac{1}{2(1+\gamma)}\frac{d}{dt}\left\|\frac{\partial v}{\partial x}(t)\right\|^2_{2,\Omega(t)}$$

$$= \varphi\frac{\partial v}{\partial t}\bigg|_{x=0} + \frac{2\gamma}{\gamma+1}\left\|\frac{\partial v}{\partial x}(t)\frac{\partial v}{\partial t}(t)|v(t)|^{\frac{\gamma-1}{\gamma+1}}\right\|_{1,\Omega(t)} + v^{\frac{2\gamma}{\gamma+1}}\frac{\partial v}{\partial t}\bigg|_{x=0}$$

$$+\frac{1}{2(\gamma+1)}\left|\frac{\partial v}{\partial x}\right|^2\bigg|_{x=R(t)} \equiv \mathcal{J}_1 + \mathcal{J}_2 + \mathcal{J}_3 + \mathcal{J}_4. \tag{6.27}$$

To estimate $\mathcal{J}_2$, we apply Hölder's inequality:

$$|\mathcal{J}_2| \le \varepsilon_1\left\||v(t)|^{-1/(1+\gamma)}\left|\frac{\partial v}{\partial t}(t)\right|^2\right\|_{1,\Omega(t)} + c_1(\varepsilon_1)\left\||v(t)|^{(2\gamma-1)/(1+\gamma)}\left|\frac{\partial v}{\partial x}(t)\right|^2\right\|_{1,\Omega(t)}$$

$$\equiv \varepsilon_1\mathcal{J}_{01} + c_1(\varepsilon_1)\mathcal{J}_{02}.$$

Making use of the identity

$$\int_{\Omega(t)}\frac{\partial}{\partial x}\left[v^{\frac{3\gamma}{1+\gamma}}\frac{\partial v}{\partial x}\right]dx = \frac{3\gamma}{\gamma+1}\int_{\Omega(t)}v^{\frac{2\gamma-1}{1+\gamma}}\left|\frac{\partial v}{\partial x}\right|^2 dx + \int_{\Omega(t)}v^{\frac{3\gamma}{1+\gamma}}\frac{\partial^2 v}{\partial x^2}dx,$$

we obtain the bound

$$|\mathcal{J}_{02}| \le \varepsilon_2\int_{\Omega(t)}v^{\frac{1}{1+\gamma}}\left|\frac{\partial^2 v}{\partial x^2}\right|^2 dx + c_2(\varepsilon_2)\int_{\Omega(t)}v^{\frac{6\gamma-1}{\gamma+1}}dx + N\left\{|v|^{\frac{5\gamma}{\gamma+1}}\bigg|_{x=0} + \left|\frac{\partial v}{\partial x}\right|\bigg|_{x=0} + 1\right\}.$$

The term $\|v(t)\|_{(6\gamma-1)/(\gamma+1),\Omega(t)}$ in the above inequality can be estimated with the use of Lemma 3, Chapter I :

$$\left\|v-1\right\|^{\frac{6\gamma-1}{\gamma+1}}_{\frac{6\gamma-1}{\gamma+1},\Omega} \le \beta\left\|\frac{\partial v}{\partial x}\right\|^{\frac{4\gamma-1}{8\gamma+4}}_{2,\Omega}\left\|v-1\right\|^{\lambda}_{\frac{2\gamma}{\gamma+1},\Omega}, \tag{6.28}$$

where $\lambda = \dfrac{6\gamma-1}{4\gamma}\left\{1 - \dfrac{(\gamma+1)(4\gamma-1)}{(2\gamma+1)(6\gamma-1)}\right\}$. Further, we have

$$v^{\frac{5\gamma}{\gamma+1}}\bigg|_{x=0} \le 1 + \frac{5\gamma}{\gamma+1}\left\|v(t)\right\|^{(1)}_{2,\Omega(t)}\left\|v(t)\right\|^{\frac{4\gamma-1}{\gamma+1}}_{\frac{8\gamma-2}{\gamma+1},\Omega(t)}.$$

We estimate the right-hand side of the latter inequality analogously to (6.28):

$$\left\|v-1\right\|_{\frac{8\gamma}{\gamma+1},\Omega} \le \beta\left\|\frac{\partial v}{\partial x}\right\|^{\lambda_1}_{2,\Omega}\left\|v\right\|^{\lambda_2}_{\frac{2\gamma}{\gamma+1},\Omega},$$

where $\lambda_1 = \dfrac{5\gamma}{8\gamma + 4} - \dfrac{1}{4}$, $\lambda_2 = \dfrac{4\gamma - 1}{4\gamma}\left\{1 - \dfrac{(\gamma - 1)(3\gamma - 1)}{(2\gamma + 1)(4\gamma - 1)}\right\}$. Hence,

$$|\mathcal{J}_2| \le N\left\{\varepsilon \int_{\Omega(t)} v^{-\frac{1}{1+\gamma}}\left|\frac{\partial v}{\partial t}\right|^2 dx + \varepsilon_2 c(\varepsilon_1) \int_{\Omega(t)} v^{\frac{1}{1+\gamma}}\left|\frac{\partial^2 v}{\partial x^2}\right| dx\right.$$

$$\left. + c(\varepsilon_1, \varepsilon_2)\left[\left\|\frac{\partial v}{\partial x}(t)\right\|^4_{2,\Omega(t)} + \left|\frac{\partial v}{\partial x}\right|^2_{x=R(t)}\right] + 1\right\}.$$

From the representations

$$v^{\frac{2\gamma}{1+\gamma}}\frac{\partial v}{\partial t}\bigg|_{x=0} = \frac{d}{dt}\left\{\frac{\gamma + 1}{3\gamma + 1}v^{\frac{3\gamma+1}{\gamma+1}}\right\}\bigg|_{x=0} = \frac{d}{dt}\left\{\frac{\gamma + 1}{3\gamma + 1} - \int_{\Omega(t)} v^{\frac{2\gamma}{\gamma+1}}\frac{\partial v}{\partial x} dx\right\},$$

$$\varphi\frac{\partial v}{\partial t}\bigg|_{x=0} = \varphi\frac{d}{dt}\left\{1 - \int_{\Omega(t)} \frac{\partial v}{\partial x} dx\right\},$$

$$\left|\frac{\partial v}{\partial x}\right|^2_{x=R(t)} = (1 + \gamma)^2\left[\varphi + v^{\frac{2\gamma}{1+\gamma}}\right]^2\bigg|_{x=0} + 2\int_{\Omega(t)} \frac{\partial v}{\partial x}\frac{\partial^2 v}{\partial x^2} dx,$$

$$v^{\frac{4\gamma}{1+\gamma}}\bigg|_{x=0} = 1 - \frac{4\gamma}{1 + \gamma}\int_{\Omega(t)} v^{\frac{2\gamma-1}{\gamma+1}}\frac{\partial v}{\partial x} dx$$

and the estimate

$$\left|\int_{\Omega(t)} v^{\frac{2\gamma}{1+\gamma}}\frac{\partial v}{\partial x} dx\right| \le \beta_1\left\|\frac{\partial v}{\partial x}(t)\right\|^{\frac{12\gamma+5}{12\gamma+8}}_{2,\Omega(t)}\left\{1 + \max_{t\in(0,T)}\left\|v(t)\right\|^{\frac{6\gamma+2}{3\gamma+1}}_{\frac{\gamma}{\gamma+1},\Omega(t)}\right\},$$

we conclude

$$\left|\int_0^t (\mathcal{J}_1 + \mathcal{J}_2 + \mathcal{J}_3)\, d\tau\right|$$

$$\le N\left\{\varepsilon_3\iint_{\Omega(t)} v^{\frac{1}{1+\gamma}}\left|\frac{\partial^2 v}{\partial x^2}\right|^2 dx d\tau + c_3(\varepsilon_3)\left(\int_0^t\left\|\frac{\partial v}{\partial x}(\tau)\right\|^4_{2,\Omega(\tau)} d\tau\right) + 1\right\}.$$

Returning to equation (6.20), we see that

$$\frac{1}{N}\int_{\Omega(t)} v^{\frac{1}{1+\gamma}}\left|\frac{\partial^2 v}{\partial x^2}\right|^2 dx \le \int_{\Omega(t)} v^{-\frac{1}{1+\gamma}}\left|\frac{\partial v}{\partial t}\right|^2 dx + \int_{\Omega(t)} v^{\frac{2\gamma-1}{\gamma+1}}\left|\frac{\partial v}{\partial x}\right|^2 dx \equiv \mathcal{J}_{01} + \mathcal{J}_{02}.$$

In view of the positiveness of $\frac{dR}{dt} + 1$ (the pressure achieves its maximum on the free boundary), taking the sum of all involved estimates and subsequently choosing small parameters $\varepsilon_1, \varepsilon_2, \varepsilon_3$, due to (6.27), (6.26) and Gronwall's lemma, we eventually arrive at the estimate

$$\max_{t\in(0,T)}\left\|\frac{\partial v}{\partial x}(t)\right\|_{2,\Omega(t)} \le N,$$

which, together with the representation

$$v(x,t) = 1 - \int_x^{R(t)} \frac{\partial v}{\partial x}(s,t)\, ds,$$

yields the assertion. $\square$

6.7. Asymptotic behaviour of the solution, as $t \to \infty$

In the simplest case where $\varphi(t) = 0$, we shall prove an exponential rate of convergence of the solution of problem (6.1) - (6.5) to a stationary solution. The general situation can be treated by means of the comparison lemma.

Lemma 14. *Let the hypotheses of Theorem 15 be satisfied and let $\varphi(t) \equiv 0$, $t \geq t_0 \geq 0$. Then*

$$|R(t) - R_\infty| \leq N \exp(-\beta t) \qquad \text{for} \quad t \geq t_0,$$

where the constants N and β depend only on M_0, and the constant R_∞ is defined by (6.10) with

$$M_\infty = Y(t_0) = \int_{\Omega(t_0)} f[p(x,t_0)]\, dx.$$

Proof. In Lagrange variables (y,t), the function $u(y,t)$ satisfies the homogeneous boundary conditions (6.13) and (6.14) for $t \geq t_0$. Hence, for $t \geq t_0$,

$$\min_{y \in G(t_0)} u(y,t_0) \leq u(y,t) \leq \max_{y \in G(t_0)} u(y,t_0) \tag{6.29}$$

and, according to Lemma 9,

$$\max_{t \in (t_0,\infty)} |u(t)|_{G(t)}^{(1)} \leq M_6(M_0). \tag{6.30}$$

Multiplying (6.12) by $u(y,t)$ and then integrating over the domain $G(t)$, we obtain the equality

$$\frac{d}{dt}\left\{ \int_0^{M_\infty} F(u,y)\, dy \right\} + \int_0^{M_\infty} f(P)\left|\frac{\partial u}{\partial y}\right|^2 dy = 0 \qquad \text{for} \quad t \geq t_0, \tag{6.31}$$

where

$$F(u,y) = \int_0^u s f'(s+y+1) f^{-2}(s+y+1)\, ds.$$

By estimate (6.29),

$$N^{-1} \leq \{f(P), f'(P), f^{-2}(P)\} \leq N, \tag{6.32}$$

$$N^{-1}u^2 \leq F(u,y) \leq Nu^2. \tag{6.33}$$

It follows from (6.33), (6.31) and the inequality $\|u\|_{2,G(t)} \leq N\|u\|_{2,G(t)}^{(1)}$ that

$$\frac{d}{dt}\left\{\int_0^{M_\infty} F(u,y)\,dy\right\} + \frac{1}{N}\int_0^{M_\infty} F(u,y)\,dy \leq 0.$$

Hence

$$\|u(t)\|_{2,G(t)}^2 \leq N\int_0^{M_\infty} F(u,y)\,dy \leq N\exp(-\beta t) \qquad \text{for} \quad t \geq t_0.$$

By (6.30) and the simplest interpolation inequality, we have

$$|u(t)|_{G(t)}^{(0)} \leq N\exp(-\beta t) \qquad \text{for} \quad t \geq t_0. \tag{6.34}$$

The position of the free boundary $x = R(t)$ is characterized by the formula

$$R(t) = \int_0^{M_\infty} \frac{1}{f\big(u(s,t)+s+1\big)}\,ds, \tag{6.35}$$

analogous to (6.10). The assertion follows from (6.35) and (6.34).   □

Lemma 15. *Let the hypotheses of Theorem 15 hold, and let*

$$|u|_{G_\infty}^{(0)} \leq N(M_0), \qquad \int_0^\infty \varphi(t)\,dt < \infty \quad \text{and} \quad \lim_{t\to\infty} \varphi(t) = 0.$$

Then

$$\lim_{t\to\infty} \big|R(t) - R_\infty\big| = 0,$$

where R_∞ is defined by (6.10) with the constant

$$M_\infty = \int_0^\infty \varphi(t)\,dt + \int_0^{R_0} f\big(p_0(x)\big)\,dx.$$

Proof. As in Lemma 14, we can construct a differential inequality for the function $F(u,y)$ (in contrast to the former case, this inequality will be non-homogeneous):

$$\frac{d}{dt}\left\{\int_{G(t)} F(u,y)\,dy\right\} + \beta\int_{G(t)} F(u,y)\,dy \leq N\varphi(t),$$

the integration of which yields

$$\int_{G(t)} F(u,y)\,dy \;\leq\; \tilde{F}(t_0)\exp\big(-\beta(t-t_0)\big) + N\int_{t_0}^t \varphi(\tau)\exp\big(-\beta(t-\tau)\big)\,d\tau$$

$$= \left(\tilde{F}(t_0) - \varphi(t_*)\frac{N}{\beta}\right)\exp\big(-\beta(t-t_0)\big) + \varphi(t_*)\frac{N}{\beta} \;\leq\; N\{\exp\big(-\beta(t-t_0)\big) + \varphi(t_*)\}.$$

Here, $\tilde{F}(t_0) = \int_{G(t_0)} F(u,y)\,dy$, and t_* is a certain point of the interval (t_0,t), specified by the mean-value theorem.

Fixing an arbitrary $\varepsilon > 0$, we choose t_0 so that $N\varphi(t_*) < \varepsilon/2$. Then

$$\int_{G(t)} F(u, y)\, dy < \varepsilon \qquad \text{for} \quad t > t_0 + \frac{1}{\beta} \log \frac{2N}{\varepsilon}.$$

The remaining arguments exactly follow the lines of Lemma 14.   □

Corollary. *The solution of problem (6.1)–(6.5) with the state function $f(p) = \exp(p-1)$ satisfies the hypotheses of Lemma 14.*

Let us now consider the case $\displaystyle\int_0^\infty \varphi(t)\, dt = \infty$. Let $\varphi_n(t)$ be regular non-negative functions such that $\varphi_n(t) = \varphi(t)$ for $t \in (0, n)$, $\varphi_n(t) = 0$ for $t \in (n+1, \infty)$ and $\varphi_n(t) \le \varphi(t)$.

By the comparison lemma,

$$R^n(t) \le R(t) \qquad \text{for} \quad t \in (0, \infty), \tag{6.36}$$

where $R^n(t)$ is the solution of problem (6.1)–(6.5) which corresponds to the data $\{p_0(x), \varphi_n(t), R_0\}$. Passing to the limit in (6.36) as $t \to \infty$, gives us the inequality

$$R^n_\infty \le \varliminf_{t \to \infty} R(t), \tag{6.37}$$

where

$$R^n_\infty = \int_0^{M^n_\infty} \frac{1}{f(s+1)}\, ds,$$

$$M^n_\infty = \int_0^{R(0)} f\big(p_0(x)\big)\, dx + \int_0^{n+1} \varphi_n(t)\, dt \ge \int_0^n \varphi(t)\, dt.$$

Providing that (6.6) holds, $\displaystyle\lim_{n \to \infty} R^n_\infty = \infty$. If $f(p)$ does not satisfy (6.6), then

$$\lim_{n \to \infty} R^n_\infty = R_*. \tag{6.38}$$

On the other hand, due to the constraints imposed on the initial data $p_0(x)$ and R_0, and by the comparison lemma, $R(t) \le R_*$. Hence,

$$\varlimsup_{t \to \infty} R(t) \le R_*. \tag{6.39}$$

Passing to the limit in (6.35) as $t \to \infty$, by making use of (6.38) and (6.39) we eventually deduce the assertion of Theorem 16.   □

Structure of the generalized solution to the one-phase Stefan problem. Existence of a mushy region

1. The inhomogeneous heat equation. Formation of the mushy region

Let us assume that only pure liquid and solid phases exist initially in the domain Ω (for simplicity we assume that only a single free boundary exists there). Assume that the temperature is prescribed non-negative on one of the components of the boundary of Ω and non-positive on the other, as well as either the heat flux condition is homogeneous or the condition

$$\frac{d\vartheta_0}{dx}(x_0 \pm 0) \neq 0 \tag{1.1}$$

holds, where $\vartheta_0(x)$ denotes the initial temperature and x_0 initially separates the different phases. Then, as already discussed, such a situation is also preserved at later times (at least for a sufficiently small interval): pure solid and liquid phases exist in the domain Ω, separated by the boundary $x = R(t)$ (see Theorem 6, Chapter II).

Still, the question arises, what will happen, if the heat flux becomes inhomogeneous and condition (1.1) is violated? Then it turns out that there exists an initial temperature distribution $\vartheta_0(x)$ with a single separation point x_0 which contributes to the formation of a mushy phase at $t > 0$, although it did not exist initially.

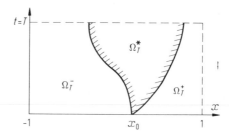

Figure 5

We shall look for a solution of the Stefan problem with the following structure: there exist two sufficiently regular functions $R^-(t)$ and $R^+(t)$, $R^-(t) \leq R^+(t)$, initially coinciding, which define the domains $\Omega^-(t) = \{x \ : \ -1 < x < R^-(t)\}$ occupied by the solid phase, $\Omega^+(t) = \{x \ : \ R^+(t) < x < 1\}$ occupied by the liquid, and

$\Omega^*(t) = \{x \ : \ R^-(t) < x < R^+(t)\}$ by the mushy phase (see Figure 5). As already proved, the temperature $\vartheta(x,t)$ satisfies the inhomogeneous nonlinear heat equation

$$a(\vartheta)\frac{\partial \vartheta}{\partial t} = \frac{\partial^2 \vartheta}{\partial x^2} + f(x,t) \qquad \text{in} \quad \Omega_T^{\pm} = \{(x,t) \ : \ x \in \Omega^{\pm}(t), \ t \in (0,T)\}. \quad (1.2)$$

In the domain $\Omega_T^* = \{(x,t) \ : \ x \in \Omega^*(t), \ t \in (0,T)\}$, the temperature $\vartheta(x,t)$ is identically zero and the specific internal energy U is unknown. U assumes values in the interval $(-1,0)$ and can be characterized as a solution of the equation

$$\frac{\partial U}{\partial t} = f(x,t) \qquad \text{for} \quad (x,t) \in \Omega_T^*. \quad (1.3)$$

On the boundaries $x = R^{\pm}(t)$ between different phases, the temperature is zero:

$$\vartheta\big(R^{\pm}(t),t\big) = 0 \qquad \text{for} \quad t \in (0,T), \quad (1.4)$$

and the energy balance equations (equations on the strong discontinuity) are of the form

$$\Big(U\big(R^-(t)+0,t\big)+1\Big)\frac{dR^-(t)}{dt} = \frac{\partial \vartheta}{\partial x}\big(R^-(t)-0,t\big) \qquad \text{for} \quad t \in (0,T), \quad (1.5)$$

$$U\big(R^+(t)-0,t\big)\frac{dR^+(t)}{dt} = \frac{\partial \vartheta}{\partial x}\big(R^+(t)+0,t\big) \qquad \text{for} \quad t \in (0,T). \quad (1.6)$$

To complete the formulation of the problem, we prescribe the initial position of the free boundary and the initial temperature distribution:

$$R^-(0) = R^+(0) = x_0, \quad \vartheta(x,0) = \vartheta_0(x) \qquad \text{for} \quad x \in \Omega, \quad (1.7)$$

as well as the conditions on the fixed boundaries:

$$\vartheta(\pm 1, t) = \vartheta^{\pm}(t) \qquad \text{for} \quad t \in (0,T). \quad (1.8)$$

In Section I.1 we proved that if $f(x_0,0) > 0$, then the condition (1.5) is equivalent to the following two conditions

$$U(R^-(t)+0,t) = -1 \qquad \text{for} \quad t \in (0,T), \quad (1.9)$$

$$\frac{\partial \vartheta}{\partial x}(R^-(t)-0,t) = 0 \qquad \text{for} \quad t \in (0,T). \quad (1.10)$$

The solution of the discussed Stefan problem may be split into three stages: (i) determine the boundary $x = R^-(t)$ of the domain Ω_T^- and the temperature $\vartheta(x,t)$ in the same domain from the conditions (1.2), (1.4), (1.7), (1.8), (1.10) (Problem (A)), (ii) determine the specific internal energy U on the boundary $x = R^+(t)$ of Ω_T^* from conditions (1.3) and (1.9), and, finally, (iii) determine the boundary $x = R^+(t)$ of Ω_T^+ and the temperature $\vartheta(x,t)$ in this domain on the basis of (1.2), (1.4), (1.6) - (1.8) (Problem (B)).
In particular, if $\frac{\partial \vartheta}{\partial x}$ is continuous on the closure of Ω_T^-, then by (1.10)

$$\frac{d\vartheta_0}{dx}(x_0 - 0) = 0.$$

Problem (A) can be solved independently and can be reduced to solving a Stefan-like problem. To this end, let us define

$$u(x,t) = \frac{\partial \vartheta}{\partial t}(x,t) \qquad \text{for} \quad (x,t) \in \Omega_T^-.$$

The function $u(x,t)$ satisfies the equation

$$a(\vartheta)\frac{\partial u}{\partial t} = \frac{\partial^2 u}{\partial x^2} - a'(\vartheta)u^2 + \frac{\partial f}{\partial t} \qquad \text{for} \quad (x,t) \in \Omega_T^-. \tag{1.11}$$

By differentiating the identity $\vartheta\big(R^-(t) - 0, t\big) = 0$ in t and making use of relation (1.10), we obtain the condition

$$u\big(R^-(t) - 0, t\big) = 0 \qquad \text{for} \quad t \in (0, T) \tag{1.12}$$

on the free boundary $x = R^-(t)$.

Since the function $x = R^-(t)$ is also to be found, there is an additional condition imposed on the unknown variables on the appropriate boundary $x = R^-(t)$. To derive such a condition, let us differentiate condition (1.10) with respect to time and, then, substitute for the second derivative with respect to x from equation (1.2), using condition (1.12):

$$\frac{\partial u}{\partial x}\big(R^-(t) - 0, t\big) = f\big(R^-(t), t\big)\frac{dR^-(t)}{dt} \qquad \text{for} \quad t \in (0, T). \tag{1.13}$$

Finally, on the known boundary $x = -1$,

$$u(-1, t) = u^-(t) \equiv \frac{d\vartheta^-(t)}{dt} \qquad \text{for} \quad t \in (0, T), \tag{1.14}$$

and, at the initial time instant,

$$u\,|_{t=0} = u_0(x) \equiv \frac{1}{a(\vartheta_0(x))}\left\{\frac{d^2\vartheta_0(x)}{dx^2} + f(x, 0)\right\} \qquad \text{for} \quad x \in \Omega^-(0). \tag{1.15}$$

Lemma 1. *Let* $a \in C^2(-\infty, 0]$, $a(s) \ge a_0 > 0$, $\vartheta^- \in C^2[0, \infty)$, $\vartheta_0 \in C^4[-1, x_0]$, $f \in C^{2,1}(\overline{\Omega}_T)$, *and suppose that the compatibility conditions*

$$\vartheta_0(-1) = \vartheta^-(0), \quad u^-(0) = u_0(-1),$$

$$\vartheta_0(x_0) = 0, \quad \frac{d\vartheta_0}{dx}(x_0 - 0) = 0, \quad u_0(x_0 - 0) = 0$$

hold at the points $x = -1$ *and* $x = x_0$. *Moreover, let*

$$\vartheta^-(t) < 0 \quad \text{for } t \in (0, \infty), \qquad \vartheta_0(x) < 0 \quad \text{for } x \in \Omega^-(0).$$

$$f(x_0, 0) > 0, \qquad \frac{du_0}{dx}(x_0 - 0) < 0.$$

Then, on a sufficiently small interval $(0, T_*)$, $0 < T_* \leq T$ *there exists a solution* $\{R^-(t), \vartheta(x, t)\}$ *of Problem (A) such that*

$$R^- \in H^{1+\gamma/2}[0, T_*], \qquad \vartheta \in C^{2,1}(\overline{\Omega_{T_*}^-}).$$

Proof. We look for functions $R^-(t)$ and $u(x, t)$ which solve the initial-boundary value problem (1.11) - (1.15), with the function

$$\vartheta(x, t) = \vartheta_0(x) + \int_0^t u(x, \tau) \, d\tau \qquad \text{for} \quad (x, t) \in \Omega_T^-, \tag{1.16}$$

entering the coefficients of equation (1.11) as an argument.

This representation defines the function $\vartheta(x, t)$ everywhere in the domain Ω_T^-, provided that $x = R^-(t)$ is a non-increasing function of t, i.e.,

$$\frac{dR^-(t)}{dt} \leq 0 \qquad \text{for} \quad t \in (0, T).$$

By (1.13) and the hypotheses of the lemma, the above derivative is strictly negative at the initial time instant. Due to the postulated regularity, at least over sufficiently small time intervals, this derivative preserves its sign.

Let us assume that the functions $R^-(t)$ and $u(x, t)$ have already been determined. Then the functions $\vartheta(x, t)$ defined by (1.16), and $R^-(t)$ represent the solution of Problem (A) we are looking for. Indeed, $u = \frac{\partial \vartheta}{\partial t}$ by construction, and (1.11) can be integrated once in time, giving rise to equation (1.2), because the integration constant vanishes (more precisely, is a function of x) at the initial time.

In view of the equalities

$$\frac{\partial \vartheta}{\partial t}(R^-(t) - 0, t) = 0, \qquad \frac{\partial^2 \vartheta}{\partial x^2}(R^-(t) - 0, t) + f(R^-(t), t) = 0,$$

the condition (1.13) can also be integrated in time:

$$\frac{\partial \vartheta}{\partial x}(R^-(t) - 0, t) = c = \text{const.}$$

Since

$$\frac{\partial \vartheta}{\partial x}(x_0 - 0, 0) = \frac{d\vartheta_0(x_0 - 0)}{dx} = 0$$

at the initial time, $c = 0$ in the above relation.

By integrating the identity

$$\frac{\partial \vartheta}{\partial x}(R^-(t) - 0, t)\frac{dR^-(t)}{dt} + \frac{\partial \vartheta}{\partial t}(R^-(t), t) = 0$$

subject to the compatibility condition $\vartheta_0(x_0) = 0$, we can conclude that $\vartheta(x, t) = 0$ on the boundary $x = R(t)$.

It remains to show that $\vartheta(x, t)$ is strictly negative in Ω_T^-. Outside a small neighbourhood of the boundary $x = R^-(t)$, this property follows from the strict negativeness of the initial function $\vartheta_0(x)$. In the neighbourhood of $\{x = R^-(t)\}$, at least at small times the derivative $u = \frac{\partial \vartheta}{\partial t}$ is strictly positive. In fact, by the hypothesis of the lemma, $\frac{du_0}{dx}(x_0 - 0) < 0$. Hence, due to the postulated regularity of the solution,

$$\frac{\partial u}{\partial x}(R^-(t) - 0, t) < 0$$

on a sufficiently small time interval. Together with (1.12) this ensures that $u(x,t)$ is strictly positive in a neighbourhood of the boundary $x = R^-(t)$. This in turn yields that $\vartheta(x,t)$ is strictly negative in a neighbourhood of $x = R^-(t)$.

To prove the existence of a solution of problem (1.11) - (1.15), we shall use a standard scheme. Let the norms of functions u_0, u^-, f, mentioned in the hypotheses of the lemma, as well as the expression

$$\left| \frac{du_0}{dx}(x_0 - 0) \right| \left| f(x_0, 0) \right|^{-1}$$

be bounded by a constant M_0. Let us fix an arbitrary constant $M > M_0$ and consider the convex set $\mathcal{M}$ which includes all pairs of functions $\{\tilde{v}(y,t), \tilde{R}(t)\}$ such that $\tilde{v}(y,t)$ is continuous in $G_T = G \times (0,T)$, $G = \{y : -1 < y < 0\}$, and is bounded in absolute value by M, and that $\tilde{R}(t)$ satisfies

$$\tilde{R} \in C^1[0,T], \qquad \text{with} \quad |\tilde{R}|^{(1)}_{[0,T]} \le M,$$

$$-1 + (1 + x_0)/2 < \tilde{R}(t) < 1 \qquad \text{for} \quad t \in (0,T),$$

$$\tilde{R}(0) = x_0, \qquad \frac{d\tilde{R}}{dt}(t) \le 0 \qquad \text{for} \quad t \in (0,T).$$

The function $\tilde{R}(t)$ defines the domain $Q_T = \{(x,t) : -1 < x < \tilde{R}(t), \ 0 < t < T\}$. In Q_T, we define $w(x,t)$ as the solution of the equation

$$a(\tilde{\vartheta}) \frac{\partial w}{\partial t} = \frac{\partial^2 w}{\partial x^2} + g(x,t), \tag{1.17}$$

which satisfies conditions (1.14), (1.15) and vanishes on the boundary $x = \tilde{R}(t)$. In equation (1.17),

$$g(x,t) = a'(\tilde{\vartheta})v^2 + \frac{\partial f}{\partial t},$$

$$\tilde{\vartheta}(x,t) = \vartheta_0(x) + \int_0^t v(x,\tau) \, d\tau,$$

$$v(x,t) = \tilde{v}\Big(\big(x - \tilde{R}(t) \big) / \big(1 + \tilde{R}(t) \big), t \Big).$$

The existence of the solution $w(x,t)$ of the above initial-boundary value problem does not follow directly from results of Section I.2, because the domain Q_T is non-cylindrical and conditions (1.14), (1.15) are inhomogeneous. However, these conditions can be easily reduced to a homogeneous form, and the transformation of variables

$$\tau = t, \quad y = \big(x - \tilde{R}(t) \big) / \big(1 + \tilde{R}(t) \big)$$

reduces the problem to an equivalent formulation over the cylinder G_T, where Theorem 5, Section I.2, holds. For the original variables, this implies that $w \in W_p^{2,1}(Q_T)$ for any $p > 2$ with

$$\|w\|_{p,Q_T}^{(2)} \le N_1(M_0, M).$$

For sufficiently large p, the space $W_p^{2,1}(Q_T)$ is embedded into $H^{\beta,\beta/2}(\bar{Q}_T)$ (see Lemma 5, Section I.2), with $\beta = 1 + \gamma$, $0 < \gamma < 1$:

$$|w|_{Q_T}^{(1+\gamma)} \le N_2(M_0, M). \tag{1.18}$$

We have to adjust the functions $\tilde{v}(y,t)$ and $\tilde{R}(t)$ so that

$$\tilde{R}(t) = x_0 + \int_0^t \frac{1}{f(\tilde{R}(\tau),\tau)} \frac{\partial w}{\partial x}\big(\tilde{R}(\tau) - 0, \tau\big)\, d\tau \equiv \Psi_1(\tilde{R}, \tilde{v}),$$

$$\tilde{v}(y,t) = w\big(\tilde{R}(t) + y\big(\tilde{R}(t) + 1\big), t\big) \equiv \Psi_2(\tilde{R}, \tilde{v}),$$

i.e., we have to determine a fixed point of the operator $\Psi = (\Psi_2, \Psi_1)$, defined on the set $\mathcal{M}$.

Estimate (1.18) implies that the operator Ψ is completely continuous. Hence, to fit hypotheses of Schauder's fixed point theorem, it is enough to show that Ψ maps $\mathcal{M}$ into itself. This can be achieved by taking a sufficiently small interval $(0, T_*)$, $0 < T_* \le T$.

By the hypothesis of the lemma,

$$\frac{\partial w}{\partial x}(x_0 - 0, 0) = \frac{du_0}{dx}(x_0 - 0) < 0.$$

Clearly, the inequality $\frac{\partial w}{\partial x}\big(\tilde{R}(t) - 0, t\big) < 0$ will remain true on a certain interval $(0, T_*)$ whose length depends only on M_0 and M :

$$\frac{\partial w}{\partial x}\big(\tilde{R}(t) - 0, t\big) = \frac{\partial w}{\partial x}(x_0 - 0, 0) + \frac{\partial w}{\partial x}\big(\tilde{R}(t) - 0, t\big) - \frac{\partial w}{\partial x}(x_0 - 0, 0)$$

$$\le \frac{du_0}{dx}(x_0 - 0) + N_2\Big(|\tilde{R}(t) - x_0|^\gamma + t^{\gamma/2}\Big) < 0.$$

Hence

$$\frac{d\Psi_1}{dt}(t) < 0 \qquad \text{for} \quad t \in (0, T_*).$$

Similarly, by taking T_* sufficiently small we can easily conclude that

$$|w(x,t)| \le |u_0(x)| + |w(x,t) - w(x,0)| \le M_0 + N_2 t^{(1+\gamma)/2} < M \quad \text{for } (x,t) \in Q_{T_*};$$

$$-1 + (1 + x_0)/2 < \Psi_1(t) < 1 \quad \text{for} \quad t \in (0, T_*);$$

$$\left|\frac{d\Psi_1}{dt}(t)\right| = \frac{1}{f(x_0,0)} \frac{du_0}{dx}(x_0 - 0)$$

$$+ \left|\frac{1}{f(\tilde{R}(t),t)} \frac{\partial w}{\partial x}\big(\tilde{R}(t) - 0, t\big) - \frac{1}{f(x_0,0)} \frac{\partial w}{\partial x}(x_0 - 0, 0)\right| < M.$$

Therefore, there exists at least one fixed point $\{\tilde{u}(y,t), R^-(t)\}$ of the operator Ψ which defines the solution $\{R^-(t), \vartheta(x,t)\}$ of Problem (A), where

$$\vartheta(x,t) = \vartheta_0(x) + \int_0^t u(x,\tau)\, d\tau,$$

$$u(x,t) = \tilde{u}\Big(\big(x - R^-(t)\big)/\big(1 + R^-(t)\big), t\Big).$$

By construction, $\frac{dR^-}{dt}$ is Hölder continuous and

$$\frac{dR^-}{dt}(0) = \frac{1}{f(x_0,0)} \frac{du_0}{dx}(x_0 - 0).$$

It is impossible to determine the domain Ω_T^* directly from the solution of Problem (A), nevertheless, by solving equation (1.3) subject to condition (1.9), we can find the internal energy function $U(x,t)$ everywhere on the curve $x = R^-(t)$:

$$U(x,t) = -1 + \int_{r^-(x)}^t f(x,\tau)\, d\tau. \tag{1.19}$$

Here, $r^-(x)$ is the function inverse to $R^-(t)$, i.e., $r^-\big(R^-(t)\big) \equiv t$.

With (1.19), Problem (B) is complete. It can be solved in exactly the same way as the problem of determining the functions $R^-(t)$ and $u = \frac{\partial \vartheta}{\partial t}$ in the domain Ω_T^-. The only requirement to be imposed on the problem data is then

$$\frac{dR^-}{dt}(0) < -\frac{d\vartheta_0}{dx}(x_0 + 0) = \frac{dR^+}{dt}(0) < 0, \tag{1.20}$$

which provides the non-degeneracy of Ω_T^- at least for small times, and the strict positiveness of $\vartheta(x,t)$ in a neighbourhood of the boundary $\{x = R^+(t)\}$ over a sufficiently small time interval. $\qquad\square$

Lemma 2. *Let* $a \in C^2[0,\infty)$, $a(s) \geq a_0 > 0$, $f \in C(\bar{\Omega}_T)$, $\vartheta^+ \in C^1[0,\infty)$, $\vartheta_0 \in C^2[x_0, 1]$, $\vartheta^+(t) > 0$ *for* $t \in (0,\infty)$, *and* $\vartheta_0(x) > 0$ *for* $x \in (x_0, 1)$. *Moreover let the compatibility conditions*

$$\vartheta_0(x_0) = 0, \quad \vartheta_0(1) = \vartheta^+(0)$$

be satisfied in the points $x = x_0$ *and* $x = 1$, *and let condition (1.20) hold.*

Then there exists a solution $\{R^+(t), \vartheta(x,t)\}$ *on a sufficiently small time interval* $(0, T_*)$, *such that* $\vartheta(x,t)$ *is strictly positive in* $\Omega_{T_*}^+$ *and*

$$R^+ \in H^{1+\gamma/2}[0, T_*], \qquad \vartheta \in C^{2,1}(\Omega_{T_*}^+) \cap H^{1+\gamma,(1+\gamma)/2}\big(\overline{\Omega_{T_*}^+}\big).$$

In choosing the interval $(0, T_*)$ it is necessary to remember that the specific internal energy $U(x,t)$ assumes values in the interval $(-1, 0)$ for $(x,t) \in \Omega_T^*$. This will be so, if T_* satisfies the inequality

$$-1 + \int_{r^-(x)}^t f(x,\tau)\, d\tau < -1 + M_0 T_* \leq -\delta < 0,$$

where $\delta > 0$ is arbitrarily fixed.

We are now ready to formulate the final result.

Theorem 1. *Let the hypotheses of Lemmas 1 and 2 be satisfied. Then the generalized solution $U(x,t)$ and $\vartheta(x,t) = \chi\big(U(x,t)\big)$ of the Stefan problem*

$$\frac{\partial U}{\partial t} = \frac{\partial^2 \vartheta}{\partial x^2} + f(x,t) \qquad for \quad (x,t) \in \Omega_T,$$

$$\vartheta(\pm 1, t) = \vartheta^{\pm}(t) \qquad for \quad t \in (0,T),$$

$$\vartheta(x,0) = \vartheta_0(x) \qquad for \quad x \in \Omega,$$

is such that, on a sufficiently small time interval $(0,T_*)$, the function $\vartheta(x,t)$ is strictly negative in the domain $\Omega_{T_*}^{-} = \big\{(x,t) : -1 < x < R^{-}(t), 0 < t < T_*\big\}$, vanishes in $\Omega_{T_*}^{*} = \big\{(x,t) : R^{-}(t) < x < R^{+}(t), 0 < t < T_*\big\}$ and is strictly positive in $\Omega_{T_*}^{+} = \big\{(x,t) : R^{+}(t) < x < 1, 0 < t < T_*\big\}$. The functions $R^{-}(t)$ and $R^{+}(t)$ are continuously differentiable, coincide at the initial time instant, and

$$R^{-}(t) < R^{+}(t) \qquad for \quad t \in (0,T_*).$$

The specific internal energy $U(x,t)$ assumes values in the interval $(-1,0)$ for $(x,t) \in \Omega_{T_*}^{*}$.

2. The homogeneous heat equation. Dynamic interactions between the mushy phase and the solid/liquid phases

In this section, we shall consider the case where the system, occupying the domain $\Omega = \{x : 0 < x < \infty\}$, is initially in the liquid state in the subdomain $\Omega^{+}(0) = \{x : 0 < x < x_0\}$ while its temperature in the rest of the domain is equal to the melting value. Thus, in $\Omega^{*}(0)$ the system may be either in solid, liquid or mushy phase at temperature equal zero, and its actual state is determined by the initial internal energy $U_0(x)$. We shall assume that the temperature on the boundary $\{x = 0\}$ of Ω is non-negative throughout.

Two different approaches may be applied to construct the generalized solution of this Stefan problem. The first of them, presented in Chapter I in the existence proof of the generalized solution to Stefan problem, exploits an approximation of the discontinuous function $\Phi(\vartheta)$ (or its inverse $\chi(U)$) by smooth functions $\Phi_n(\vartheta)$, solving the resulting initial-boundary problem for a uniformly parabolic equation and, subsequently, a passage to the limit.

Such an approach exhibits two essential shortcomings: it turns out to be difficult to characterize the structure of solution, in particular, to recover the free boundary, and it is necessary to solve an initial-boundary value problem over the whole domain $\Omega_T = \Omega \times (0,T)$ for each approximation. Let us recall the definition: a pair $\{U, \vartheta\}$ is the generalized solution of the Stefan problem if, for each smooth bounded function φ, vanishing at $x = 0$ and $t = T$,

$$\iint_{\Omega_T} \left\{ -U \frac{\partial \varphi}{\partial t} + \frac{\partial \vartheta}{\partial x} \frac{\partial \varphi}{\partial x} \right\} dx dt = \int_{\Omega} U_0(x)\, \varphi(x, 0)\, dx. \qquad (2.1)$$

The corresponding identity for the approximate solutions ϑ_n can be written in the form

$$\iint_{\Omega_T} \left\{ -\Phi_n(\vartheta_n) \frac{\partial \varphi}{\partial t} + \frac{\partial \varphi}{\partial x} \frac{\partial \vartheta_n}{\partial x} \right\} dx dt = \int_{\Omega} U_0(x)\, \varphi(x, 0)\, dx.$$

To obtain the differential equation for ϑ_n, it is necessary to apply the appropriate inverse transformations to the latter identity:

$$\iint_{\Omega_T} \left\{ \frac{\partial \Phi_n(\vartheta_n)}{\partial t} - \frac{\partial^2 \vartheta_n}{\partial x^2} \right\} \varphi\, dx dt = \int_{\Omega} \left\{ U_0(x) - \Phi_n(\vartheta_n(x, 0)) \right\} \varphi(x, 0)\, dx.$$

It is natural to assume that at the initial time $\Phi_n(\vartheta_n(x, 0)) = U_0(x)$ or $\vartheta_n(x, 0) = \vartheta_0^n(x) \equiv \chi_n(U_0(x))$.

Hence, due to the arbitrariness of φ, the approximate solutions ϑ_n satisfy the nonlinear heat equation

$$\frac{\partial \Phi_n(\vartheta_n)}{\partial t} = \frac{\partial^2 \vartheta_n}{\partial x^2} \qquad \text{for} \quad (x, t) \in \Omega_T.$$

For Φ_n, χ_n appropriately coinciding with Φ, χ for ϑ and U positive, the functions $\vartheta_0^n = \chi_n(U_0)$ are negative as long as U_0 assumes values in the interval $(-1, 0)$. Then, the sign of $\vartheta_n(x, t)$ is not fixed.

Similarly, if $\Phi_n = \Phi$ at negative temperatures, then ϑ_0^n is positive, and the corresponding approximate solution $v_n(x, t)$ is strictly positive in the interior of Ω_T. Hence, the approximate solutions do not propagate perturbations with finite speed in contrast to the original Stefan problem.

The second approach, more effective in our opinion, is based on reducing the original Stefan problem to a one-phase Stefan-like problem that consists of determining the domain Ω_T^+, of all those points of Ω_T where the temperature is strictly positive. Such an approach preserves the finite propagation speed of perturbations: for each time instant τ, there exists $x = x_\tau$ such that the temperature vanishes everywhere in the domain $\{(x, t) : x_\tau < x < \infty,\ 0 < t < \tau\}$.

Assume the liquid phase occupies the domain $\Omega^+(t) = \{x : 0 < x < R(t)\}$ for all time, and the mushy phase occupies the domain $\Omega^*(t) = \{x : R(t) < x < \infty\}$, where $R(t)$ is a non-decreasing, continuously differentiable function. Let $U_0(x)$ be the initial distribution of the internal energy for $x \in (x_0, \infty)$. If $U_0(x) < 0$ everywhere (i.e., no liquid subdomains enter the mushy region), then

$$\vartheta(R(t), t) = 0, \qquad \frac{\partial \vartheta}{\partial x}(R(t) - 0, t) = U_0(R(t)) \frac{dR(t)}{dt} \qquad (2.2)$$

on the phase separation line. Under the above hypotheses, we have already obtained condition (2.2) in Section I.1. In the domain $\Omega_T^+ = \{(x, t) : x \in \Omega^+(t),\ t \in (0, T)\}$, the temperature $\vartheta(x, t)$ satisfies the heat equation

$$a(\vartheta) \frac{\partial \vartheta}{\partial t} = \frac{\partial^2 \vartheta}{\partial x^2}, \qquad a(\vartheta) = \Phi'(\vartheta). \tag{2.3}$$

There is no need to approximate the function $\Phi(s)$ in the above equation, it is enough to extend this function by continuity, defining

$$a(0) = \lim_{s \to +0} a(s).$$

To complete the problem formulation, the temperature on the boundary $\{x = 0\}$ and at the initial time instant is to be prescribed, together with the initial position of the free boundary:

$$\vartheta(0, t) = \vartheta^+(t) \qquad \text{for} \quad t \in (0, T),$$
$$R(0) = x_0, \quad \vartheta(x, 0) = \vartheta_0(x) \qquad \text{for} \quad x \in \Omega^+(0). \tag{2.4}$$

For the initial internal energy $U_0(x)$ continuously differentiable and strictly negative, the problem (2.2) - (2.4) can be easily solved as in Section V.1 in the case of the one-phase Stefan problem.

We now formulate the summarizing result.

Lemma 3. *Let* $a \in C^1[0, \infty)$, $U_0 \in C^1[x_0, \infty)$, $a(s) \geq a_0 = \text{const.} > 0$, $U_0(x) \leq -\delta = \text{const.} < 0$; *let* $\vartheta^+(t)$, $t \in (0, \infty)$, *and* $\vartheta_0(x)$, $x \in (0, x_0)$ *be non-negative bounded functions.*

Then there exists a unique classical solution $\{R(t), \vartheta(x, t)\}$ *of problem (2.2) - (2.4) on an infinite time interval, such that*

$$R \in C[0, \infty) \cap C^1(0, \infty), \qquad \frac{dR}{dt}(t) > 0 \quad \text{for} \quad t \in (0, \infty);$$

$$\vartheta \in C^{2,1}(\Omega_\infty^+) \cap (\overline{\Omega_\infty^+ \cap \Omega_\lambda^\lambda}), \qquad \Omega_\lambda^\lambda = \{(x, t) : x > \lambda, t > \lambda\}, \ \lambda > 0;$$

$$0 < \vartheta(x, t) \leq M_0 = \max\{|\vartheta_0|_{[0, x_0]}^{(0)}, |\vartheta^+|_{[0, \infty)}^{(0)}\}.$$

Lemma 3 indicates how to construct the generalized solution of the Stefan problem with an arbitrary measurable initial function $U_0(x)$ which assumes values in the interval $[-1, 0]$ for all $x \in \Omega^*(0)$. To this end, we take a sequence of strictly negative functions $U_0^n(x)$, pointwise convergent to $U_0(x)$, and then select a convergent subsequence from the sequence of solutions $\{R^n(t), \vartheta_n(x, t)\}$ to problem (2.2) - (2.4), corresponding to the initial data $\{\vartheta^+(t), \vartheta_0(x), U_0^n(x)\}$.

Theorem 2. *Let* $a \in C^1[0, \infty)$, $a(s) \geq a_0 > 0$, *and let* $\vartheta^+(t)$, $t \in (0, \infty)$ *and* $\vartheta_0(x)$, $x \in (0, x_0)$, *be non-negative measurable functions such that*

$$\text{ess max } \vartheta_0(x) > 0 \qquad \text{for} \quad x \in (0, x_0).$$

Let $U_0(x)$ be a measurable function defined for $x \in (x_0, \infty)$, with values in the interval $[-1, 0]$. Then the generalized solution $\vartheta(x, t)$ of the Stefan problem

$$\begin{cases} \dfrac{\partial U}{\partial t} = \dfrac{\partial^2 \vartheta}{\partial x^2} & \text{for} \quad (x, t) \in \Omega_\infty, \\ \vartheta(0, t) = \vartheta^+(t) & \text{for} \quad t \in (0, \infty), \\ U(x, 0) = U_0(x) & \text{for} \quad x \in \Omega, \end{cases} \tag{2.5}$$

where $U_0(x) = \Phi(\vartheta_0(x))$ for $x \in (0, x_0)$, is strictly positive in the domain $\Omega_\infty^+ = \{(x, t) : 0 < x < X_*, h(x) < t < \infty\}$ and vanishes identically in its complement $\Omega_\infty^* = \{(x, t) : 0 < x < X_*, 0 < t < h(x)\} \cup \{(x, t) : x > X_*, t > 0\}$ (see Figure 6). The continuous function h(x), defined on the interval $[0, X_*)$, vanishes everywhere for $x \in (0, x_0)$, is monotone increasing on the interval (x_0, X_*) and

$$\lim_{x \to X_*} h(x) = \infty, \quad \text{if} \quad X_* < \infty.$$

If $\displaystyle\int_0^\infty \vartheta^+(t)\, dt = \infty$, then $X_* = \infty$.

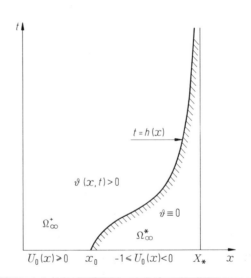

Figure 6

Proof. Let us consider a sequence of infinitely differentiable functions $U_0^n(x)$, strictly negative and pointwise convergent to $U_0(x)$ in the domain $\{x : x_0 < x < \infty\}$, such that

$$U_0^1(x) = -1, \quad U_0^n(x) \leq U_0^{n+1}(x) \leq \ldots \leq U_0(x),$$

with the equality $U_0^n(x) = U_0(x)$ holding only at the points x where $U_0(x) = -1$.

For each set of data $\{\vartheta^+(t), \vartheta_0(x), U_0^n(x)\}$, according to Lemma 3 there exists a solution $\{R_n(t), \vartheta_n(x,t)\}$ of problem (2.2)–(2.4). We shall extend ϑ_n by zero onto the domain $\{(x,t) : R_n(t) < x < \infty, 0 < t < \infty\}$ and take $U_n(x,t) = U_0^n(x)$ in this domain. Whenever $\vartheta_n > 0$, we set $U_n = \Phi(\vartheta_n)$. The function $U_n(x,t)$, with a first-kind discontinuity on the curve $\{x = R_n(t)\}$, is the generalized solution of Stefan problem (2.5) that corresponds to the initial distribution of the specific internal energy $U_0^n(x) = \Phi(\vartheta_0(x))$ for $x \in (0, x_0)$.

The comparison theorem (cf. Theorem 11, Section I.3) and the special choice of initial data provide monotonicity of the sequences $\{R_n(t)\}$ and $\{\vartheta_n(x,t)\}$:

$$R_n(t) \le R_{n+1}(t), \quad \vartheta_n(x,t) \le \vartheta_{n+1}(x,t). \tag{2.6}$$

Due to the uniform boundedness and continuity of functions $\vartheta_n(x,t)$ in the interior of Ω_∞, there exists the limit function

$$\vartheta(x,t) = \lim_{n\to\infty} \vartheta_n(x,t),$$

bounded in the closed domain $\bar{\Omega}_\infty$ and continuous in its interior Ω_∞. Moreover, it follows easily that the function $\vartheta(x,t)$ belongs to the Hölder space $H^{\alpha,\alpha/2}(\bar{Q})$ for some $\alpha > 0$ on each bounded subdomain Q such that $\bar{Q} \subset \Omega_\infty$ (i.e., $\bar{Q} \cap \partial\Omega_\infty = \emptyset$).

The latter property can be derived for each approximation $\vartheta_n(x,t)$ in the same way as in Chapter V, with constants that depend only on the domain Q and are independent of index n. We now recall an outline of the proof. Let $\delta > 0$ be taken so that the domain Q is contained in the interior of the square $G = \{(x,t) : \delta < x < \delta^{-1}, \delta < t < \delta^{-1}\}$ and $\delta < x_0$.

It follows by local parabolic estimates that the derivatives of $\vartheta_n(x,t)$ are bounded on the line $\{(x,t) : x = \delta, \delta < t < \delta^{-1}\}$:

$$\left| \frac{\partial \vartheta_n}{\partial t}, \frac{\partial \vartheta_n}{\partial x} \right| \le N_1(M_0, \delta), \quad M_0 = \max_n |v_n|^{(0)}_{\Omega_\infty}. \tag{2.7}$$

Let us now multiply the equation (2.3) for ϑ_n by $\vartheta_n(x,t)$, then integrate over $\{(x,t) : \delta < x < R_n(t), \delta < t < \delta^{-1}\}$, to obtain the inequality

$$\int_\delta^{\delta^{-1}} \int_\delta^{R_n(t)} \left| \frac{\partial \vartheta_n}{\partial x} \right|^2 dx dt \le N_2(M_0, N_1, a_0). \tag{2.8}$$

Assume $\zeta(t)$ to be an infinitely differentiable function which vanishes at $t = 0$ and is equal to 1 for $t > \delta$. After multiplying equation (2.3) by $\zeta \frac{\partial \vartheta_n}{\partial t}$ and integrating the result with respect to x over $[\delta, R_n(t)]$, we get the equality

$$\int_\delta^{R_n(t)} \zeta a(\vartheta_n) \left| \frac{\partial \vartheta_n}{\partial t} \right|^2 dx + \frac{d}{dt} \int_\delta^{R_n(t)} \frac{\zeta}{2} \left| \frac{\partial \vartheta_n}{\partial x} \right|^2 dx + \frac{\xi}{2} \frac{dR_n}{dt} \left| \frac{\partial \vartheta_n}{\partial x} \right|^2 \Bigg|_{x=R_n(t)-0}$$

$$= \frac{1}{2} \int_\delta^{R_n(t)} \zeta' \left| \frac{\partial \vartheta_n}{\partial x} \right|^2 dx - \frac{\partial \vartheta_n}{\partial t} \frac{\partial \vartheta_n}{\partial x} \Bigg|_{x=\delta}. \tag{2.9}$$

In deriving the above equality, we have used the identity

$$\frac{\partial \vartheta_n}{\partial t}(R_n(t) - 0, t) = -\frac{\partial \vartheta_n}{\partial x}(R_n(t) - 0, t)\frac{dR_n(t)}{dt}.$$

Since

$$\left.\frac{\partial \vartheta_n}{\partial x}\right|_{x=R_n(t)-0} \leq 0 \quad \text{and} \quad U_0^n(R_n(t)) < 0,$$

it follows that

$$\frac{dR_n(t)}{dt} \geq 0,$$

and by the estimates (2.7), (2.8) and equality (2.9),

$$\int_\delta^{\delta^{-1}} \int_\delta^{R_n(t)} \left|\frac{\partial \vartheta_n}{\partial t}\right|^2 dx\, dt + \max_{t \in (\delta, \delta^{-1})} \int_\delta^{R_n(t)} \left|\frac{\partial \vartheta_n}{\partial x}\right|^2 dx \leq N_3(M_0, \delta).$$

An analogous estimate also holds for the function $\vartheta_n(x, t)$ in the part of G where it vanishes. Therefore,

$$\iint_G \left|\frac{\partial \vartheta_n}{\partial t}\right|^2 dx\, dt + \max_{t \in (\delta, \delta^{-1})} \int_{G(t)} \left|\frac{\partial \vartheta_n}{\partial x}\right|^2 dx \leq N_3.$$

This inequality implies the Hölder continuity of the function $\vartheta_n(x, t)$ in G, as shown in Section I.3. Obviously, the limit function $\vartheta(x, t)$ is thus Hölder continuous in G, hence also in Q.

For proving the convergence of the sequences $\{R_n(t)\}$, it is more convenient to consider the sequence $\{h_n(x)\}$ of the inverse functions to $R_n(t)$:

$$h_n(R_n(t)) \equiv t.$$

Since $R_n(t) \geq R_n(0) = x_0$, the functions $h_n(x)$ are defined for $x \geq x_0$. We extend these functions onto $(0, x_0)$ by zero.

Returning to (2.6), we can see that the function sequence $h_n(x)$ is non-increasing:

$$h_1(x) \geq h_2(x) \geq \ldots \geq h_n(x) \geq \ldots. \tag{2.10}$$

Denote

$$X_n = \lim_{t \to \infty} R_n(t).$$

The functions $h_n(x)$ are defined on the interval $(0, X_n)$, where

$$x_0 < X_1 \leq X_2 \leq \ldots \leq X_n \leq \ldots \leq X_*,$$

with $X_* = \lim_{n \to \infty} X_n$.

Clearly, the limit function $h(x) = \lim_{n \to \infty} h_n(x)$, the existence of which follows from (2.10), is defined on the interval $[0, X_*)$. In particular, if

$$\int_0^\infty \vartheta^+(t)\, dt = \infty,$$

then the results of Section V.2 on the asymptotic behaviour of solutions as $t \to \infty$ hold for the solution $\{R_1(t), \vartheta_1(x,t)\}$ of problem (2.2) - (2.4). We have

$$X_1 = \lim_{t \to \infty} R_1(t) = \infty.$$

By construction, the function $h(x)$ is increasing in x, bounded on the interval $(0, X_*)$ and

$$\lim_{x \to X_*} h(x) = \infty \quad \text{if} \quad X_* < \infty.$$

Suppose this does not hold; then $X_* < \infty$ and $\lim_{x \to X_*} h(x) = t_* < \infty$. The domain $Q^+ = \{(x,t) : 0 < x < X_*, \ t_* < t < \infty\}$ represents the limit of

$$Q_n^+ = \{(x,t) : 0 < x < R_n(t), t_* < t < \infty\},$$
$$Q_n^+ \subset Q_{n+1}^+ \subset \ldots \subset Q^+,$$

with $\vartheta(x,t)$ strictly positive as a solution of the nonlinear heat equation, provided that

$$\text{ess} \max_{x \in (0,x_0)} \vartheta_0 > 0.$$

This follows from the strong maximum principle ([139], Theorem 2.4, p. 175) applied to $\vartheta(x,t)$ in $\Omega_{t_*}^+$.
By hypothesis,

$$\lim_{n \to \infty} R_n(t) = X_* \qquad \text{for} \quad t \in (t,\infty),$$

hence $\vartheta(X_*,t) = 0$ for $t \in (t_*,\infty)$. Thus, $\vartheta(x,t)$ attains its maximal value on the boundary $\{x = X_*\}$ of the domain Q^+, and by Hopf's theorem ([92], p. 69),

$$\frac{\partial \vartheta}{\partial x}(X_* - 0, t) < 0 \qquad \text{for} \quad t \in (t_*,\infty). \tag{2.11}$$

On the other hand (see Figure 7),

$$\vartheta \equiv 0, \quad \frac{\partial U}{\partial t} \equiv 0 \qquad \text{for} \quad x \in (X_*,\infty), \ t \in (t_*,\infty).$$

Consider the integral identity (2.1) with an arbitrary function φ whose support is concentrated in a small neighbourhood of the point (X_*,t), $t > t_*$. By the above equalities and the heat equation, which is satisfied by ϑ in Q^+, we obtain the condition

$$\frac{\partial \vartheta}{\partial x}(X_* - 0, t) = 0 \qquad \text{for} \quad t \in (t_*,\infty),$$

contradicting (2.11). Thus, the function $h(x)$ increases to ∞ as $x \to X_*$.
As a byproduct, we have obtained the following result.

Lemma 4. *Under the hypotheses of Theorem 2, for all x_1, t_0, t_1, $0 < l_0 < l_1 < \infty$, $x_0 < x_1 < X_*$, the set $\{(x,t) : x = x_1, t \in (t_0,t_1)\}$ is not contained in $\Omega_\infty^+ \cap \Omega_\infty^*$.*

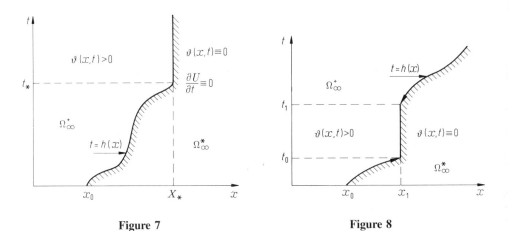

Figure 7   **Figure 8**

Lemma 4 implies the continuity of $h(x)$. Suppose this to be false and let $x_1 > x_0$ such that (see Figure 8)

$$\lim_{x \to x_1 - 0} h(x) = t_0 < t_1 = \lim_{x \to x_1 + 0} h(x).$$

By construction, $\vartheta(x, t)$ is strictly positive for $x < x_1$ and $t \in (t_0, t_1)$, and vanishes for $x > x_1$ and $t \in (t_0, t_1)$. Hence, the segment $\{(x, t) : x = x_1, t \in (t_0, t_1)\}$ is contained in $\Omega_\infty^+ \cap \Omega_\infty^*$, contradicting Lemma 4. This completes the proof of the theorem. □

If $a(s)$ is a sufficiently regular function, then the further regularity of the function $h(x)$ and its asymptotic behaviour depend on the differential properties of the initial specific internal energy.

Lemma 5. *Let* $a \in C^\infty[0, \infty)$, $U_0 \in C^k(x_1, x_2)$, $U_0(x) < 0$ *for* $x \in [x_1, x_2]$, *where* $x_0 \leq x_1 < x_2 \leq X_*$ *in the hypotheses of Theorem 2.*
   Then $h \in C^{k+1}(x_1, x_2)$ *and* $\frac{dh(x)}{dx} > 0$. *If* $U_0(x) \equiv 0$ *for* $x \in (x_1, x_2)$, *then* $h(x) \equiv$ const. $= h(x_1)$ *there.*
   The reverse statement is also true: if $h(x) \equiv$ const. *on the interval* (x_1, x_2), *then* $U_0(x) \equiv 0$.

Proof. The first assertion follows by a local analysis of the problem in the domain $\Omega_{t_2}^+ \setminus \Omega_{t_1}^+$, where $t_i = h(x_i)$. Indeed, the solution of problem (2.2) - (2.4) on the interval (t_1, t_2), with $U(x, t_1)$ taken as the initial specific internal energy, due to the proved uniqueness of the generalized solution of the Stefan problem, coincides with the solution of problem (2.5). The function $R(t)$ is thus at least once continuously differentiable in the interior of the interval (t_1, t_2). Its further regularity (and that of the function $h(x)$) can be shown as in Section V.1.

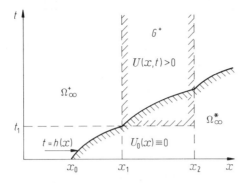

Figure 9

Let $U_0(x) \equiv 0$ on the interval $J = (x_1, x_2)$. We shall consider the following Stefan problem: find a function $\tilde{U}(x, t)$ defined in the domain $G^+ = \{(x, t) : x \in J, t \in (t_1, \infty)\}$, $t_1 = h(x_1)$, such that

$$\frac{\partial \tilde{U}}{\partial t} = \frac{\partial^2 \tilde{\vartheta}}{\partial x^2}, \qquad \tilde{\vartheta} = \chi(\tilde{U});$$

$$\tilde{\vartheta}(x_i, t) = \vartheta(x_i, t), \ i = 1, 2; \qquad \tilde{U}(x, t_1) = 0 \qquad \text{for} \quad x \in J.$$

Since $\vartheta(x_1, t) > 0$ (the segment $\{(x, t) : x = x_1, \ t > t_1\}$ is contained in the domain Ω_∞^+), the function $\tilde{U}(x, t)$ is strictly positive in the interior of G^+ (see Figure 9).

On the other hand, the segment $\{(x, t) : x \in J, \ t = t_1\}$ is contained in the closure of Ω_∞^* (recall that $h(x)$ is an increasing function), where $\frac{\partial U}{\partial t} = 0$. Hence

$$U(x, t_1) = U_0(x) = 0 \qquad \text{for} \quad x \in J.$$

By the uniqueness of the generalized solution of the Stefan problem,

$$U(x, t) = \tilde{U}(x, t) \qquad \text{in} \quad G^+.$$

By construction,

$$\tilde{U}(x, t) > 0 \qquad \text{for} \ (x, t) \in G^+, \qquad \text{and}$$
$$U(x, t) = 0 \quad \text{for } x \in J, \ t < t_1,$$

hence $h(x) \equiv t_1$ for $x \in J$.

If $h(x) \equiv t_1$ for $x \in J$, then according to the integral identity (2.1) with bounded test functions φ having supports concentrated in a small neighbourhood of the points (x, t_1), $x \in J$, we can conclude that

$$U(x, t_1 - 0) = U(x, t_1 + 0) = 0 \qquad \text{for } x \in J.$$

The rectangle $\{(x, t) : x \in J, \ t \in (0, t_1)\}$ is contained in the domain Ω_∞^*, where $U(x, t) = U_0(x)$. Hence, $U_0(x) = U(x, t_1 - 0) = 0$ for $x \in J$. $\square$

The results of this section indicate that, for the same data $\vartheta^+(t)$, $\vartheta_0(t)$, the smaller the specific internal energy in the mushy phase (roughly speaking, the more "solid" is the mushy phase), the slower is the movement of the boundary separating the liquid and mushy phases. If there are parts of the mushy phase completely filled with the liquid, i.e., with $U_0(x) \equiv 0$, then the corresponding boundary moves with infinite speed.

The case where the solid phase, occupying $\Omega^-(0) = \{x : 0 < x < x_0\}$, contacts the mushy phase that fills the domain $\Omega^*(0) = \{x : x_0 < x < \infty\}$, can be considered similarly.

3. The homogeneous heat equation. Coexistence of different phases

Assume that the system considered occupies the domain $\Omega = \{x : -1 < x < 1\}$, on the boundary $x = -1$ the temperature is maintained negative throughout, it is positive on the boundary $x = 1$, and at the initial time the mushy phase coexists with the liquid and solid phases. In this section, we shall study the simplest case where the domain Ω admits an initial decomposition into three subdomains $\Omega^-(0) = \{x : -1 < x < x_0^-\}$ $\Omega^*(0) = \{x : x_0^- < x < x_0^+\}$ and $\Omega^+(0) = \{x : x_0^+ < x < 1\}$, occupied by the solid, mushy and liquid phases, respectively. Such an initial distribution of the specific internal energy allows the treatment of the discussed Stefan problem to be split into the local in time treatment (on some interval $(0, t_*)$) of two one-phase problems as considered in the preceding section, and a two-phase Stefan problem with two-phase initial condition on the interval (t_*, ∞). The time instant t_* at which the mushy phase disappears may even be infinite. Its value depends on the initial distribution of the specific internal energy and on the behaviour of the prescribed temperature on the boundary of Ω.

As already noted, there are several alternative forms of the integral identity and the generalized solution of the Stefan problem. In particular, the test function φ does not have to vanish at $t = T$. We can formulate the following Stefan problem: determine the specific internal energy $U(x, t)$ and temperature $\vartheta(x, t) = \chi(U(x, t))$ which in the domain $\Omega_\infty = \Omega \times (0, \infty)$ satisfy the integral identity

$$\iint_{\Omega_T} \left\{ -U\frac{\partial\varphi}{\partial t} + \frac{\partial\vartheta}{\partial x}\frac{\partial\varphi}{\partial x} \right\} dxdt + \int_\Omega \left\{ U(x, T)\varphi(x, T) - U_0(x)\varphi(x, 0) \right\} dx = 0 \quad (3.1)$$

for any continuously differentiable function φ, such that $\varphi|_{\partial\Omega} = 0$ for all $T \in (0, \infty)$.

The temperature ϑ is assumed to be prescribed on the boundary of Ω by

$$\vartheta(\pm 1, t) = \vartheta^\pm(t) \qquad \text{for } t \in (0, \infty). \tag{3.2}$$

We introduce the following notation:

$$A^- = \int_0^\infty \left|\vartheta^-(t)\right| dt - \int_{-1}^{x_0^-} (1+x)\left|1 + U_0(x)\right| dx,$$

$$A^+ = \int_0^\infty \left|\vartheta^+(t)\right| dt + \int_{x_0^+}^1 (1-x)U_0(x)\, dx,$$

$$f^-(X) = \int_{x_0^-}^X (1+x)\big|1 + U_0(x)\big|\,dx,$$

$$f^+(X) = \int_X^{x_0^+} (1-x)\big|U_0(x)\big|\,dx.$$

Let us also define

$$X_\infty^- = \min\{X \ : \ f^-(X) = A^-\}, \quad X_\infty^+ = \max\{X \ : \ f^+(X) = A^+\}.$$

We shall assume that:

(I) $\vartheta^\pm \in W^1_{2,\mathrm{loc}}(0,\infty)$, $(\pm 1)\vartheta^\pm(t) > 0$ for $t \in (0,\infty)$.

(II) $U_0(x) \le -1$ for $x \in \Omega^-(0)$, $U_0(x) \ge 0$, for $x \in \Omega^+(0)$, $-1 \le U_0(x) \le 0$ for $x \in \Omega^*(0)$, $\vartheta_0 = \chi(U_0) \in W^1_2(\Omega)$.

(III)   In a sufficiently small neighbourhood of the point x_0^-, the set of points to its right, where $U_0 = -1$, has measure zero; analogously, in a sufficiently small neighbourhood of the point x_0^+, the set of points to its left, where $U_0 = 0$, has measure zero.

(IV)   If $A^+ + A^- < \infty$, then $\lim \vartheta^\pm(t) = 0$ for $t \to \infty$.

Theorem 3. *Let hypotheses (I) - (IV) be satisfied and let $a \in C^1(-\infty,0] \cap C^1[0,\infty)$, $a(s) \ge a_0 > 0$.*

If $A^+ + A^- = \infty$ or $A^+ + A^- < \infty$, but $X_\infty^- > X_\infty^+$, then there exist real numbers x_* and t_*, $0 < t_* < \infty$, $x_0^- < x_* < x_0^+$, and continuous functions $h^-(x)$, $h^+(x)$, such that $h^-(x) = 0$ on the interval $(-1, x_0^-)$ and is increasing on the interval (x_0^-, x_*); $h^+(x) = 0$ on the interval $(x_0^+, 1)$ and is decreasing on the interval (x_*, x_0^+), with $h^-(x_*) = h^+(x_*) = t_*$. The continuous function $R(t)$, with values in $(-1,1)$, is defined on the half-line (t_*, ∞) and is continuously differentiable at all its interior points, $R(t_*) = x_*$ (see Figure 10).

The generalized solution $\vartheta(x,t)$ of Stefan problem (3.1), (3.2) is strictly positive in the domain $\Omega_\infty^+ = \{(x,t) \ : \ x_* < x < 1, \ h^+(x) < t < t_*\} \cup \{(x,t) \ : \ R(t) < x < 1, \ t_* < t < \infty\}$, strictly negative in the domain $\Omega_\infty^- = \{(x,t) \ : \ -1 < x < x_*, \ h^-(x) < t < t_*\} \cup \{(x,t) \ : \ -1 < x < R(t), \ t_* < t < \infty\}$ and vanishes everywhere in the domain $\Omega_{t_*}^* = \{(x,t) \ : \ x_0^- < x < x_*, \ 0 < t < h^-(x)\} \cup \{(x,t) \ : \ x_* < x < x_0^+, \ 0 < t < h^+(x)\}$.

If $A^+ + A^- < \infty$ and $X_\infty^- \le X_\infty^+$, then the functions $h^-(x)$, $h^+(x)$ are defined on the intervals $[-1, X_\infty^-)$ and $(X_\infty^+, 1]$, and $\lim h^-(x) = \infty$ as $x \to X_\infty^-$, $\lim h^+(x) = \infty$ as $x \to X_\infty^+$ (see Figure 11).

The generalized solution $\vartheta(x,t)$ of Stefan problem (3.1) - (3.2) is strictly positive in the domain $\Omega_\infty^+ = \{(x,t) \ : \ X_\infty^+ < x < 1, \ h^+(x) < t < \infty\}$, is strictly negative in the domain $\Omega_\infty^- = \{(x,t) \ : \ -1 < x < X_\infty^-, \ h^-(x) < t < \infty\}$ and vanishes everywhere in $\Omega_\infty^* = \Omega_\infty \setminus (\Omega_\infty^- \cup \Omega_\infty^+)$.

Proof. For small times T, the solution of the problem (3.1), (3.2) on the interval $(0,T)$ can be decomposed into the solution of two independent problems as considered in the preceding section: the problem of determining the domain $\Omega_T^- = \{(x,t) \ : \ -1 < x <$

$x_T,\ h^-(x) < t < T\}$, $h^-(x_T) = T,$ and the function $\vartheta(x,t)$ there, and the problem of determining the appropriate domain Ω_T^+ and $\vartheta(x,t)$ on that domain.

The main question thus reduces to finding conditions which ensure that the graphs of functions $t = h^-(x)$ and $t = h^+(x)$ intersect (or, alternatively, do not intersect). Since the functions $h^-(x)$ and $h^-(x)$ are monotone, the mushy phase diminishes with time; if the appropriate graphs intersect at a finite time t_*, the mushy phase completely disappears at this time instant. Then the extension of the solution of problem (3.1), (3.2) onto the interval (t_*, ∞) is uniquely defined by the solution of the standard two-phase Stefan problem with a two-phase initial state, as considered in Chapter V.

If $A^- = \infty$, then Theorem 2 implies that the function $h^-(x)$ is defined on the interval $[-1, x_0^+]$ and its graph intersects the graph of function $t = h^+(x)$ at some time $t_* \le h^-(x_0^+)$.

Similarly, if $A^+ = \infty$, then the function $h^+(x)$ is defined on $[x_0^-, 1]$ and its graph intersects the graph of the function $t = h^-(x)$ at $t_* \le h^+(x_0^-)$.

Let us specify the domains of the functions $h^+(x)$ and $h^-(x)$ for $A^+ + A^- < \infty$. More precisely, we specify the conditions which guarantee that the domains of $h^+(x)$ and $h^-(x)$ do not intersect, hence the mushy phase exists for all finite time. It will be more convenient to consider the regular case where $U_0(x)$ is continuously differentiable in $\Omega^*(0)$ and there assumes values from $(-1, 0)$. To fit the general situation, it is sufficient to approximate the function $U_0(x)$ by smooth functions $U_0^n(x)$, $-1 < U_0^n(x) < 0$ for $x \in \Omega^*(0)$, and then pass to the limit. The functions U_0^n can be chosen so that the sequence $\{R_n^-(t)\}$ is increasing, where the curves $x = R_n^-(t)$ approximate $t = h^-(x)$, and the sequence $\{R_n^+(t)\}$ is decreasing. To this end, it suffices to choose the sequence $\{U_0^n\}$ decreasing in the former case, and increasing in the latter.

Let now $U_0(x)$ be continuously differentiable, $-1 < U_0(x) < 0$ for $x \in \overline{\Omega^*(0)}$, and let the mushy phase exist for all finite time. The right boundary $x = R^-(t)$ of Ω_∞^- and the temperature $\vartheta(x,t)$ in this domain are defined as the solution of problem (2.2) - (2.4) of the preceding section. Similarly, the left boundary $x = R^+(t)$ of Ω_∞^+ and the temperature $\vartheta(x,t)$ in this domain are defined as the solution of problem (2.2) - (2.4) subject to the appropriate initial and boundary conditions. By hypothesis,

$$R^-(t) < R^+(t) \qquad \text{for} \quad t \in (0, \infty).$$

Clearly, the function $U(x,t)$, equal to $\Phi(\vartheta(x,t))$ for $-1 < x < R^-(t)$ and equal to $U_0(x)$ for $R^-(t) < x < R_\infty^-$, $t > 0$, where

$$R_\infty^- = \lim_{t\to\infty} R^-(t)$$

is the solution of the Stefan problem, analogous to (3.1), (3.2), formulated in the domain $\{(x,t) : -1 < x < R_\infty^-,\ 0 < t < \infty\}$, with the conditions

$$\vartheta(-1,t) = \vartheta^-(t), \quad \vartheta(R_\infty^-, t) = 0 \qquad \text{for} \quad t \in (0, \infty).$$

Let $T > 0$ be arbitrary. By the strict monotonicity of $R^-(t)$, we have $R^-(T) < R_\infty^-$. Therefore, we can choose a function $\varphi_T = \varphi_T(x)$ so that $\varphi_T = 1 + x$, $x \in (-1, R^-(T))$ and $\varphi_T = 0$ for $x = R_\infty^-$. We substitute this function φ_T into (3.1). Since $\vartheta = \chi(U)$

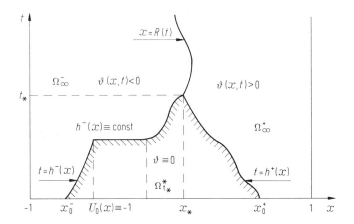

Figure 10

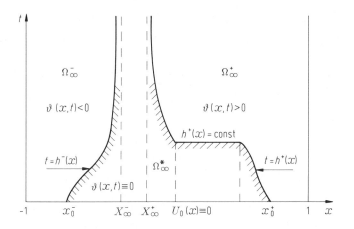

Figure 11

vanishes everywhere outside the domain Ω_T^- for $x < R_\infty^-$, we easily obtain

$$\int_{-1}^{R^-(T)} (x+1) U \, dx = \int_0^T \vartheta^-(t) \, dt + \int_{-1}^{R^-(T)} (x+1) U_0(x) \, dx.$$

The latter equality can be transformed to the form

$$\int_{-1}^{R^-(T)} (x+1)|U(x,T)+1|\,dx - \int_{x_0^-}^{R^-(T)} (x+1)|U_0(x)+1|\,dx$$

$$= \int_0^T \vartheta^-(t)\,dt + \int_{-1}^{x_0^-} (x+1)|U_0(x)+1|\,dx. \tag{3.3}$$

By the hypothesis of the theorem, the temperature ϑ on the boundary $x = -1$ converges to zero as $t \to \infty$. In this case, as shown in Section V.2,

$$\lim_{t\to\infty} \max_{x\in\Omega^-(t)} |\vartheta(x,t)| = 0.$$

Since, further,

$$\lim_{\vartheta\to-0} U = \lim_{\vartheta\to-0} \Phi(\vartheta) = -1,$$

we have $\displaystyle\lim_{t\to\infty} \max_{x\in\Omega^-(t)} |U(x,t)+1| = 0$.

Passing in (3.3) to the limit as $T \to \infty$, by the latter relation we get the equality $f^-(R_\infty^-) = A^-$. Thus, $X_\infty^- = R_\infty^-$. Similarly it follows that $X_\infty^+ = R_\infty^+$, where $R_\infty^+ = \lim_{t\to\infty} R^+(t)$.

Therefore, the inequality $X_\infty^- \le X_\infty^+$ is precisely the condition that ensures the existence of the mushy phase for all finite time.   □

4. The case of an arbitrary initial
distribution of specific internal energy

Let us consider the Stefan problem treated as a model of phase change in a system that occupies the domain $\Omega = \{x : |x| < 1\}$. We shall assume that the temperature is prescribed on the boundaries $\{x = \pm 1\}$ of Ω and preserves its sign there, whereas the initial distribution of the internal energy, although arbitrary, provides the existence of a generalized solution. Under these hypotheses, we shall show that subject to some restrictions on the input data, there exists a finite time T_* such that for $t > T_*$ either only the solid and liquid phases will exist in the domain Ω and the generalized solution of the Stefan problem will coincide with the classical solution of the two-phase single-front Stefan problem, or only one phase, solid or liquid, will exist for $t > T_*$, and then the generalized solution of the Stefan problem will coincide with the classical solution of the heat equation.

Thus, in the domain $\Omega_T = \Omega \times (0, T)$ we seek a solution $U(x,t)$, $\vartheta(x,t) = \chi\big(U(x,t)\big)$ of the equation

$$\frac{\partial U}{\partial t} = \frac{\partial^2 \vartheta}{\partial x^2} \tag{4.1}$$

in the sense of distributions. On the boundaries $\{x = \pm 1\}$ of Ω, the temperature $\vartheta(x, t)$ is prescribed by

$$\vartheta(\pm 1, t) = \vartheta^{\pm}(t) \qquad \text{for} \quad t \in (0, T), \tag{4.2}$$

and, at the initial time instant the specific internal energy distribution is known:

$$U(x, 0) = U_0(x). \tag{4.3}$$

The functions $\vartheta^{\pm}(t), U_0(x)$ are assumed to be bounded. Let

$$\Lambda = \max \left\{ 1, \max_{x \in \Omega} |U_0(x)|, \max_{t \in (0, \infty)} |\Phi(\vartheta^{\pm}(t))| \right\}$$

and denote by T^+, T^- the solutions of the equations

$$\int_0^T |\vartheta^+(t)| \, dt = 4\Lambda \quad \text{and} \quad \int_0^T |\vartheta^-(t)| \, dt = 4\Lambda,$$

respectively. If the former equation does not have a solution, we set $T^+ = \infty$ and we proceed similarly with the latter equation. We shall use the notation

$$T_* = \min\{T^+, T^-\}. \tag{4.4}$$

The main result of this section is formulated below.

Theorem 4. *Let $\Phi \in C^2(-\infty, 0] \cap C^2[0, \infty)$, $\vartheta^{\pm} \in C[0, \infty] \cap W_{2,\text{loc}}^1(0, \infty)$. Let $U_0(x)$ be a bounded measurable function such that $\vartheta_0 = \chi(U_0) \in W_2^1(\Omega)$, and the compatibility conditions $\vartheta^{\pm}(0) = \vartheta_0(\pm 1)$ are satisfied at the points $x = \pm 1$. Assume further that the functions $\vartheta^{\pm}(t)$ have constant sign and the time T_*, defined by (4.4), is bounded.*

Under these hypotheses, if $\vartheta^+(t)\vartheta^-(t) > 0$, then for $t > T_*$ there exists only one phase (solid or liquid) in the domain Ω, and the solution of the Stefan problem for $t > T_*$ coincides with the classical solution of a nonlinear heat equation. If $\vartheta^+(t)\vartheta^-(t) < 0$, then for $t > T_*$ there exist both solid and liquid phases, and the generalized solution of the Stefan problem coincides with the classical solution of two-phase single-front Stefan problem.

There follows an outline of the proof of Theorem 4. First the Stefan problem with a finite number of components of liquid, solid and mushy phases is considered, i.e., the domain Ω decomposes into a finite number of subdomains each containing just one phase. Theorem 4 is proved for such a problem.

Secondly, an arbitrary initial distribution of specific internal energy is approximated by data that correspond to the above situation, and the corresponding approximate solutions are constructed. The solution of the original problem is then the appropriate limit.

The consideration of the first step exploit the following Stefan problems in turn:

– the problem on half-line $\mathbb{R}^+ = \{x : -1 < x < \infty\}$, with prescribed temperature $\vartheta^-(t)$ on the boundary $x = -1$, a finite number of liquid and solid components occupying bounded domains, and the mushy phase occupying the unbounded, extreme right domain (Problem (A));

– the problem on the whole line $\mathrm{IR} = \{x : |x| < \infty\}$, with a finite number of solid and liquid components occupying bounded domains, and two components of the mushy phase, occupying the extreme left and right, unbounded subdomains (Problem (B));
– the multi-front two-phase Stefan problem in Ω, with a finite number of the components of the solid and liquid phases (Problem (C)).

Let $\vartheta^-(t) < 0$, for definiteness.

Definition 1. A generalized solution of the Stefan problem on the half-line IR^+, satisfying condition (4.2) on the boundary $x = -1$, *exhibits* (A, m)-*structure at time* τ, if there exist real numbers $x_1, \ldots, x_m$, $-1 = x_0 < x_1 < \ldots < x_m < x_{m+1} = \infty$, such that the domains $\Omega^{(k)}(\tau) = \{x : x \in (x_{k-1}, x_k)\}$, $k = 1, \ldots, m$, are occupied by the liquid and solid phases at even and odd k, respectively, and $\Omega^{(m+1)}(\tau)$ is occupied by the mushy phase.

Definition 2. A generalized solution of the Stefan problem *preserves* on the interval $(0, T)$ *the* (A, m)-*structure imposed at the initial time*, if there exist continuous functions $r_k(t)$, $k = 1, \ldots, m$, $-1 = r_0(t) < r_1(t) < \ldots < r_m(t) < r_{m+1}(t) = \infty$ for $t \in [0, T)$, such that the domains $\Omega_T^{(k)} = \{(x, t) : x \in (r_{k-1}(t), r_k(t)), t \in (0, T)\}$, $k = 1, \ldots, m$, are occupied by the liquid and solid phases, respectively, at even and odd k, and $\Omega_T^{(m+1)}$ is occupied by the mushy phase.

Analogous definitions can be introduced for Problems (B) and (C). Let us now consider Problem (A). To be determined is a solution of that problem, satisfying in $\mathrm{IR}_T^+ = \mathrm{IR}^+ \times (0, T)$ equation (4.1) in sense of distributions, condition (4.2) on the boundary $\{x = -1\}$, the initial condition

$$U(x, 0) = U_0(x) \qquad \text{for} \quad x \in R^+, \tag{4.5}$$

and exhibiting (A, m)-structure at the initial time moment.

Lemma 6. *Assume that the hypotheses of Theorem 4 hold and let $\vartheta_0 \in W_2^1(R^+)$. At the initial time let the solution exhibit (A, m)-structure, and let the function $U_0(x)$ be continuously differentiable in the domain $\Omega^{(m+1)}(0)$, without attaining the values 0 and -1.*

Then the whole time interval $(0, \infty)$, where there exists a generalized solution of the Stefan problem, can be decomposed into at most m subintervals (T_{k-1}, T_k), $0 = T_0 < T_1 < \ldots < T_p < T_{p+1} = \infty$, $p < m$, such that on each of the intervals (T_{k-1}, T_k) $k = 1, \ldots, p+1$, the generalized solution of the Stefan problem preserves (A, m_{k-1})-structure, $m_0 = m > m_1 > \ldots > m_p$, imposed at the time $t = T_{k-1}$ (see Figure 12).

Proof. The intervals (T_{k-1}, T_k) are maximal because at least one of the components of the liquid or solid phase disappears at $t = T_k$. Furthermore, on (T_k, T_{k+1}) the solution of Stefan problem preserves its (A, m_k)-structure imposed at $t = T_k$, although with a reduced number of liquid and solid phase components compared with the preceding time interval.

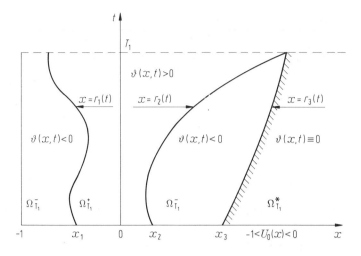

Figure 12

We shall at first assume that $\vartheta_0 \in H^{2+\alpha}(\overline{\Omega^{(k)}}(0))$, $\vartheta^- \in H^{(2+\alpha)/2}[0,\infty)$. One can expect that, at least on a sufficiently small interval $(0,T)$, the generalized solution of the Stefan problem preserves the (A,m)-structure imposed at the initial time, and the functions $r_k(t)$, $k = 1, \ldots, m$, belong to the space $H^{(2+\alpha)/2}[0,T]$.

We can prove in a standard way that the temperature $\vartheta(x,t)$ fulfils the nonlinear heat equation

$$a(\vartheta)\frac{\partial\vartheta}{\partial t} = \frac{\partial^2\vartheta}{\partial x^2} \tag{4.6}$$

on each of the domains $\Omega_T^{(k)}$, $k = 1, \ldots, m$; this function satisfies for $m > 1$ two strong discontinuity conditions on $x = r_k(t)$, $k = 1, \ldots, m-1$, for $t \in (0,T)$:

$$\vartheta(r_k(t), t) = 0, \tag{4.7}$$

$$\left[\frac{\partial\vartheta}{\partial x}(x,t)\right]_{x=r_k(t)+0}^{x=r_k(t)-0} = (-1)^{k+1}\frac{dr_k(t)}{dt}. \tag{4.8}$$

The temperature ϑ vanishes everywhere in the domain $\Omega_T^{(m+1)}$ and the specific internal energy is defined by its initial distribution,

$$U(x,t) = U_0(x) \qquad \text{for} \quad (x,t) \in \Omega_T^{(m+1)}.$$

On the boundary $x = r_m(t)$, the temperature $\vartheta(x,t)$ fulfils condition (4.6) and

$$\frac{\partial\vartheta}{\partial x}(r_m(t) - 0, t) = U_0(r_m(t))\frac{dr_m(t)}{dt} \qquad \text{for} \quad t \in (0,T), \tag{4.9}$$

if the domain $\Omega_T^{(m)}$ is filled with the liquid, and

$$\frac{\partial\vartheta}{\partial x}(r_m(t) - 0, t) = (U_0(r_m(t)) + 1)\frac{dr_m(t)}{dt} \qquad \text{for} \quad t \in (0,T), \tag{4.10}$$

if the latter domain is occupied by the solid. As in the above equations, we postulate that the initial data at $x = x_k$, $k = 0, 1, \ldots, m$, satisfy first-order compatibility conditions.

As in Chapter II, where smooth approximations to the solutions of the multidimensional Stefan problem have been constructed, we can prove that a classical solution of Problem (A) exists on a sufficiently small interval. To this end, we consider the set $\mathcal{M}$ of all vector functions $\{\bar{r}_1(t), \ldots, \bar{r}_m(t)\} = \bar{r}(t)$ which coincide with a given vector $\bar{r}(0) = \{x_1, \ldots, x_m\}$ at the initial time and have all components $\bar{r}_k(t)$ bounded in $H^{(2+\alpha)/2}[0,T]$:

$$|\bar{r}_k - x_k|_{[0,T]}^{(2+\alpha)/2} \le 1, \quad d = 1, \ldots, m.$$

Let

$$\kappa_m(0, T; \bar{r}) = \min_{k=1,\ldots,m} \left\{ \min_{t \in (0,T)} |\bar{r}_{k-1}(t) - \bar{r}_k(t)| \right\}, \qquad \bar{r}_0(t) = -1.$$

The necessary condition for each element $\bar{r}(t)$ of the set $\mathcal{M}$ to define the domains $\Omega_T^{(k)}$, $k = 1, \ldots, m+1$, is that $\kappa_m(0, T; \bar{r}) > 0$. We shall assume further that $\kappa_m(0, T; \bar{r}) \ge \frac{1}{2}\kappa_m(0, T; \bar{r}(0))$ for all $\bar{r} \in \mathcal{M}$.

In each of the domains $\Omega_T^{(k)}$, $k = 1, \ldots, m$, constructed for a fixed $\bar{r} \in \mathcal{M}$, let us consider the initial-boundary value problem (4.5) - (4.7), (4.2). According to Theorem 3, Section I.2, there exists a unique solution $\bar{\vartheta} \in H^{2+\alpha,(2+\alpha)/2}\overline{\Omega_T^{(k)}})$, $k = 1, \ldots, m$, such that

$$|\bar{\vartheta}|_{\Omega_T^{(k)}}^{(2+\alpha)} \le M_1, \quad k = 1, \ldots, m,$$

where M_1 depends only on the norms $|\vartheta^-|_{[0,T]}^{(2+\alpha)/2}$, $|\vartheta_0|_{\Omega^{(k)}(0)}^{(2+\alpha)}$, $k = 1, \ldots, m$.

The conditions (4.8), (4.9) (or (4.10)) define the operator $\Re(\bar{r})$ whose fixed points determine the solution $\Re(\bar{r}) = \{r_1, \ldots, r_m\}$ of the problem considered,

$$r_k(t) = x_k + (-1)^{k+1} \int_0^t \left[\frac{\partial \bar{\vartheta}}{\partial x}(x, \tau) \right]_{x=\bar{r}_k(\tau)+0}^{x=\bar{r}_k(\tau)-0} d\tau, \quad k = 1, \ldots, m-1,$$

$$r_m(t) = x_m + \int_0^t (U_0(r_m(\tau)))^{-1} \frac{\partial \bar{\vartheta}}{\partial x}(\bar{r}_m(\tau) - 0, \tau) d\tau.$$

We have

$$|r_k - x_k|_{[0,T]}^{(0)} \le TM_2(M_1, \beta), \quad k = 1, \ldots, m, \tag{4.11}$$

where

$$\beta = \min_{x \in \Omega^{(m+1)}(0)} \{\min\{|U_0(x)|, |1 + U_0(x)|\}\}$$

and $\Omega^{(m+1)}(0)$ is the set occupied by the mushy phase at the initial time,

$$|r_k - x_k|_{[0,T]}^{(3+\alpha)/2} \le M_3(M_1, \beta). \tag{4.12}$$

By this estimate, the operator $\Re$ is completely continuous on the set $\mathcal{M}$. The existence of a fixed point of $\Re$ follows by Schauder's theorem, provided that $\Re$ maps $\mathcal{M}$ into itself.

This is a consequence of the estimate

$$\left|r_k - x_k\right|_{[0,T]}^{(2+\alpha)/2} \leq \delta\left|r_k - x_k\right|_{[0,T]}^{(3+\alpha)/2} + \delta^{-2-\alpha}\left|r_k - x_k\right|_{[0,T]}^{(0)}$$

for arbitrarily small $\delta > 0$, and the bounds (4.11), (4.12), provided that T has been chosen sufficiently small.

The function $r_m(t)$, defining the extreme right boundary $x = r_m(t)$, increases not faster than the function $r^+(t)$, where $\{U^+(x,t), r^+(t)\}$ is the solution of the Stefan problem in R_T^+, coinciding with $|\vartheta^-(t)|$ on the boundary $x = -1$ and, at the initial time, with the function $U_0^+(x)$ defined by

$$U_0^+(x) = \begin{cases} |\vartheta_0|_{R^+}^{(0)} & \text{for } x \in (-1, x_m + 1), \\ -\beta & \text{for } x \in (x_{m+1}, \infty). \end{cases}$$

The latter property is implied by the comparison lemma applied to generalized solutions of the Stefan problem.

As in Chapter V, where we have considered the single-front two-phase Stefan problem, the norms of functions $r_k(t)$, $k = 1, \ldots, m$, in the space $W_4^1[0, T]$ and the norms of $\vartheta(x, t)$ in the space $L_\infty(0, T; W_2^1(\mathrm{I\!R}^+))$ can be proved to be bounded by a constant dependent only on the value

$$M_0(T) = \max\left\{\|\vartheta^-\|_{2,(0,T)}^{(1)}, \|\vartheta_0\|_{2,\mathbf{R}+}^{(1)}\right\}.$$

Let $G \subset \Omega_T^{(k)}$, $k = 1, \ldots, m$, be the domain whose distance from the boundaries $\{x = -1\}$ and $\{t = 0\}$ is larger than δ. Then, the norm of $\vartheta(x, t)$ in $H^{2+\alpha,(2+\alpha)/2}(\bar{G})$ depends only on $\delta, M_0(T)$ and $\kappa_m(0, T; r)$, where $\kappa_m(0, T; r)$ which characterizes the minimal distance between the boundaries $x = r_{k-1}(t)$ and $x = r_k(t)$, $k = 1, \ldots, m$, has been defined above.

Because the space $W_4^1[0, T]$ is contained in $H^{3/4}[0, T]$, the functions $r_k(t)$, $k = 1, \ldots, m$, are Hölder continuous with Hölder norm dependent only on $M_0(T)$. Therefore, the interval $(0, T)$ can be chosen so as to ensure that

$$\kappa_m(0, T; r) \geq \frac{1}{2}\kappa_m(0, 0; r) = \kappa,$$

with the norms of the functions ϑ independent of $\kappa_m(0, T; r)$. The above set of constraints on functions will be referred to as the *property* $N(M_0(T), T, \kappa)$.

We are now ready to consider the case of arbitrary initial data

$$\vartheta^- \in C[0, \infty] \cap W_{2,\text{loc}}^1(0, \infty) \quad \text{and} \quad \vartheta_0 \in W_2^1(\mathrm{I\!R}^+),$$

such that the corresponding solution has (A, m)-structure at the initial time. We shall construct the appropriate smooth approximations in such a way that the resulting solution not only has (A, m)-structure initially but also determines a smooth solution of Problem (A) on a sufficiently small time interval $(0, T)$. These approximations obviously have the property $N(M_0(T), T, \kappa)$ and the corresponding solutions preserve the initial (A, m)-structure.

Hence, there exists a subsequence of approximate solutions, convergent to the classical solution of Problem (A) which preserves the (A, m)-structure, imposed at the initial time instant on the whole interval $(0, T)$. We have

$$\vartheta^-(T) = \vartheta(-1, T), \quad \|\vartheta(T)\|_{2, R^+}^{(1)} < \infty, \quad \kappa_m(0, T; r) > 0.$$

The latter property enables us to extend the solution of Problem (A) onto a larger interval. Since the norm $\|\vartheta(T)\|_{2, R^+}^{(1)}$ depends only on $M_0(T)$ and T, and is finite for finite T, the above procedure can be continued as long as no component of either liquid or solid phase disappears, i.e., until $\kappa_m(0, T_1; r) = 0$.

At $t = T_1$, the solution of the Stefan problem has (A, m_1)-structure and, treating T_1 as new initial time, we can construct a solution of Problem (A) that preserves on (T_1, T_2) the structure imposed at $t = T_1$. Due to the finiteness of the total number of components of the liquid and solid phases, the entire time interval $(0, \infty)$ will be covered in a finite number of steps.   □

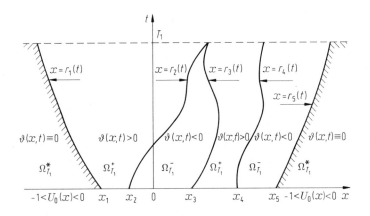

Figure 13

The existence of the classical solution of Problem (B) can be shown in a completely analogous way.

Lemma 7. *Let the hypotheses of Theorem 4 hold, $\vartheta_0 \in W_2^1(\mathbb{R})$, let the function $\vartheta_0(x)$ vanish everywhere outside of $\Omega^+(0) \cup \Omega^-(0)$, and let $U_0(x)$ be a continuously differentiable function which defines the initial distribution of the specific internal energy and assumes values from the interval $(-1 + \beta, -\beta)$, $0 < \beta < 1$, in the mushy phase. Assume that the generalized solution of the Stefan problem has initially (B, m)-structure. Then the whole time interval $(0, \infty)$ can be decomposed into at most m subintervals (T_{k-1}, T_k), $0 = T_0 < T_1 < \ldots < T_p < T_{p+1} = \infty$, $p < m$, where the generalized solution of the Stefan problem on each interval (T_{k-1}, T_k), $k = 1, \ldots, p$ preserves the (B, m_{k-1})-structure imposed at $t = T_{k-1}$, with $m = m_0 > m_1 > \ldots > m_p$ (see Figure 13).*

Problem (C) has already been considered in Section V.5.

We shall now study problem (4.1) - (4.3), with the set $\Omega = \{x \; : \; |x| < 1\}$ decomposed into a finite number of subintervals $\Omega^{(k)}(0)$, each of them occupied either by solid, liquid or mushy phase. We assume that the specific internal energy function is continuously differentiable on the domain $\Omega^*(0)$ occupied by the mushy phase, and has there values from the interval $(-1 + \beta, -\beta)$.

For a given initial distribution of the specific internal energy, on a sufficiently small time interval the solution of problem (4.1) - (4.3) admits splitting into the solution of Problem (A) on the half-line $\mathbb{R}^+$, and on the half-line $\mathbb{R}^- = \{x \; : \; -\infty < x < 1\}$, where the temperature coincides with a given function $\vartheta^+(t)$ at $x = 1$, and, possibly, some problems (B). If there is no mushy phase, the solution of problem (4.1) - (4.3) coincides with that of Problem (C).

Let, for instance, $x = r_m(t)$ be the extreme right boundary in Problem (A) on half-line $\mathbb{R}^+$, separating the liquid or solid phase in $\{x < r_m(t)\}$ from the mushy phase. Further, let $x = r_{m+1}(t)$ be the free boundary nearest to the above curve, extreme left in Problem (B) or in Problem (A) on the half-line $\mathbb{R}^-$ and separating the mushy phase from either liquid or solid. The domain $\Omega^{(m+1)}(t) = \{x \; : \; r_m(t) < x < r_{m+1}(t)\}$ is occupied by the mushy phase; its measure decreases in time. If, for some $t = T_1$, $r_m(T_1) = r_{m+1}(T_1)$, then either one Problem (B) less (the number of connected components of the mushy phase is larger than one) or Problem (C) (the mushy phase has entirely disappeared) is to be solved.

The generalized solution of problem (4.1) - (4.3) can be extended onto the whole half-line $(0, \infty)$. This half-line can be partitioned into a finite number of intervals (T_{k-1}, T_k), $T_0 < T_1 < \ldots < T_p < T_{p+1}$, such that on each of them the solution represents a composition of the solutions of Problems (A), (B) and (C), preserving on (T_{k-1}, T_k) the structure imposed at $t = T_{k-1}$. If, at any $t = T_n$, the mushy phase disappears, then the solution of problem (4.1) - (4.3) coincides for $t > T_n$ with the solution of Problem (C).

Clearly, the number of connected components of each phase may only decrease in time.

We shall now prove Theorem 4 for the initial distribution of the specific internal energy specified above. For definiteness, let $\vartheta^-(t) < 0$, $\vartheta^+(t) > 0$ and $T^- < \infty$.

Suppose the converse is true. For $t > T^-$, let there exist a domain $\Omega^{(k)}(t) = \{x \; : \; r_{k-1}(t) < x < r_k(t)\}$, $-1 < r_{k-1}(t) < r_k(t) < 1$, occupied by the liquid or mushy phase, and let the boundary $x = r_k(t)$ separate the phases. If this domain is occupied by the liquid, then there is at least one domain $\Omega^{(k+1)}(t)$, occupied by the solid or mushy phase and separating the above component of the liquid phase from the component of the liquid adjacent to the boundary $x = 1$.

There is a continuous curve Γ_T in the domain Ω_T, $T > T^-$, containing the point $(r_k(T), T)$ and separating different phases so that the liquid remains always to its left. Such a curve may be non-unique, but it does not matter which of them is chosen in our construction.

It follows from an analysis of the structure of solutions to Problems (A), (B) and (C) that the curve Γ_T admits definition by the function $r_k(t) \in W_4^1(0, T) : \; \Gamma_T = \{(x, t) \; : \; x = r_k(t), \; t \in (0, T)\}$. Hence, the Gauss formula holds for the domain

$\{(x,t)\ :\ -1 < x < r_k(t),\ t \in (0,T)\}$ and equation (4.1) can transformed to the equivalent integral identity

$$\int_0^T \int_0^{r_k(t)} \left\{ \frac{\partial \varphi}{\partial x} \frac{\partial \vartheta}{\partial x} - U \frac{\partial \varphi}{\partial t} \right\} dx dt$$

$$+ \int_{-1}^{r_k(t)} \varphi(x,t) U(x,t)\, dx \Big|_{t=0}^{t=T} - \int_0^T \varphi(x,t) \frac{\partial \vartheta}{\partial x}(x,t)\, dt \Big|_{x=-1}^{x=r_k(t)-0} = 0,$$

for any smooth function φ. In deriving the above identity we have used the equality $U(r_k(t) - 0, t) = 0$ which follows from the fact that the liquid phase remains all the time to the left of $x = r_k(t)$.

Let us take $\varphi = 1 + x$ in the latter identity, to obtain

$$\mathcal{J}_1(0) - \mathcal{J}_1(T) + \mathcal{J}_2(T) = \int_0^T \left| \vartheta^-(t) \right| dt, \tag{4.13}$$

where

$$\mathcal{J}_1(t) = \int_{-1}^{r_k(t)} (1+x) U(x,t)\, dx,$$

$$\mathcal{J}_2(t) = \int_0^T (1 + r_k(t)) \frac{\partial \vartheta}{\partial x} (r_k(t) - 0, t)\, dt.$$

Let us estimate the left- and right-hand sides of (4.13). By the hypothesis of the theorem and due to the assumption $T > T^- \geq T_*$, the right-hand side of (4.13) is larger than 4Λ. On the other hand, the derivative of ϑ in the term $\mathcal{J}_2(T)$ of (4.13) is non-positive on the boundary Γ_T, because the temperature is positive to the left of Γ_T and vanishes on this boundary. Hence,

$$\mathcal{J}_1(0) - \mathcal{J}_1(T) > 4\Lambda. \tag{4.14}$$

It follows easily that the maximum principle holds for function $U(x,t)$:

$$|U(x,t)| \leq \Lambda.$$

Therefore,

$$\left| \mathcal{J}_1(t) \right| \leq \Lambda \int_{-1}^{r_k(t)} (1+x)\, dx \leq 2\Lambda.$$

The above relations imply the upper bound for the left-hand side of inequality (4.14):

$$4\Lambda < \left| \mathcal{J}_1(0) \right| + \left| \mathcal{J}_1(T) \right| \leq 4\Lambda.$$

This contradiction shows that no liquid phase may exist in $\Omega^{(k)}(t)$ for $t > T_-$.

Suppose now that for $t > T^-$ the domain $\Omega^{(k)}(t)$ is occupied by the mushy phase. There will then exist a unique curve $\Gamma_T = \{(x,t)\ :\ x = r_k(t),\ t \in (0,T)\}$, including the point $(r_k(T), T)$ and separating different phases, such that the mushy phase remains to its left throughout. The function $r_k(t)$ is continuously differentiable there and decreasing for $t \in [0,T]$.

Taking the test function $\varphi(x) = 1 + x$ in the integral identity for the solution $\{U, \vartheta\}$ of problem (4.1) - (4.3), as previously we get the equality

$$\mathcal{J}_1(0) - \mathcal{J}_1(T) - \mathcal{J}_3(T) = \int_0^T |\vartheta^-(t)| \, dt, \tag{4.15}$$

where $\mathcal{J}_3(T) = \displaystyle\int_{r_k(T)}^{r_k(0)} (1 + x) U_0(x) \, dx.$

Because

$$\left| \mathcal{J}_1(0) - \mathcal{J}_3(T) \right| = \left| \int_{-1}^{r_k(T)} (1 + x) U_0(x) \, dx \right| \le 2\Lambda,$$

by estimating the right-hand side of (4.15) from below and its left-hand side from above, we arrive at a contradiction.

The remaining cases can be treated in a similar way. So, for $t > T_*$, there exist either single connected components of the liquid and solid phases in Ω, separated by a smooth boundary $x = r(t)$ or the whole of Ω is occupied by one phase, liquid or solid.

Let us now pass to the proof of Theorem 4 in a general situation. Since $\vartheta_0 \in W_2^1(\Omega)$, the function $\vartheta_0(x)$ is Hölder continuous in Ω, hence it is continuous everywhere. Thus, the set $\Omega^+(0) = \{x : \vartheta_0(x) > 0\}$ is the sum of at most a countable number of open non-intersecting intervals Ω_m^+ of length a_m. We shall order these intervals so that $a_{m+1} \le a_m$.

Analogously, the set $\Omega^-(0) = \{x : \vartheta_0(x) < 0\}$ is the sum of at most a countable number of open non-intersecting intervals Ω_k^- of length b_k. As above, we shall assume that $b_{k+1} \le b_k$.

The series $\displaystyle\sum_{m=1}^{\infty} a_m$ and $\displaystyle\sum_{k=1}^{\infty} b_k$ are convergent, hence for each positive integer n there exists N such that

$$\sum_{m=N+1}^{\infty} a_m + \sum_{k=N+1}^{\infty} b_k \le \frac{1}{n}.$$

We shall define the approximate solutions $\{U^{(n)}, \vartheta^{(n)}\}$ of problem (4.1) - (4.3) as the solution of the same problem subject to the initial distribution of the specific internal energy $U_0^{(n)}(x)$.

The function $U_0^{(n)}(x)$ will be defined as follows. On the intervals Ω_m^+ and Ω_k^-, $m, k = 1, \ldots, N$, we take $U_0^{(n)}(x) = U_0(x)$. The appropriate complements $\Omega_{(N)}^*$ in Ω represent the sum of a finite number of closed intervals. We put there

$$\tilde{U}_0(x) = \begin{cases} U_0(x), & \text{if } -1 + 2/n \le U_0(x) \le -2/n, \\ -1 + 2/n, & \text{if } U_0(x) < -1 + 2/n, \\ -2/n, & \text{if } U_0 > -2/n. \end{cases}$$

We define $U_0^{(n)}(x)$ as a continuously differentiable function such that

$$\|U_0^{(n)}(x) - \tilde{U}_0(x)\|_{L_2(\Omega_{(N)}^*)} \leq \frac{1}{n},$$

with values in the interval $(-1 + 1/n, -1/n)$.

Clearly,

$$\lim_{n \to \infty} \|\vartheta_0 - \vartheta_0^{(n)}\|_{2,\Omega}^{(1)} = \lim_{n \to \infty} \|U_0 - U_0^{(n)}\|_{2,\Omega} = 0,$$

$$\vartheta_0^{(n)} = \chi(U_0^{(n)}),$$

and the number T_*, corresponding to the functions $U_0^{(n)}$, is the same as that taken for the function U_0.

As we have already proved, each approximate solution of problem (4.1) - (4.3), corresponding to the initial specific internal energy $U_0^{(n)}(x)$, is for $t > T_*$ the classical solution either of a single-front two-phase Stefan problem or of the heat equation.

As in Theorem 1, Section I.3, the family $\{\vartheta^{(n)}\}$ can be proved to be uniformly bounded in the norm of the space $W_2^{1,1}(\Omega_T) \cap H^{\gamma, \gamma/2}(\bar{\Omega}_T)$, $T > T_*$, and thus to contain a subsequence $\{\vartheta^{(n_k)}\}$, convergent together with $\{U^{(n_k)}\}$, to the solution $\{\vartheta, U\}$ of Stefan problem (4.1) - (4.3), which corresponds to the initial specific internal energy $U_0(x)$. Up to renumbering, we can assume that the above convergence takes place for the entire sequences $\{\vartheta^{(n)}\}$ and $\{U^{(n)}\}$.

Let $T > T_*$ and, for definiteness, $\vartheta^-(t) < 0$, $\vartheta^+(t) > 0$. By the results of Chapter V, the boundary $x = r^{(n)}(t)$ separating the liquid and solid phases is continuously differentiable on the closed interval $[T_*, T]$, and the function $\vartheta^{(n)}(x, T_*)$ is continuously differentiable on each of the closed intervals $[-1, r^{(n)}(T_*)]$ and $[r^{(n)}(T_*), 1]$. By definition, the specific internal energy is smoothly dependent on the temperature on each of the above intervals, hence is continuously differentiable there. The constant estimating the modulus of the derivative of the specific internal energy on these intervals is independent of n.

Without loss of generality, we can assume that the sequences of numbers $\{r^{(n)}(T_*)\}$ converge to $r(T_*)$, the sequence $\{\tilde{U}_S^{(n)}(y, T_*)\}$, given by

$$\tilde{U}_S^{(n)}(y, T_*) = U^{(n)}(x, T_*), \ y = \left[(1 + r(T_*))x + r(T_*) - r^{(n)}(T_*)\right]\left[1 + r^{(n)}(T_*)\right]^{-1},$$

converges uniformly to $\tilde{U}_S(y, T_*)$ on the interval $[-1, r(T_*)]$, and the sequence $\{\tilde{U}_L^{(n)}(y, T_*)\}$, where

$$\tilde{U}_L^{(n)}(y, T_*) = U^{(n)}(x, T_*), \ y = \left[(1 - r(T_*))x - r^{(n)}(T_*) - r(T_*)\right]\left[1 - r^{(n)}(T_*)\right]^{-1}$$

converges to $\tilde{U}_L(y, T_*)$ on the interval $[r(T_*), 1]$.

Clearly, $\tilde{U}_S$ and $\tilde{U}_L$ coincide with the function $U(x, T_*)$ for $x < r(T_*)$ and $x > r(T_*)$, respectively, where $U(x, t)$ is the solution of problem (4.1) - (4.3), corresponding to the initial distribution $U_0(x)$ of the specific internal energy.

Because $U^{(n)}(x, T_*) \leq 1$ for $x \leq r^{(n)}(T_*)$ and $U^{(n)}(x, T_*) \geq 0$ for $x \geq r^{(n)}(T_*)$, we have $U(x, T_*) \leq -1$ for $x \leq r(T_*)$ and $U(x, T_*) \geq 0$ for $x \geq r(T_*)$.

For $t > T_*$, the only generalized solution of problem (4.1) - (4.3) for which the specific internal energy coincides with the function $U(x, T_*)$ at $t = T_*$, is the classical solution of the two-phase single-front Stefan problem. We have already constructed such a solution $\{U(x, t), \vartheta(x, t)\}$ as the limit of approximate solutions $\{U^{(n)}, \vartheta^{(n)}\}$. Therefore, this solution will remain classical also for $t > T_*$. The remaining cases can be treated analogously, hence the proof of Theorem 4 has been completed.   □

Remark. It is obvious that if the solution of the Stefan problem has structure (A), (B) or (C) at the initial time instant, then this property is preserved on the whole time interval $(0, \infty)$, even if the function $\vartheta^+(t)$ changes sign. The only requirement, as in Section V.5, is that $\vartheta^+(t)$ has a finite number of zeroes on each bounded interval (its zeroes can accumulate only at infinity). Then, each connected component of the mushy phase has the form of the set

$$\{(x, t) \in \Omega_\infty \, : \, x^- < x < x_\infty^-, \, 0 < t < h^-(x)\}$$
$$\cup \{(x, t) \in \Omega_\infty \, : \, x_\infty^+ < x < x^+, \, 0 < t < h^+(x)\}.$$

The continuous functions $h^-(x)$ and $h^+(x)$ are increasing or decreasing, respectively, and $x_\infty^- \leq x_\infty^+$.

Hence, the intersections of each connected component of the mushy phase with the line $\{t = \text{const.}\}$ strictly decrease in time.

Chapter VII

Time-periodic solutions of the
one-dimensional Stefan problem

Let us consider a two-phase Stefan problem formulated on the infinite strip $\Omega_\infty = \Omega \times (-\infty, \infty)$, $\Omega = \{x : |x| < 1\}$, in the case where the temperature $\vartheta(x, t)$ is a given periodic function with period $T > 0$ on the boundaries $x = \pm 1$ of Ω_∞. By the general theory, the problem admits solutions in the case when the initial condition is replaced by the condition of periodicity

$$U(x, t) = U(x, t + T) \qquad \text{for} \quad (x, t) \in \Omega_\infty.$$

The periodic solution has no physical realization, nevertheless it represents an asymptotic solution to which the solution of the Stefan problem with arbitrary initial specific internal energy and with periodic boundary temperature converges as $t \to \infty$.

The following is an outline of the analysis of the periodic solution. One proves theorems on the existence and uniqueness of the generalized time-periodic solution of the Stefan problem (see Section 1). In Section 2, for the strictly negative function $\vartheta^-(t)$ and the strictly positive function $\vartheta^+(t)$, the structure of the solution obtained is explored with the use of the Kronrod-Landis theorem and the domain Ω_∞ is shown to decompose into three subdomains: $\Omega_\infty^- = \{(x, t) : -1 < x < R^-(t), -\infty < t < \infty\}$, where the solution ϑ is strictly negative, $\Omega_\infty^* = \{(x, t) : R^-(t) < x < R^+(t), -\infty < t < \infty\}$ where the solution vanishes, and Ω_∞^+ where it is strictly positive. By the results of Chapter VI and due to the periodicity of the determined solution it can be proved that the set Ω_∞^* is empty, i.e., $R^-(t) = R^+(t)$ for all $t \in (-\infty, \infty)$. In the last section we shall consider the case when $\vartheta^+(t)$ has variable sign on the interval $(0, T)$. It can be proved that the solution will be classical.

1. Construction of the generalized solution

Let us denote $\Omega_T = \Omega \times (-T, T)$ and let $\tilde{W}_2^{1,1}(\Omega_T)$ be the set of functions which belong to $W_2^{1,1}(\Omega_T)$ and are time-periodic with period T. In a similar way, we also define the spaces $\tilde{L}_q(\Omega_T)$, $\tilde{W}_q^{1,0}(\Omega_T)$, $\tilde{W}_p^s[-T, T]$, $s \geq 0$, $p \geq 0$, $q > 0$.

Definition 1. The measurable functions $U \in \tilde{L}_\infty(\Omega_T)$ and $\vartheta \in \tilde{W}_2^{1,0}(\Omega_T)$, satisfying the constitutive relations $\vartheta = \chi(U)$, $U = \Phi(\vartheta)$ and the boundary conditions

$$\vartheta(-1,t) = \vartheta^-(t), \quad \vartheta(1,t) = \vartheta^+(t) \qquad \text{for} \quad t \in (-\infty,\infty), \tag{1.1}$$

with T-periodic in time functions $\vartheta^-(t)$ and $\vartheta^+(t)$, will be referred to as a *time-periodic generalized solution of the Stefan problem,* if

$$\iint_{\Omega_T} \left\{ \frac{\partial \vartheta}{\partial x} \frac{\partial \varphi}{\partial x} - U \frac{\partial \varphi}{\partial t} \right\} dx\, dt = 0 \tag{1.2}$$

is satisfied for all test functions $\varphi \in \tilde{W}_2^{1,1}(\Omega_T)$, vanishing on the boundary $\{x = \pm 1\}$.

Theorem 1. *Let $\Phi \in C^3(-\infty,0] \cap C^3[0,\infty)$, $\Phi'(s) = a(s) \geq a_0 = \text{const.} > 0$, ϑ^-, $\vartheta^+ \in \tilde{W}_1^2[-T,T]$.*
   Then there exists at least one time-periodic generalized solution of the Stefan problem, with

$$U \in \tilde{L}_\infty(\Omega_T), \qquad \vartheta \in \tilde{W}_2^{1,1}(\Omega_T).$$

Proof. Let us consider a sequence $\{\Phi_n(s)\}$ of functions which are three times continuously differentiable, converge pointwise to the function $\Phi(s)$ at $s \neq 0$ and are such that $\Phi'_n(s) = a_n(s) \geq a_0$ for $|s| < \infty$, $\Phi_n(s) = \Phi(s)$ for $|s| > 1/n$.
   By repeating the above procedure, we shall construct the generalized solution of the Stefan problem as a limit as $n \to \infty$ of time-periodic solutions ϑ_n of the equation

$$a_n(\vartheta_n)\frac{\partial \vartheta_n}{\partial t} = \frac{\partial^2 \vartheta_n}{\partial x^2} \qquad \text{for} \quad (x,t) \in \Omega_\infty, \tag{1.3}$$

subject to the boundary conditions (1.1).
   The existence of solutions ϑ_n of problem (1.1), (1.3) follows from results due to Shmulev [210] on the time-periodic solutions of the first boundary value problem for quasilinear parabolic equations. Clearly, the functions ϑ_n and $U_n = \Phi_n(\vartheta_n)$ satisfy identity (1.2) for all n.
   We now derive estimates uniform in n, which justify passing to the limit in the appropriate integral identity.
   The estimate

$$|\vartheta_n|_{\Omega_T}^{(0)} \leq M_0 = \max\left(|\vartheta^-|_{[-T,T]}^{(0)}, |\vartheta^+|_{[-T,T]}^{(0)}\right) \tag{1.4}$$

follows from the maximum principle. The inequality

$$\left\|\frac{\partial \vartheta_n}{\partial x}\right\|_{2,\Omega_T}^2 \leq 4|U_n|_{\Omega_T}^{(0)} \max\|\vartheta^-, \vartheta^+\|_{1,[-T,T]}^{(1)} + 4TM_0^2 = M_2^2 \tag{1.5}$$

results from multiplying equation (1.3) by the difference $\vartheta_n - \tilde{\vartheta}$, where

$$\tilde{\vartheta}(x,t) = \vartheta^-(t) + \frac{x+1}{2}[\vartheta^+(t) - \vartheta^-(t)],$$

and by the corresponding partial integration in Ω_T.

According to the choice of Φ_n, the functions U_n are uniformly bounded:

$$|U_n|_{\Omega_T}^{(0)} \leq M_1,$$

hence the constant M_2 in (1.5) does not depend on n.

The basic estimate can be concluded as in Section I.3, after multiplication of (1.3) by the difference $(\partial \vartheta_n / \partial t - \partial \tilde{\vartheta} / \partial t)$ and the integration by parts in Ω_T :

$$a_0 \left\| \frac{\partial \vartheta_n}{\partial t} \right\|_{2,\Omega_T}^2 \leq M_1 \left\| \frac{\partial \tilde{\vartheta}}{\partial t} \right\|_{1,\Omega_T} + M_2 \left\| \frac{\partial^2 \tilde{\vartheta}}{\partial t \partial x} \right\|_{2,\Omega_T} = a_0 M_3. \qquad (1.6)$$

We shall prove that the sequence $\{\vartheta_n\}$ is uniformly bounded in the space $H^{\alpha,\alpha/2}(\overline{\Omega}_T)$:

$$|\vartheta_n|_{\Omega_T}^{(\alpha)} \leq M_4. \qquad (1.7)$$

To prove this, as shown in Section I.3, it is enough that, together with the bounds (1.4), (1.6),

$$\max_{t \in (-T,T)} \left\| \frac{\partial \vartheta_n}{\partial x}(t) \right\|_{2,\Omega} \leq M_5. \qquad (1.8)$$

The estimate (1.8) can be derived in the same way as (1.6), up to integrating the equality

$$\left(\frac{\partial U_n}{\partial t} - \frac{\partial^2 \vartheta_n}{\partial x^2} \right) \left(\frac{\partial \vartheta_n}{\partial t} - \frac{\partial \tilde{\vartheta}}{\partial t} \right) = 0$$

first over Ω and then in time over $(t_*^{(n)}, t)$, $t_*^{(n)} < t < t_*^{(n)} + T$, provided that

$$\left\| \frac{\partial \vartheta_n}{\partial x}(t_*^{(n)}) \right\|_{2,\Omega}^2 \leq M_6. \qquad (1.9)$$

The existence of such a value $t_*^{(n)} \in (-T, T)$ and the estimate (1.9) follow from (1.5) and the mean-value theorem:

$$\left\| \frac{\partial \vartheta_n}{\partial x}(t_*^{(n)}) \right\|_{2,\Omega}^2 = \frac{1}{2T} \int_{-T}^{T} \left\| \frac{\partial \vartheta_n}{\partial x}(t) \right\|_{2,\Omega}^2 \leq \frac{M_2^2}{2T}.$$

Taking the sum of all these estimates, we can conclude the existence of some subsequences of $\{\vartheta_n\}$ and $\{U_n\}$ (we use the same indices for them) such that $U_n(x,t)$ converge weakly in $\tilde{L}_2(\Omega_T)$ to a bounded measurable function $U \in \tilde{L}_\infty(\Omega_T)$; $\vartheta_n(x,t)$ converge strongly in $H^{\gamma,\gamma/2}(\overline{\Omega}_T)$ and weakly in $\tilde{W}_2^{1,1}(\Omega_T)$ to a certain $\vartheta \in H^{\gamma,\gamma/2}(\overline{\Omega}_T) \cap \tilde{W}_2^{1,1}(\Omega_T)$.

Passing to the limit in the integral identity (1.2) for functions (U_n, ϑ_n), we infer that the pair of functions (U, ϑ) satisfies identity (1.2), with $\vartheta = \chi(U)$ in Ω_T. $\square$

Theorem 2. *The bounded time-periodic generalized solution of the Stefan problem is uniquely defined.*

The proof differs from that of the uniqueness theorem formulated in Section I.3 for non-periodic solutions in the construction of approximate solutions to an adjoint problem (we refer to the appropriate results by Shmulev, [210]).

2. Structure of the mushy phase for temperature on the boundary of Ω_∞ with constant sign

In this section we shall consider the case $\vartheta^-(t) < 0$, $\vartheta^+(t) > 0$. Our main result will state that the mushy phase occupies the domain $\Omega_\infty^* = \{(x,t) \; : \; R^-(t) < x < R^+(t), -\infty < t < \infty\}$, where $-1 < R^-(t) \le R^+(t) < 1$, the solid phase is to the left of $x = R^-(t)$ and the liquid is to the right of $x = R^+(t)$.

Suppose we have already proved this property. Let us take $U_0(x) = U(x,0)$, where $U(x,t)$ is the time-periodic generalized solution of the Stefan problem.

We shall consider the Stefan problem

$$
\begin{cases}
\dfrac{\partial \tilde{U}}{\partial t} = \dfrac{\partial^2 \tilde{\vartheta}}{\partial x^2} & \text{for} \quad x \in \Omega, \quad t \in (0,\infty), \\[2mm]
\tilde{\vartheta}(\pm 1, t) = \vartheta^\pm(t) & \text{for} \quad t \in (0,\infty), \\[2mm]
\tilde{U}(x,0) = U_0(x) & \text{for} \quad x \in \Omega.
\end{cases}
\tag{2.1}
$$

On the one hand, because the generalized solution of the Stefan problem is unique (see Theorem 10, Section I.3), this solution coincides with the time-periodic generalized solution of the Stefan problem constructed in Section 1 of the present chapter.

On the other hand, by the results of Chapter VI we can show the existence of a time t_* such that for $t > t_*$ the solution of the Stefan problem will be classical regardless the behaviour of $U_0(x)$ on the interval $(R^-(0), R^+(0))$. At the same time, since the solution is T-periodic in t and the mushy phase existed at $t = 0$, it will exist also at $t = T, 2T, \ldots, kT$, where $kT > t_*$. The latter property contradicts the results we obtained earlier. Thus, the mushy phase does not exist at any time instant and the time-periodic generalized solution of the Stefan problem is classical everywhere: there exists a continuously differentiable T-periodic function $R(t)$ such that the domain $\Omega_\infty^- = \{(x,t) \; : \; -1 < x < R(t), |t| < \infty\}$ is occupied by the solid phase and $\Omega_\infty^+ = \{(x,t) \; : \; R(t) < x < 1, |t| < \infty\}$ by the liquid.

It is therefore enough to prove the following result.

Theorem 3. *Let $\vartheta^+(t) \ge 2\beta = \text{const.} > 0$, $\vartheta^-(t) \le -2\beta < 0$, for $t \in (-\infty, \infty)$ in the hypotheses of Theorem 1. Then there exist two T-periodic functions $R^-(t)$ and $R^+(t)$ such that the domain $\Omega_\infty^- = \{(x,t) \; : \; -1 < x < R^-(t), |t| < \infty\}$ is occupied by the solid phase, the domain $\Omega_\infty^* = \{(x,t) \; : \; R^-(t) < x < R^+(t), |t| < \infty\}$ by the mushy phase and $\Omega_\infty^+ = \{(x,t) \; : \; R^+(t) < x < 1, |t| < \infty\}$ by the liquid.*

In the proof of Theorem 3 we shall use the following result.

Theorem 4. (cf., [131], [139]) *For almost all ε, the level set $\{(x,t) \ : \ \vartheta(x,t) = \varepsilon\}$ of a function ϑ twice continuously differentiable in the rectangle Ω_T consists of a finite number of closed or open smooth rectifiable non-intersecting curves such that*

$$\left|\frac{\partial \vartheta}{\partial x}(x,t)\right|^2 + \left|\frac{\partial \vartheta}{\partial t}(x,t)\right|^2 > 0 \tag{2.2}$$

along them, only their end-points belong to the boundary $\partial\Omega_T$, and they separate the sets $\{(x,t) \ : \ \vartheta > \varepsilon\}$ and $\{(x,t) \ : \ \vartheta < \varepsilon\}$.

The above theorem is analogous to the famous Sard theorem [177] which states that the folding of a mapping $f : \mathbb{R}^n \to \mathbb{R}^m$ (i.e., of the set of points $a \in \mathbb{R}^m$ such that the differential of f at least in one point of the set $\{x \in \mathbb{R}^n \ : \ f(x) = a\}$ has rank less than $\min(m,n)$) has measure zero.

Proof of Theorem 3. Let $\Omega_\infty^- = \{(x,t) \in \Omega_\infty \ : \ \vartheta(x,t) < 0\}$. In an analogous way we define the domain Ω_∞^+. Because $\vartheta \in H^{\gamma,\gamma/2}(\bar\Omega_\infty)$, each of the domains Ω_∞^- and Ω_∞^+ contains a strip of width δ that is adjacent to the appropriate boundary. The constant δ is dependent on the Hölder norm of the function ϑ and on β given by

$$2\beta = \min\left\{ \min_{t\in(0,\infty)} |\vartheta^-(t)|, \ \min_{t\in(0,\infty)} \vartheta^+(t)\right\}.$$

At the interior points of the domains $\Omega_T^\pm = \Omega_\infty^\pm \cap \Omega_T$, the differential properties of the function ϑ are only dependent on those of Φ. Thus, ϑ is at least twice continuously differentiable in every closed subdomain of $\Omega_T^\pm$.

To apply Theorem 4, for each n let us consider an infinitely differentiable monotone function $\Psi(s)$, equal s if $s < -1/n$ and vanishing whenever $s > -1/2n$. The function $\Psi(\vartheta(x,t))$ meets all hypotheses of Theorem 4 and its level set $\{(x,t) \ : \ \Psi(\vartheta(x,t)) = -\varepsilon\}$, $\varepsilon > 0$, coincides with the level set $\{(x,t) \ : \ \vartheta(x,t) = -\varepsilon\}$, if $\varepsilon > 1/n$.

By Theorem 4, there exists $\varepsilon_n \in (1/n, 2/n)$ such that the corresponding level set $\{(x,t) \ : \ \vartheta(x,t) = -\varepsilon_n\}$ comprises a finite number of rectifiable curves for which inequality (2.2) holds. Without loss of generality, we can assume that the set $\{(x,t) \ : \ \vartheta(x,t) = \varepsilon_n\}$ in Ω_T^+ also comprises a finite number of rectifiable curves for which (2.2) holds.

We shall choose numbers n large enough to provide that $\varepsilon_n < \beta$. Then all open level sets cannot touch the boundary of Ω_∞. They can begin or end on the lines $\{t = -T\}$ and $\{t = T\}$.

Everywhere in the interior of Ω_T^-, the temperature ϑ satisfies the heat equation

$$a(\vartheta)\frac{\partial \vartheta}{\partial t} = \frac{\partial^2 \vartheta}{\partial x^2}, \qquad a(\vartheta) = \Phi'(\vartheta). \tag{2.3}$$

By the maximum principle ([139], p. 175), it then follows that the level set $L = \{(x,t) \ : \ \vartheta(x,t) = -\varepsilon_n\}$ cannot be closed, otherwise we would have $\vartheta(x,t) \equiv -\varepsilon_n$ everywhere in its interior, in contradiction with (2.2).

For the same reason, L cannot begin and end on the line $\{t = T\}$. Indeed, if L begins and ends on the line $\{t = T\}$, then the temperature ϑ would be constant everywhere inside the domain bounded by this line and by the curve L.

We shall prove that each level set $L = \{(x, t) : \vartheta(x, t) = -\varepsilon_n\}$ which begins on the line $\{t = -T\}$ and ends on the line $\{t = T\}$ represents a curve $x = R^-(t)$, $t \in (-T, T)$, with the function $R^- \in W_1^1(-T, T)$. It suffices to note that $\bar{t}(s)$ strictly increases in the parametrization L:

$$x = \bar{x}(s), \quad t = \bar{t}(s) \qquad \text{for} \quad s \in (0, S).$$

Because L is rectifiable, $R^- \in W_1^1(-T, T)$, with

$$\int_{-T}^{T} \left(1 + \left|\frac{dR^-(t)}{dt}\right|^2\right)^{1/2} dt = S < \infty.$$

Indeed, suppose the opposite. Then there exists $s_1 < s_2$ such that $\bar{t}(s_1) = \bar{t}(s_2)$. Two cases can occur:
(i) the curve $\ell = \{(x, t) \in L : s \in J = (s_1, s_2)\}$ includes points below the line $\{t = \bar{t}(s_1) = \text{const.}\}$, i.e., $\bar{t}(s_1) < \bar{t}(s_2)$ at least for some $s \in J$;
(ii) $\bar{t}(s) = \bar{t}(s_1)$ for all $s \in J$.

A third case is also possible, with the curve ℓ situated completely above the line $\{t = \bar{t}(s_1) = \text{const.}\}$. But because $\bar{t}(s_1) < T$ and the end-point of L belongs to the line $\{t = T\}$, there are two other points s_3 and s_4, $s_3 < s_2 < s_4$, such that $\bar{t}(s_3) = \bar{t}(s_4)$ and the curve $\{(x, t) \in L : s \in (s_3, s_4)\}$ includes points below the line $\{t = \bar{t}(s_3) = \text{const.}\}$ (in particular, the point $x = \bar{x}(s_2)$, $t = \bar{t}(s_2)$). Then, relabelling s_3 and s_4, we again come to the case (i).

Suppose the first case occurs. By the general theorem on the structure of open subsets of the line [130], we can assume that $\bar{t}(s) < \bar{t}(s_1) = \bar{t}(s_2)$ for all $s \in J$. Indeed, the set of points s of the line, where the continuous function $f(s) = \bar{t}(s_1) - \bar{t}(s)$ is positive, is open and consists of at most a countable number of open intervals. We can take J in the form of one of those intervals.

The curve ℓ on which $\vartheta = -\varepsilon_n$, which we have chosen, and the line $\{t = \bar{t}(s_1) = \text{const.}\}$ bound a domain G such that ϑ satisfies equation (2.3) in its interior. ϑ can thus attain its extremal values only on the lower "parabolic" boundary ℓ, where $\vartheta(x, t) = -\varepsilon_n$. Hence, $\vartheta(x, t)$ is identically constant everywhere in the domain G, but this is impossible since then all the first derivatives of ϑ would vanish everywhere in G, including ℓ.

In the second case,

$$\vartheta(x, t) = -\varepsilon_n, \quad \frac{\partial \vartheta}{\partial x}(x, t) = \frac{\partial^2 \vartheta}{\partial x^2}(x, t) = 0 \qquad \text{for} \quad (x, t) \in \ell.$$

Since ϑ satisfies the heat equation with strictly positive function $a(\vartheta)$, we can conclude that

$$\frac{\partial \vartheta}{\partial t}(x, t) = 0 \qquad \text{for} \quad (x, t) \in \ell.$$

The equalities we have obtained contradict (2.2).

Now let L be the level set that begins on the line $\{t = -T\}$ and ends on $\{t = T\}$, with a function $R^-(t)$ non-periodic in time. We shall consider the curve $L^{(1)}$ obtained by shifting by one period in time the part of L located in the strip $\{(x,t) : -T < t < 0\}$. One of the end-points of $L^{(1)}$ meets then the line $\{t = 0\}$ and the other belongs to $\{t = T\}$. Since $L^{(1)}$ is contained in the level set $\{(x,t) : \vartheta(x,t) = -\varepsilon_n\}$, by Theorem 4 it should be a part of the level curve $L^{(2)}$ whose end-points are on the boundary of Ω_T. Because one of the end-points of $L^{(2)}$ belongs to the line $\{t = T\}$ and both end-points cannot be on the same line, the only possibility is that the second end-point belongs to $\{t = -T\}$. This procedure admits infinite continuation which contradicts the finite number of components of the level set.

Let now L be the level set that begins and ends on the line $\{t = -T\}$. As in the previous case, we can show that L consists of two branches L_1 and L_2, each of them described by the equation $x = h_i(t)$ on the interval $(-T, t_*)$ and $h_1(t_*) = h_2(t_*)$. If $t_* < 0$, then the curve $L^{(1)}$, obtained from L by the shift of T, should be a part of the curve $L^{(2)}$ whose end-points are located on the line $\{t = -T\}$.

In the opposite case, the curve $L^{(2)}$ which begins on the line $\{t = -T\}$ and ends on $\{t = T\}$, would have local maxima and minima inside Ω_T and, consequently, would not admit description in the form $x = R^-(t)$. The latter property contradicts the former result.

The curve $L^{(2)}$, as L, comprises two branches $L_i^{(2)}$, $i = 1, 2$, each of them described by the equation $x = h_i^{(2)}(t)$, $-T \leq t \leq t_*^{(2)}$, where $t_*^{(2)}$, is already positive, unlike t_*. But then the curve $L^{(3)}$, obtained by the shift of T from the part of $L_1^{(2)}$ contained in the strip $\{(x,t) : -T < t < 0\}$, is a part of the non-periodic curve $L^{(4)}$ which begins on the line $\{t = -T\}$ and ends on $\{t = T\}$. As already shown, there exist no such level curves. Therefore, all level curves L admit on $(-T, T)$ the description $x = R_n^-(t)$ with the function $R_n^-(t)$ periodic in time.

Each level set actually comprises just one component. If there were two curves for which $\vartheta = -\varepsilon_n$, then the function ϑ would be constant everywhere in the domain Q bounded by these curves and the lines $\{t = -T\}$ and $\{t = T\}$. This behaviour is impossible as it would imply, due to the continuity of the first derivatives, that the equality

$$\left|\frac{\partial\vartheta}{\partial x}(x,t)\right|^2 + \left|\frac{\partial\vartheta}{\partial t}(x,t)\right|^2 = 0 \qquad \text{for} \quad (x,t) \in Q,$$

also holds on the level curves, contradicting (2.2).

It can be proved in exactly the same way that the level set $\{(x,t) : \vartheta(x,t) = \varepsilon_n\}$ is the curve $x = R_n^+(t)$, with a periodic function $R_n^+ \in \tilde{W}_1^1[-T, T]$.

We can always assume that the sequence $\{\varepsilon_n\}$ is strictly decreasing. Then the sequence of functions $\{R_n^-\}$ will be strictly increasing, while $\{R_n^+\}$ will be strictly decreasing. Let

$$R^\pm(t) = \lim_{n\to\infty} R_n^\pm(t) \qquad \text{for} \quad t \in (-T, T).$$

Obviously,

$$-1 < R^-(t) \leq R^+(t) < 1 \qquad \text{for} \quad t \in (-T, T).$$

By construction, to the left of the curve $x = R_n^-(t)$ the temperature is less than $-\varepsilon_n$, to the right of the curve $x = R_n^+(t)$ it is greater than ε_n. Since there exist no other level sets in Ω_∞, the temperature in the domain $\{(x,t) : R_n^-(t) < x < R_n^+(t), |t| < \infty\}$ has absolute value smaller than ε_n. Hence, the temperature vanishes everywhere in the domain $\Omega_\infty^* = \{(x,t) : R^-(t) < x < R^+(t), |t| < \infty\}$.

Therefore, the set of points where the time-periodic generalized solution ϑ of the Stefan problem is strictly negative, consists of one connected component Ω_∞^-, whose right boundary is the curve $x = R^-(t)$. Analogously, the set where ϑ is strictly positive consists of one connected component Ω_∞^+, whose left boundary is the curve $x = R^+(t)$.

It can be shown that the functions $R^-(t)$ and $R^+(t)$ are continuous, but, as follows from the considerations at the beginning of this section, what one actually needs to know is that there exist points $x_0^-, x_0^+, -1 < x_0^- \leq x_0^+ < 1$ such that $\vartheta(x,0) < 0$ for $x \in (-1, x_0^-)$, $\vartheta(x,0) = 0$ for $x \in (x_0^-, x_0^+)$ and $\vartheta(x,0) > 0$ for $x \in (x_0^+, 1)$. $\quad\square$

Remark. As in Chapter V, the regularity hypotheses for the functions $\vartheta^-(t)$ and $\vartheta^+(t)$ can be relaxed. It suffices to assume that $\vartheta^-(t), \vartheta^+(t) \in \tilde{W}_2^1[-T, T]$.

3. The case of $\vartheta^+(t)$ with variable sign

Let all hypotheses of Theorem 1 be fulfilled. Then, as proved in Section 1 of the present chapter, there exists a unique time-periodic generalized solution $\{U, \vartheta\}$ of the Stefan problem, satisfying boundary conditions (1.1) and the integral identity

$$\int_{\Omega_T} \int \left\{ \frac{\partial \vartheta}{\partial x} \frac{\partial \varphi}{\partial x} - U \frac{\partial \varphi}{\partial t} \right\} dx dt = 0$$

for arbitrary function $\varphi \in \tilde{W}_2^{1,1}(\Omega_T)$, vanishing on the boundary of Ω. In the above formulation, $U \in \tilde{L}_\infty(\Omega_T)$, $\vartheta \in \tilde{W}_2^{1,1}(\Omega_T) \cap L_\infty(0, T; W_2^1(\Omega)) \cap H^{\gamma,\gamma/2}(\bar\Omega_T)$.

In this section we shall consider the case where $\vartheta^-(t) < 0$ for all $t \in [0, T]$, and $\vartheta^+(t) > 0$ for $t \in (0, t_0)$ and $\vartheta^+(t) < 0$ for $t \in (t_0, T)$. As in Section 2, we shall prove that the generalized solution is also the classical solution, i.e., there exists a finite number of curves in Ω_T, separating the liquid and solid phases.

To this end, let us first analyze the structure of the sets $\{(x,t) : \vartheta(x,t) = \pm\varepsilon\}$, where $\varepsilon > 0$ is sufficiently small. As in Section 2, by Theorem 4 (due to Kronrod and Landis) there exists a positive number $\varepsilon < \beta$ such that each of the subsets $\{\vartheta(x,t) = \varepsilon\}$ and $\{\vartheta(x,t) = -\varepsilon\}$ of the domain Ω_T comprise a finite number of non-intersecting smooth rectifiable curves on which

$$\left| \frac{\partial \vartheta}{\partial x}(x,t) \right|^2 + \left| \frac{\partial \vartheta}{\partial t}(x,t) \right|^2 > 0.$$

These curves cannot be closed and they begin or end on the lines $\{t = \pm T\}$ or on the boundary $\{x = 1\}$. The number ε can be assumed small enough so as to ensure that the end-points of the above curves on the boundary $\{x = 1\}$ are located in a small neighbourhood of the points $t = -T, -T + t_0, 0, t_0, T$.

In contrast to the previous section, there now exist non-periodic curves which begin on the line $\{t = -T\}$ and end either on the line $\{t = T\}$ or on the boundary $\{x = 1\}$ or on the line $\{t = -T\}$. There also exist curves which begin on the boundary $\{x = 1\}$ and end on the lines $\{t = \pm T\}$ or on the boundary $\{x = 1\}$.

Each of these curves in the domain Ω_T is contained in a level curve in Ω_∞, which begins on the boundary $\{x = 1\}$ at a certain point t_1 and ends at some t_2 on the same boundary, where $|t_2 - t_1| < T$; all non-periodic level curves in Ω_∞ are obtained by shifting a certain curve by kT, $k = \pm 1, \pm 2, \ldots$

Indeed, let us consider the domain Ω_{2T}. Due to the time-periodicity of the solution $\vartheta(x, t)$, the structure of level sets in the strip $\{(x, t) \in \Omega_{2T} : 0 < t \leq 2T\}$ is the same as in the domain Ω_T. In particular, none of the level curves can have the end-point in the interior of the above strip. Expressed in an alternative way, if a non-periodic level curve L in Ω_T ends on the line $\{t = T\}$, then it represents a subset of some level curve $L^{(1)}$ in Ω_{2T} which may begin on the line $\{t = -2T\}$ or on the boundary $\{x = 1\}$ and end either on the lines $\{t = \pm 2T\}$ or on the boundary $\{x = 1\}$.

Suppose $L^{(1)}$ begins on the line $\{t = -2T\}$ and ends on the line $\{t = -(k+1)T\}$, or begins on the boundary $\{x = 1\}$ and ends in the domain Ω_{3T}. At a certain step, the procedure breaks off, i.e., L represents a part of some level curve $L^{(k)}$ in $\Omega_{(k+1)T}$, with starting point either on the line $\{t = -(k+1)T\}$ or on the boundary $\{x = 1\}$ and the end-point on the same line. Otherwise, there would exist at least one non-periodic level curve L^∞ in Ω_∞, whose end-point does not belong to the boundary of Ω_∞. Because the function $\vartheta(x, t)$ is non-periodic in time, there would then exist an even number of non-intersecting level curves in the domain Ω_T, leading to a contradiction with Theorem 4.

If the curve $L^{(k)}$ begins or ends on the line $\{t = -(k+1)T\}$, then by considering the expanding domains $\Omega_{(k+1)T} \subset \Omega_{(k+2)T} \subset \ldots$, we can conclude that $L^{(k)}$ is a subset of the level curve $L^{(m)}$ in $\Omega_{(m+1)T}$ which begins and ends on the boundary $\{x = 1\}$.

We now describe the above level curves in more detail. We shall consider only the "extreme" level curves, i.e., the curves whose end-points are located in a sufficiently small neighbourhood of the points $t = -T, +t_0 - T, 0, t_0$ and T on the boundary $\{x = 1\}$ and which are the most distant from these points. Three cases may occur.

In the first case, call it (a), the level curve $\{\vartheta(x, t) = \varepsilon\}$ in Ω_∞ begins at the point t_1^+ on the boundary $\{x = 1\}$, where $t_1^+ < t_0$ is close to t_0, and this curve ends at a certain t_2^+ on the same boundary, such that $t_2^+ > T$ and is close to T. The above curve L^+ together with the boundary $\{x = 1\}$ bound the domain Q^+. The level curve $\{\vartheta(x, t) = -\varepsilon\}$, which begins at some point t_1^- on the boundary $\{x = 1\}$, where $t_1^- > t_0$, and ends on the same boundary at t_2^-, $t_2^- < T$, is contained inside the domain Q^+. The curve L^- and boundary $\{x = 1\}$ bound the domain Q^-. The notion "extreme" applied to L^- means that everywhere in the interior of Q^- the temperature is strictly less than $-\varepsilon$.

All remaining non-periodic level curves can be constructed by an appropriate shift in time of kT, $k = \pm 1, \pm 2, \ldots$ The notion "extreme" applied to L^+ reflects the fact that everywhere outside the domain Q^+ and outside the domains obtained by its time-shifting by a multiple period, the temperature is strictly greater than ε. Because the temperature was strictly negative on the boundary $\{x = -1\}$, for the set $\{\vartheta(x, t) = \varepsilon\}$ exactly one periodic level curve L_0^+ should exist which separates the boundary $\{x = -1\}$ from these sets. In turn, because the temperature on the boundary $\{x = -1\}$ is strictly less than $-\varepsilon$, there exists exactly one level curve L_0^- of the set $\{\vartheta(x,t) = -\varepsilon\}$ which separates the line L_0^+ and the boundary $\{x = -1\}$. To the left of L_0^-, the temperature is strictly less than $-\varepsilon$, and $-\varepsilon < \vartheta(x,t) < \varepsilon$ for all points (x,t) situated between the lines L_0^- and L_0^+.

In the second case, call it (b), the line L^- of the set $\{\vartheta(x,t) = -\varepsilon\}$ in the domain Ω_∞ begins at a certain t_1^- on the boundary $\{x = 1\}$, with negative t_1^- close to 0, and it ends at some t_2^- on the same boundary, where $t_2^- > t_0$ and is close to t_0. The line L^- together with the boundary $\{x = 1\}$ bound the domain Q^-. Inside Q^-, there exists a level curve $\{\vartheta(x,t) = \varepsilon\}$ that begins at a certain t_1^+ on the boundary $\{x = 1\}$, where $t_1^+ > 0$, and ends at some t_2^+, $t_2^+ < t_0$, on the same boundary. The line L^+ together with the boundary $\{x = 1\}$ bound the domain Q^+.

All remaining non-periodic level curves in Ω_∞ can be constructed by time-shifting the curves $L^\pm$ by a multiple period. In the domain Q^+, the temperature is everywhere greater than ε, while outside Q^- and the domains obtained by time-shifting Q^- by a multiple period, the temperature is strictly less than $-\varepsilon$. Therefore, the sets $\{\vartheta(x,t) = \pm\varepsilon\}$ cannot have periodic level curves $L_0^\pm$.

In the third case, call it (c), the line L^+ is the same as in (b), whereas L^- is the same as in (a). In this case, the domains $Q^\pm$, bounded by the curves $L^\pm$ and the boundary $\{x = 1\}$, have no common points. Outside the domains $Q^\pm$ and their time-shifts by a multiple period, the temperature is between $-\varepsilon$ and ε. There exists then exactly one periodic curve L_0^- of the set $\{\vartheta(x,t) = -\varepsilon\}$ which separates the domains $Q^\pm$ and their shifted counterparts from the boundary $\{x = 1\}$.

In all situations different from (a) - (c) there would exist level curves intersecting each other, in contradiction with Theorem 4.

What is the structure of the curves $L_0^\pm$ and $L^\pm$? It follows from the results of the preceding section that $L_0^\pm$ is defined by the equation $x = R_0^\pm(t)$ with time-periodic continuous function $R_0^\pm(t)$.

It can be shown in exactly the same way that if $t_*^\pm = \max\{t : (x,t) \in L^\pm\}$, then $L^\pm$ consists of two curves. One of them begins at $t = t_1^\pm$ on the boundary $\{x = 1\}$ and is defined by the equality $x = R_1^\pm(t)$ on the interval $(t_1^\pm, t_*^\pm)$. The second curve begins at the point $t = t_2^\pm$ on the same boundary $\{x = 1\}$ and is defined by the equality $x = R_2^\pm(t)$ on the interval $(t_2^\pm, t_*^\pm)$. At the point $t = t_*^\pm$, $R_1^\pm = R_2^\pm$.

Assume now that a sequence $\{\varepsilon_n\}$ of positive real numbers has been chosen according to Theorem 4 and

$$\lim \varepsilon_n = 0 \quad \text{as} \quad n \to \infty.$$

It is not surprising that the structure of the level curves $\{(x,t_1) : \vartheta(x,t) = \pm\varepsilon_n\}$ for sufficiently large n can be described by either of the above cases (a), (b) or (c), and this property is preserved for further increase of n. We shall denote by $\Omega_{n,\infty}^*$ the subset of Ω_∞ where the temperature has absolute value less than ε_n. In the case (a), this will be the set Q_n^0 located between two periodic level curves $L_{0,n}^\pm = \{(x,t) : \vartheta(x,t) = \pm\varepsilon_n\}$, the set $Q_n^+ \setminus Q_n^-$ contained between the level curves $L_n^\pm = \{(x,t) : \vartheta(x,t) = \pm\varepsilon_n\}$

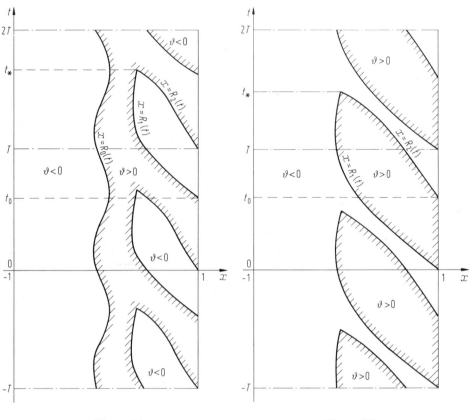

Figure 14   **Figure 15**

and the boundary $\{x = 1\}$, and its copies obtained by time-shifting by a multiple period. In the case (b), it will be the set $Q_n^- \setminus Q_n^+$ and its copies obtained by time-shifting of multiple period. And finally, in the case (c), the set $\Omega_{n,\infty}^*$ is the subset of Ω_∞, is located to the right of the unique time-periodic level curve $L_{0,n}^- = \{(x,t) \; : \; \vartheta(x,t) = -\varepsilon_n\}$ and it does not intersect either the sets Q_n^+, Q_n^- or their time-shifted copies.

In each of the cases (a) - (c), the set $\Omega_{n,\infty}^*$ intersects the line $\{t = 0\}$ on a finite number of connected intervals $\mathcal{J}_1^n, \ldots, \mathcal{J}_{k_n}^n$. Clearly, the number of these intervals depends on the particular case. For n increasing, the sets $\Omega_{n,\infty}^*$ define a decreasing sequence: $\Omega_{n,\infty}^* \supset \Omega_{n+1,\infty}^*$. The same is true for the sets $\mathcal{J}^n = \bigcup_{j=1}^{k_n} \mathcal{J}_j^n$, $\mathcal{J}^n \supset \mathcal{J}^{n+1}$. In this construction, the total number of intervals $\mathcal{J}_j^n$ may differ depending on the specific case, nevertheless it remains bounded from above by a finite constant independent of n. Indeed, let us fix some $n = n_0$ for the cases (a) and (b). Suppose the number of intervals $\mathcal{J}_j^{n_0}$ is bounded by k_0. Then it is easy to see that the total number of the intervals $\mathcal{J}_j^n$ is for $n > n_0$ bounded from above by $2k_0$ and does not depend on n. The case (c) is intermediate between (a) and (b), hence the above estimate of the total number of the intervals $\mathcal{J}_j^n$ holds also there. By construction, on the interval $\mathcal{J}_j^n$, $j = 1, \ldots, k_n$, $k_n \le k_0$, the absolute value of the temperature is below ε_n. In the limit, the temperature is equal to 0 everywhere on each of the intervals $\mathcal{J}_j = \bigcap_{n=1}^{\infty} \mathcal{J}_j^n$.

Therefore, at $t = 0$ there exists a finite number of connected components of the liquid, solid and mushy phases in the domain Ω. Let us take $U_0(x) = U(x,0)$ and consider the solution $\{\tilde{U}, \tilde{\vartheta}\}$ of the Stefan problem, where $\tilde{U}$ coincides with the function $U_0(x)$ at the initial time and $\tilde{\vartheta}$ coincides with the function $\vartheta^\pm(t)$ on the boundary $\{x = \pm 1\}$. It follows from the results of Section VI.4 that there exists a unique generalized solution of the same problem, with a finite number of connected components of the liquid, solid and mushy phases in the domain $\{(x,t) : x \in \Omega, t \in (0,T)\}$.

Referring to the uniqueness theorem, we conclude that the periodic solution $\{U, \vartheta\}$ coincides with $\{\tilde{U}, \tilde{\vartheta}\}$. Thus, the time-periodic generalized solution of the Stefan problem in Ω_T has a finite number of connected components of the liquid, solid and mushy phases.

Let $\mathcal{J}(t)$ be the section of any connected component of the mushy phase. We have proved in Chapter VI that the interval $\mathcal{J}(t)$ decreases as t grows: $\mathcal{J}(t) \supset \mathcal{J}(\tau)$, if $t < \tau$. On the other hand, due to the periodicity of the solution, $\mathcal{J}(t) = \mathcal{J}(t+T)$. In particular, $\mathcal{J}(0) = \mathcal{J}(T)$. The contradiction, at which we have arrived, shows that no mushy phase can exist for time-periodic generalized solutions of the Stefan problem and that the solution is in fact classical.

What is the structure of the constructed time-periodic classical solution of the Stefan problem? As above, three cases may occur.

In the case (a), there exists one connected component of the liquid phase, separated to the left from a connected component of the solid phase adjacent to the boundary $\{x = -1\}$ by the periodic curve $x = R_0(t)$. This component of the liquid is separated from the connected components of the solid phase adjacent to the boundary $\{x = 1\}$ by the continuous curves $x = R_1(t+kT)$ and $x = R_2(t+kT)$. The curve $x = R_1(t)$ begins there at the point $t = t_0$ on the boundary $\{x = 1\}$, the curve $x = R_2(t)$ begins on the same boundary $\{x = 1\}$ at the point $t = T$ and both these curves intersect at the point (x_*, t_*), where $x_* = R_i(t_*)$, $i = 1, 2$ (see Figure 14). The functions $R_i(t)$, $i = 1, 2$, are continuously differentiable for all $t \in (-\infty, \infty)$.

In the case (b), each connected component $\Omega_{(k)}^+$ of the liquid phase is adjacent to the boundary $\{x = 1\}$ and is separated from the unique connected component of the solid phase by two continuous curves $x = R_1(t+kT)$ and $x = R_2(t+kT)$. The curve $x = R_1(t)$ begins at the point $t = 0$ on the boundary $\{x = 1\}$ while the curve $x = R_2(t)$ begins at the point $t = t_0$ of the same boundary; both these curves intersect at the point (x_*, t_*), where $x_* = R_i(t_*), i = 1, 2$ (see Figure 15). The functions $R_i(t)$, $i = 1, 2$, are continuously differentiable everywhere except for the points $0, t_*$ and t_0, t_*, respectively.

In the case (c) (see Figure 16), there exists one time-periodic curve $x = R_0(t)$, continuously differentiable everywhere except for the points $t = t_* + kT$, $k = 0, \pm 1, \pm 2, \ldots,$, and separating the component of solid adjacent to the boundary $\{x = -1\}$ from the components of liquid and solid that are next to the boundary $\{x = 1\}$. What is the structure of the component adjacent to the latter boundary ? The connected liquid component, adjacent to the segment $(0, t_0)$ on the boundary $\{x = 1\}$, is bounded by this boundary, the curve $x = R_0(t)$ and two curves $x = R_i(t)$, $i = 1, 2$, defined over time intervals $(0, t_* - T)$ and (t_0, t_*), respectively. The connected solid component, adjacent to the boundary $\{x = 1\}$ along the segment (t_0, T), is bounded by this boundary, the curve $x = R_2(t)$, where $t \in (t_0, t_*)$, and the line $x = R_1(t - T)$ for $t \in (T, t_*)$. The functions $R_i(t), i = 1, 2$, are continuously differentiable everywhere on their domains except for the points $0, t_* - T$ and t_0, t_*, respectively. All remaining components of the liquid and solid phases, adjacent to the boundary $\{x = 1\}$, can be obtained there by time-shifting by an appropriate multiple of the period.

It turns out that if the integral

$$P = \int_0^T \left\{ \vartheta^+(t) - \vartheta^-(t) \right\} dt \tag{3.1}$$

is different from zero, then it is possible to estimate the size of the connected component adjacent to the boundary $\{x = 1\}$. To this end, we need some detailed insight into the differential properties of the solutions $\vartheta(x, t)$.

We show at first that in all three cases the derivative $\frac{\partial \vartheta}{\partial x}$ is summable at $x = \pm 1$ on the interval $(0, T)$. For definiteness, let us consider the case (b). The domain Q^+, bounded by the curve $x = R_1(t)$ and the boundary $\{x = 1\}$, contained in the strip $\{(x, t) : 0 < t < \delta_0\}$, is occupied by the liquid phase, and we have there

$$\frac{\partial}{\partial t} \Phi(\vartheta) = \frac{\partial^2 \vartheta}{\partial x^2}, \tag{3.2}$$

$$\left\{ \left\| \frac{\partial \vartheta}{\partial t}, \frac{\partial^2 \vartheta}{\partial x^2} \right\|_{2, Q^+}, \max_{t \in (0, \delta_0)} \left\| \frac{\partial \vartheta}{\partial x}(t) \right\|_{2, Q^+(t)} \right\} \leq M. \tag{3.3}$$

The constant δ_0 can be selected small enough to ensure that the entire domain $Q^- = \{(x, t) : R_1(2\delta_0) < x < R_1(t), \ 0 < t < \delta_0\}$ is occupied by the solid phase. The temperature is then strictly negative on the boundary $\{x = R_1(2\delta_0)\}$ and, thus, continuously differentiable:

$$\left| \frac{\partial \vartheta}{\partial x}(R_1(2\delta_0), t) \right| \leq M \qquad \text{for} \quad t \in (0, \delta_0). \tag{3.4}$$

In Q^-, the temperature satisfies equation (3.2) and estimates analogous to (3.3). Obviously, the Stefan condition

$$\frac{\partial \vartheta}{\partial x}(R_1(t) - 0, t) - \frac{\partial \vartheta}{\partial x}(R_1(t) + 0, t) = \frac{dR_1(t)}{dt} \tag{3.5}$$

holds on the separating boundary $x = R_1(t)$.

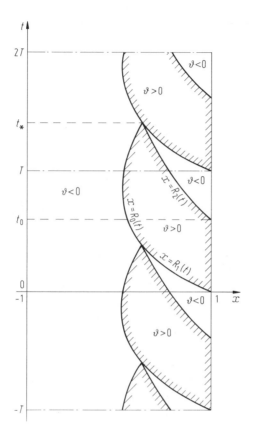

Figure 16

The function $\vartheta(x,t)$ attains its minimal value in Q^+ on the boundary $x = R_1(t)$. Hence,

$$\frac{\partial \vartheta}{\partial x}(R_1(t) + 0, t) \geq 0 \qquad \text{for} \quad t \in (0, \delta_0).$$

The latter inequality, together with the representation

$$\frac{\partial \vartheta}{\partial x}(1,t) = \frac{\partial \vartheta}{\partial x}(R_1(t) + 0, t) + \int_{R_1(t)}^{1} \frac{\partial^2 \vartheta}{\partial x^2}(x,t)\, dx, \tag{3.6}$$

the Stefan condition (3.5) and the bounds (3.3), (3.4), gives us the estimate

$$\int_0^{\delta_0} \left| \frac{\partial \vartheta}{\partial x}(1,t) \right| dt \leq \int_0^{\delta_0} \frac{\partial \vartheta}{\partial x}(R_1(t) + 0, t)\, dt + \iint_{Q^+} \left| \frac{\partial^2 \vartheta}{\partial x^2} \right| dx dt$$

$$\leq -\int_0^{\delta_0} \frac{dR_1(t)}{dt}\, dt + \int_0^{\delta_0} \left| \frac{\partial \vartheta}{\partial x}(R_1(t) - 0, t) \right| dt + M \left| \text{meas}(Q^+) \right|^{1/2}.$$

A representation similar to (3.6) is true in Q^-, hence it follows that

$$\int_0^{\delta_0} \left| \frac{\partial \vartheta}{\partial x}(R_1(t) - 0, t) \right| dt \leq M + M \left| \text{meas } (Q^-) \right|^{1/2}. \tag{3.7}$$

Taking the sum of these two inequalities, we eventually obtain the estimate

$$\int_0^{\delta_0} \left| \frac{\partial \vartheta}{\partial x}(1, t) \right| dt \leq 1 - R_1(\delta_0) + M \left\{ 1 + \left| \text{meas } (Q^+) \right|^{1/2} + \left| \text{meas } (Q^-) \right|^{1/2} \right\}.$$

To estimate the derivative $\frac{\partial \vartheta}{\partial x}(1, t)$ on the interval (δ_0, t), we shall use the representation

$$\left| \frac{\partial \vartheta}{\partial x}(1, t) \right|^2 = \left| \frac{\partial \vartheta}{\partial x}(x_0, t) \right|^2 + 2 \int_{x_0}^1 \frac{\partial \vartheta}{\partial x}(x, t) \frac{\partial^2 \vartheta}{\partial x^2}(x, t) \, dx. \tag{3.8}$$

If $R_1(t_0) < 1$, then the desired estimate follows by integrating (3.8) with respect to x_0 on the interval $(R_1(t), 1)$ and with respect to t on (δ_0, t_0).

If $R_1(t_0) = 1$, then let us integrate (3.8) with respect to x_0 from $R_1(t)$ to 1 and with respect to t from t_0 to $t_0 - \delta_1$, where δ_1 is a sufficiently small positive number. Because the function $(1 - R_1(t))$ is strictly positive on the interval $(\delta_0, t_0 - \delta_1)$, the derivative $\frac{\partial \vartheta}{\partial x}(1, t)$ is summable on this interval. The summability of the latter derivative on $(t_0 - \delta_1, t_0)$ can be shown in the same way as for the interval $(0, \delta_0)$.

The interval (t_0, T) and the cases (a), (c) admit a similar treatment. The temperature is strictly negative on the boundary $\{x = -1\}$, hence the derivative $\frac{\partial \vartheta}{\partial x}(-1, t)$ belongs to $L_2(0, T)$ by local estimates of the solutions of linear parabolic equations.

The derivatives $\frac{\partial \vartheta}{\partial x}(\pm 1, t)$ are defined and summable on $(0, T)$, hence, the test functions in the definition of the generalized solution can be permitted to be different from zero on the boundary of Ω. To prove this, it is sufficient to take the product $\varphi(x, t) h_\varepsilon(x)$, with an arbitrary $\varphi(x, t) \in \tilde{W}_2^{1,1}(\Omega_T) \cap C(\bar{\Omega}_T)$, where $h_\varepsilon(x)$ vanishes at $x = \pm 1$, is equal to 1 for $|x| < 1 - \varepsilon$, is continuous in $\bar{\Omega}$ and linear with respect to x outside the interval $[-1 + \varepsilon, 1 - \varepsilon]$. The appropriate integral identity will then be obtained after passing to the limit in (2.2) as $\varepsilon \to 0$ and applying the mean-value theorem:

$$\iint_{\Omega_T} \left\{ \frac{\partial \vartheta}{\partial x} \frac{\partial \varphi}{\partial x} - U \frac{\partial \varphi}{\partial t} \right\} dx dt = \int_{-T}^{T} \left[\varphi \frac{\partial \vartheta}{\partial x} \right]_{x=-1}^{x=1} dt. \tag{3.9}$$

By substituting $\varphi = 1 + x$ and $\varphi = 1 - x$ into (3.9), we get the equality

$$\frac{1}{2} P = \int_0^T \frac{\partial \vartheta}{\partial x}(-1, t) \, dt = \int_0^T \frac{\partial \vartheta}{\partial x}(1, t) \, dt. \tag{3.10}$$

The integral identity (3.9) can be extended in an obvious way onto the domain $\Omega \times (0, \tau)$, where τ is an arbitrary positive number, and to the case of non-periodic test functions. For this, it suffices to take as a test function $\psi(x, t) = \varphi(x, t) h_\varepsilon(t)$, with an arbitrary $\varphi(x, t) \in W_2^{1,1}(\Omega_\tau) \cap C(\bar{\Omega}_\tau)$. Here $h_\varepsilon(t)$ is a continuous function, vanishing at $t < 0$ and $t > \tau$, equal to 1 on the interval $[\varepsilon, \tau - \varepsilon]$ and linear in t elsewhere.

Substituting ψ into (3.9) and passing to the limit as $\varepsilon \to 0$, we obtain

$$\int_0^\tau \int_\Omega \left\{ \frac{\partial \vartheta}{\partial x} \frac{\partial \varphi}{\partial x} - U \frac{\partial \varphi}{\partial t} \right\} dx dt + \int_\Omega U(x,t) \varphi(x,t) \, dx \Big|_{t=0}^{t=\tau}$$

$$= \int_0^\tau \left[\varphi(x,t) \frac{\partial \vartheta}{\partial x}(x,t) \right]_{x=-1}^{x=1} dt. \tag{3.11}$$

Let us consider the case (b) and suppose $P < 0$. If $x = R_1(t)$ is the left boundary of the connected liquid component adjacent to the segment $(0, t_0)$ on the boundary $\{x = 1\}$, then, due to condition (3.5) and estimate (3.7), the derivative $\frac{\partial \vartheta}{\partial x}(R_1(t) + 0, t)$ is strictly positive on the interval $(0, t_*)$ and summable there.

Assume $(k-1)T < t_* \leq kT$. Let us take $\tau = (k-1)T$ in the integral identity (3.11), and the test function $\varphi_\varepsilon(x,t)$ at any fixed $t \in (0, \tau)$ equal to 0 for $x < R_1(t)$, equal to 1 for $x > R_1(t) + \varepsilon$, continuous in Ω and linear with respect to x in the open interval $(R_1(t), R_1(t) + \varepsilon)$. By passing to the limit as $\varepsilon \to 0$, we get the identity

$$\int_{R_1(\tau)}^1 U(x,\tau) \, dx = \int_0^{(k-1)\tau} \frac{\partial \vartheta}{\partial x}(1,t) \, dt - \int_0^{(k-1)T} \frac{\partial \vartheta}{\partial x}(R_1(t) + 0, t) \, dt.$$

From the periodicity of the functions involved and by equality (3.10), the first term on the right-hand side of the latter relation is equal to $(k-1)\, P/2$ and the second term is negative. Thus,

$$\left| \int_{R_1(\tau)}^1 U(x,\tau) \, dx \right| \geq \frac{1}{2}(k-1)\, |P|.$$

On the other hand, $U(x,t)$ has its absolute value bounded from above by the quantity

$$M_0 = \max_{t \in (0,T)} \max \left| \Phi\big(\vartheta^-(t)\big), \Phi\big(\vartheta^+(t)\big) \right|, \tag{3.12}$$

and, finally, $2M_0 \geq (1/2)(k-1)|P|$. Correspondingly, $t_* < T(4M_0/|P| + 1)$.

If $P > 0$, then all the above arguments apply to the domain $\{(x,t) : -1 < x < R_2(t), t_0 < t < (k-1)T\}$. The cases (a) and (c) permit analogous treatment.

We are now ready to formulate the following summarizing result.

Theorem 5. *Assume that the hypotheses of Theorem 1 hold with $\vartheta^+(t) > 0$ for $t \in (0, t_0)$ and $\vartheta^+(t) < 0$ for $t \in (t_0, T)$. Then the time-periodic generalized solution of the Stefan problem is also classical and there are three different cases:*

(a) The unique connected component of the liquid phase is separated by a periodic curve $x = R_0(t)$, with T-periodic continuously differentiable function $R_0(t)$, from the connected component of the solid, on the left adjacent to the boundary $\{x = -1\}$. The above liquid component is separated by continuous curves $x = R_1(t + kT)$ and $x = R_2(t + kT)$, $k = 0, \pm 1, \pm 2, \ldots$, from the connected components $\Omega_{(k)}^-$, adjacent to the boundary $\{x = 1\}$. The function $R_1(t)$ is defined on the interval $[t_0, t_*]$ and function $R_2(t)$ on the interval $[T, t_*]$, $t_* \in [T, \infty)$. Moreover, $R_1(t_0) = 1$, $-1 < R_1(t) < 1$ for $t \in (t_0, t_*)$, $R_2(T) = 1$, $-1 < R_2(t) < 1$ for $t \in (T, t_*)$, $R_1(t) < R_2(t)$ for $t \in$

(T, t_*), $R_1(t_*) = R_2(t_*)$. *The functions* $R_i(t), i = 1, 2$, *are continuously differentiable on their domains except for the points* t_0, t_* *and* T, t_*, *respectively.*

If the quantity P, *given by (3.1), is different from zero, then*

$$t_* \leq T(4(M_0 + 1)/|P| + 2),$$

with M_0 *defined in (3.12).*

(b) The unique connected component of the solid phase is separated by two continuous curves, $x = R_1(t + kT)$, $x = R_2(t + kT)$, $k = 0, \pm 1, \pm 2, \ldots$, *from connected components* $\Omega_{(k)}^+$ *of the liquid, adjacent to the boundary* $\{x = 1\}$. *The function* $R_1(t)$ *is defined on the interval* $[0, t_*]$ *whereas* $R_2(t)$ *is defined on* $[t_0, t_*]$, $t_* \in [t_0, \infty)$.

Moreover, $R_1(0) = 1$, $-1 < R_1(t) < 1$ *for* $t \in (0, t_*)$, $R_2(t_0) = 1$, $-1 < R_2(t) < 1$ *for* $t \in (t_0, t_*)$, $R_1(t) < R_2(t)$ *for* $t \in (t_0, t_*)$ *and* $R_1(t_*) = R_2(t_*)$. *The functions* $R_i(t), i = 1, 2$, *are continuously differentiable on their domains except for the points* $0, t_*$ *and* t_0, t_*, *respectively.*

If $P \neq 0$, *then*

$$t_* \leq T(4M_0/|P| + 1).$$

(c) The connected component of the solid phase, adjacent to the boundary $\{x = -1\}$, *is separated from connected components of the liquid and solid phases, adjacent to the boundary* $\{x = 1\}$ *by a continuous* T-*periodic curve* $x = R_0(t)$. *The function* $R_0(t)$ *is continuously differentiable everywhere except for the points* $t = t_* + kT$, $k = 0, \pm 1, \pm 2, \ldots$. *The connected component of the liquid phase, adjacent to the segment* $(0, t_0)$ *on the boundary* $\{x = 1\}$, *is bounded by this boundary, the curve* $x = R_0(t)$ *and two curves* $x = R_i(t), i = 1, 2$, *defined for* t *in* $[0, t_* - T]$ *and* $[t_0, t_*]$, *respectively. The connected component of the solid, adjacent to the boundary* $\{x = 1\}$ *on the segment* (t_0, T), *is bounded by this boundary, the curve* $x = R_2(t)$, *where* $t \in [t_0, t_*]$, *and by the curve* $x = R_1(t - T)$ *for* $t \in [T, t_*]$. *Moreover,* $R_1(0) = 1$, $R_2(t_0) = 1$, $R_0(t_*) = R_1(t_* - T) = R_2(t_*)$. *The functions* $R_i(t), i = 1, 2$, *are continuously differentiable everywhere in their domains except for the points* $0, t_* - T$ *and* t_0, t_*, *respectively. All remaining components of the liquid and solid phases, adjacent to the boundary* $\{x = 1\}$, *can be constructed from those introduced above, by time-shifting by a multiple period.*

If $P \neq 0$, t_* *can be estimated as in the case (a).*

Remark. As in the preceding section, the regularity hypotheses assumed for $\vartheta^\pm(t)$ can be relaxed; what we actually need is the square integrability of their first derivatives.

Chapter VIII

Approximate approaches to the two-phase Stefan problem

1. Problem statement. Formulation of the results

In the present chapter, we study a few models for the crystallization processes of cylindrical ingots. First we consider a model of the crystallization or melting of a cylindrical ingot occupying a domain Ω in $\mathbb{R}^3$ that takes heat conduction in the surrounding medium Q into account (Problem (A_ε)). The second approximate model we study describes the same process, but subject to the hypothesis of infinite thermal conductivity of the ingot material in directions orthogonal to the ingot axis (Problem (A_0)). We also introduce an exact model that describes phase transitions in the same ingot Ω, this time without accounting for heat conduction in the environment Q, postulating instead that the heat flux across the common part Π of the boundaries of Q and Ω is proportional to the difference of the ingot temperature on Π and the temperature in Q, assumed to be known (Problem (B_ε)). Analogously to Problem (A_0), the corresponding approximate model (Problem (B_0)) is constructed in the form of a standard two-phase Stefan problem for non-homogeneous heat equation. In the latter case, only the quasi-steady situation (Problem (C)) is studied in detail.

The formulations treated in the present chapter differ from the usual approach where a sequence of auxiliary models which in some sense approximate the exact model is introduced. In contrast, here we have a sequence of exact models dependent on a parameter $\varepsilon > 0$, which in a certain sense are themselves approximated as $\varepsilon \to 0$ by an approximate model ($\varepsilon = 0$). A similar set-up is known in the theory of surface waves [173], where a sequence of exact equations is approximated by the shallow water equations.

We now formulate the problem. As before, let $\Phi(\vartheta)$ be a function discontinuous at $\vartheta = 0$, equal to $\vartheta - 1$ for $\vartheta < 0$, equal to ϑ for $\vartheta > 0$ and assuming values in the segment $[-1, 0]$ at $\vartheta = 0$. By Ω and Q denote cylindrical domains of a given height in $\mathbb{R}^3$, with common part Π of the boundaries and with axis parallel to the axis x_3 (see Figure 17). More precisely, the boundary of the intersection of Ω and the plane $\{x_3 = \text{const.}\}$ is a closed curve γ (intersection of the surface Π and the same plane $\{x_3 = \text{const.}\}$), which coincides with the inner boundary of the intersection of Q and the plane $\{x_3 = \text{const.}\}$, which in turn is a annular domain. Denote also $G = \Omega \cup \Pi \cup Q$, $F = \partial G$, and let ℓ be the projection of the domain G onto the x_3-axis. In Problem (A_0) $\ell = (-1, 1)$ while $\ell = \mathbb{R}$ in Problem (C).

Let $\eta(x)$ be the characteristic function of Ω, equal to 1 on Ω and 0 elsewhere. We shall assume that the specific internal energy of the ingot material coincides with the function $\Phi(\vartheta)$, where ϑ represents the temperature of the ingot, and the specific internal energy of

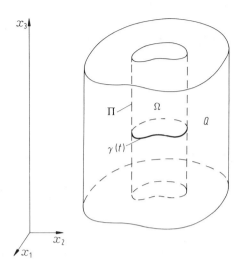

Figure 17

the surrounding medium coincides with its temperature. We thus assume that the melting temperature of the ingot is equal to 0 and no phase change may occur outside the ingot. From the above we infer that the specific internal energy of the medium occupying the domain G is

$$u \equiv \Phi_0(x, \vartheta) = \eta \Phi(\vartheta) + (1 - \eta)\vartheta. \tag{1.1}$$

Let $\vartheta = \chi(u)$ be the correspondence inverse to the relation $u = \Phi(\vartheta)$. Then

$$\vartheta = \eta \chi(u) + (1 - \eta)u \equiv \chi_0(x, u) \quad \text{in} \quad G. \tag{1.2}$$

We shall assume the medium occupying G to be isotropic, with the thermal conductivity normalized to 1. In contrast, the material of the ingot will be treated as anisotropic, with different thermal conductivities κ_i in various directions x_i:

$$\kappa_1 = \kappa_2 = 1/\varepsilon, \quad \kappa_3 = 1, \quad \varepsilon \ll 1.$$

The thermal conductivities in G are then

$$\kappa_1 = \kappa_2 = (1/\varepsilon)\eta + 1 - \eta, \quad \kappa_3 = 1. \tag{1.3}$$

Problem (A_ε). In the domain $G_T = G \times (0, T)$, determine a solution $u(x, t)$ of the equation

$$D_t u = \sum_{i=1}^{3} D_i(\kappa_i D_i \vartheta), \quad \vartheta = \chi_0(x, u) \tag{1.4}$$

in sense of distributions, satisfying the condition

$$\frac{\partial \vartheta}{\partial n} = 0, \tag{1.5}$$

on the boundary $\mathcal{F}_T = \mathcal{F} \times (0, T)$, where $n = (n_1, n_2, n_3)$ is the normal vector to surface $\mathcal{F}$, and subject to the initial condition

$$u(x, 0) = u_0(x) \qquad \text{for} \quad x \in G. \tag{1.6}$$

If $\psi(x, t)$ is a smooth function, vanishing at $t = T$, then Problem (A_ε) can be equivalently formulated as an integral identity.

Definition 1. By the *generalized solution of problem (1.4) - (1.6)* we shall mean $u_\varepsilon \in L_\infty(G_T)$ such that the function ϑ_ε, defined by (1.2), belongs to $W_2^{1,0}(G_T)$ and the integral identity

$$\int_{G_T} \left\{ \sum_{i=1}^3 \kappa_i D_i \vartheta_\varepsilon D_i \psi - u_\varepsilon D_t \psi \right\} dx dt = \int_G u_0^\varepsilon(x) \psi(x, 0) \, dx \tag{1.7}$$

holds for any function $\psi \in W_2^{1,1}(G_T)$ which vanishes at $t = T$.

We now introduce (only formally at first) the approximate model (A_0) by passing to the limit as $\varepsilon \to 0$.

As noted in Section I.1, the solution ϑ_ε satisfies the equation

$$D_t u_\varepsilon = \frac{1}{\varepsilon}(D_1^2 \vartheta_\varepsilon + D_2^2 \vartheta_\varepsilon) + D_3^2 \vartheta_\varepsilon$$

in the interior of Ω_T. By letting $\varepsilon \to 0$, we deduce that the limit function

$$\vartheta(x, t) = \lim_{\varepsilon \to 0} \vartheta_\varepsilon(x, t)$$

depends only on x_3 and t :

$$\vartheta(x, t) = \Theta(x_3, t) \qquad \text{for} \quad (x, t) \in \Omega_T. \tag{1.8}$$

Let us substitute $\vartheta(x, t)$ into (1.7), to get the identity

$$\int_{Q_T} \left\{ D\vartheta D\psi - \vartheta D_t \psi \right\} dx dt + \int_{\Omega_T} \left\{ D_3 \Psi D_3 \Theta - U D_t \Psi \right\} dx dt$$

$$= \int_\Omega U_0(x_3) \Psi(x_3, 0) \, dx + \int_Q \vartheta_0(x) \psi(x, 0) \, dx, \tag{1.9}$$

which holds for all functions $\psi \in W_2^{1,1}(G_T)$, vanishing at $t = T$ and independent of variables (x_1, x_2) in Ω_T:

$$\psi(x, t) = \Psi(x_3, t) \qquad \text{for} \quad (x, t) \in \Omega_T.$$

In the above identity,

$$U = \Phi(\Theta).$$

Definition 2. By the *generalized solution of Problem* (A_0) we mean a function $u \in L_\infty(G_T)$, independent of the variables (x_1, x_2) in Ω_T, such that the function $\vartheta = \chi_0(x, u)$ belongs to the space $W_2^{1,0}(G_T)$ and functions u, ϑ satisfy the integral identity (1.9).

Problem (A_0) represents the desired approximate model which describes phase transitions inside the cylindrical ingot Ω subject to heat conduction in the surrounding medium Q. In this model, ϑ can be interpreted as the temperature, u as the specific internal energy. Inside the ingot, both the temperature $\vartheta = \Theta$ and the specific internal energy $u = U$ depend only on t and x_3.

In (1.9), ψ is an arbitrary function. In particular, ψ can be finite in Q_T. If so, then ϑ satisfies the integral identity

$$\int_{Q_T} \{D\vartheta D\psi - \vartheta D_t \psi\} \, dx \, dt = 0,$$

which, for sufficiently smooth $\vartheta(x, t)$, is equivalent to the heat equation

$$D_t \vartheta = \Delta \vartheta \qquad \text{for} \quad (x, t) \in Q_T. \tag{1.10}$$

And what happens in the domain Ω_T? Suppose that the normal derivative $\frac{\partial \vartheta}{\partial n} \in L_1(\Pi_T)$ is defined everywhere on Π_T, where $n = (n_1, n_2, n_3)$ is the normal vector to Π, outward with respect to Q. Let us consider the identity (1.9) with functions ψ vanishing at $t = 0, T$, to deduce by equation (1.10) and partial integration that

$$\int_{\Omega_T} \{D_3 \Psi D_3 \Theta - U D_t \Psi\} \, dx \, dt = - \int_{\Pi_T} \frac{\partial \vartheta}{\partial n} \Psi \, d\sigma \, dt.$$

Let the closed planar curve γ be the intersection of the surface Π and the plane $\{x_3 = \text{const.}\}$. Then, because Ψ in Ω_T depends only on (x_3, t), the integral on surface Π_T in the latter equality can be transformed into a double integral and, further, into an integral over the domain Ω_T :

$$\int_{\Pi_T} \frac{\partial \vartheta}{\partial n} \Psi(x_3, t) \, d\sigma \, dt = \int_0^T \left\{ \int_{-1}^1 \left\{ \oint_\gamma \frac{\partial \vartheta}{\partial n} \, ds \right\} \Psi(x_3, t) \, dx_3 \right\} dt$$

$$= a \int_0^T \left\{ \int_\Omega \left(\oint_\gamma \frac{\partial \vartheta}{\partial n} \, ds \right) \Psi(x_3, t) \, dx \right\} dt,$$

where a is one over the area of the intersection of the domain Ω with the plane $\{x_3 = \text{const.}\}$. Hence, the identity

$$\int_{\Omega_T} \left\{ D_3 \Theta D_3 \Psi - U D_t \Psi + a \oint_\gamma \frac{\partial \vartheta}{\partial n} \, ds \, \Psi \right\} dx \, dt = 0 \tag{1.11}$$

holds in Ω_T, for any function Ψ, finite in ℓ_T and dependent only on (x_3, t).

Therefore, provided that the normal derivative $\frac{\partial \vartheta}{\partial n}$ of the solution to Problem (A_0) is summable on the surface Π_T, each generalized solution of Problem (A_0) will satisfy equation (1.10) in the domain Q_T and the integral identity (1.11) on the surface Π_T (or in the domain Ω_T).

The classical solution of Problem (A_0) can be defined in the usual way, as the solution with a strong discontinuity in Ω_T. Suppose that there exists a phase transition surface $\Gamma_T = \{(x,t) \ : \ x \in \Omega, \ x_3 = R(t), \ t \in (0,T)\}$ in Ω which represents the strong discontinuity surface of functions U and $D_3\Theta$. Everywhere in $\Omega_T \setminus \Gamma_T$, the functions U and $D_3\Theta$ are smooth and satisfy the non-homogeneous heat equation

$$D_t U = D_3^2 \Theta - a \oint_\gamma \frac{\partial \vartheta}{\partial n} \, ds, \tag{1.12}$$

where $\Theta = \chi(U)$.

On the surface Γ_T, not only the isothermal condition

$$\Theta = 0 \qquad \text{for} \quad (x,t) \in \Gamma_T, \tag{1.13}$$

but also the standard Stefan condition is imposed:

$$\frac{dR}{dt} = \left[D_3\Theta\right]_{\Theta \to +0}^{\Theta \to -0}. \tag{1.14}$$

At the initial time,

$$u(x,0) = u_0(x) \qquad \text{for} \quad x \in G, \tag{1.15}$$

where $u_0(x) = U_0(x_3)$ for $x \in \Omega$.

The above initial-boundary value Problem (A_0) is non-standard both in the classical and the generalized formulation. In this problem, a solution of the heat equation is to be determined in the domain Q_T, but its value on the boundary Π_T is unknown and itself represents a solution of a certain parabolic equation formulated in variables tangent to Π_T. The equation on Π_T is treated in the sense of distributions and its right-hand side represents a functional with respect to the normal derivative of the solution on the boundary Π_T.

If the temperature ϑ does not change sign in Ω_T (no phase transitions in the ingot), the above strong discontinuity does not occur and identity (1.11) is equivalent to an inhomogeneous heat equation understood in the standard sense. We have already faced a similar situation in Chapter II, in constructing approximate solutions of multidimensional Stefan problem. As in Chapter II, here also a proof of the correctness for Problem (A_0) is quite simple.

Everything becomes more complicated when both liquid and solid phases exist in Ω_T. First, due to the non-homogeneity of the heat inflow equation a mushy phase may appear on the boundary Π_T (in Ω_T). Secondly, for a motion with strong discontinuity the functional containing the normal derivative of the solution in condition (1.12) may be unbounded on approaching the discontinuity curve Γ_T. Thus, most probably the generalized solution will exist globally in time whereas the existence of the classical solution will be restricted to a small time interval.

Remark 1. Condition (1.12) (or (1.11)) can be derived in a different way, with considerations restricted to the domain Ω where the energy balance is set up under the hypothesis that the temperature is constant on all sections $\{x_3 = \text{const.}\}$ and there is a heat flux from the domain Q. According to [197], the boundary Π (or domain Ω) is called a "concentrated capacity".

Theorem 1. *Let the boundary Π be of the class $H^{2+\alpha}$ and $u_0(x)$ be a bounded function such that $u_0(x) = U_0(x_3)$ for $x \in \Omega$,*

$$\vartheta_0 = \chi(u_0) \in W_2^1(G), \quad \Theta_0(x_3) = \chi(U_0(x_3))$$

and let ϑ_0 satisfy boundary condition (1.5).

Then there exists a unique generalized solution of Problem (A_0). This solution can be constructed as the limit as $\varepsilon \to 0$ of the corresponding solutions of Problem (A_ε).

Theorem 2. *Assume that the hypotheses of Theorem 1 are satisfied, surface Π is in the class $H^{5+\alpha}$, $\vartheta_0 \in H^{5+\alpha,(5+\alpha)/2}(\bar{Q})$ and that there exists a real number R_0, $R_0 \in \ell = \{x_3 : |x_3| < 1\}$ such that $\Theta_0(x_3) < 0$ for $x_3 < R_0$, $\Theta_0(x_3) > 0$ for $x_3 > R_0$, and*

$$\left| D_3 \Theta_0(R_0 \pm 0) \right| \geq d > 0. \tag{1.16}$$

Furthermore, on Π, let the first-order compatibility conditions

$$\Delta \vartheta_0 = D_3^2 \Theta - a \oint_\gamma \frac{\partial \vartheta_0}{\partial n} \, ds \quad \text{for} \quad x \in \Pi, \tag{1.17}$$

$$\Delta \vartheta_0(x_1, x_2, R_0 \pm 0) = 0 \quad \text{for} \quad (x_1, x_2, R_0) \in \Pi, \tag{1.18}$$

be satisfied. Due to the smoothness of the function $\Theta_0(x_3)$ these follow from equation (1.10) and condition (1.12), and from equation (1.10) and condition (1.14), respectively. Outside the curve $\mathring{\gamma} = \{x_3 = R_0\}$, let the second-order compatibility conditions which correspond to the heat equation (1.10) in Q_T be satisfied on the surface Π and let the boundary condition (1.12) hold on Π_T.

Then, on a sufficiently small time interval $(0, T_*)$, there exists a classical solution $\{R(t), \vartheta(x,t)\}$ of Problem (A_0) such that $R \in W_2^2[0, T_*]$ and $\vartheta(x,t)$ has continuous derivatives entering equation (1.10) and boundary condition (1.12) everywhere except the curves Γ_{T_*} on Π_{T_*}. Moreover, the boundary condition (1.14) is satisfied in the standard sense with all the derivatives appearing Hölder continuous in t with Hölder index $1/2$.

In Problem (B_ε), equation (1.4) is considered only in the domain Ω_T, with the condition

$$\frac{1}{\varepsilon} \frac{\partial \vartheta}{\partial n} = f(x_3, t) - \vartheta \tag{1.19}$$

for the solution $\vartheta(x,t)$ on the boundary Π_T, where $n = (n_1, n_2, n_3)$ is the normal vector to the surface Π outward with respect to Ω.

The problem is completed by imposing the condition of thermal insulation on the remaining part of the boundary of Ω (on sections $\{x_3 = \pm 1\}$) and prescribing the initial specific internal energy.

Without formulating existence and uniqueness theorems for the generalized solution of Problem (B_ε) (this could be done in the same way as in Section I.3), we immediately proceed to the approximate model (B_0).

Let ϑ_ε be the solution of Problem (B_ε). Then it satisfies the integral identity

$$\int_{\Omega_T} \left\{ D_3\vartheta_\varepsilon D_3\psi + \frac{1}{\varepsilon}(D_1\vartheta_\varepsilon D_1\psi + D_2\vartheta_\varepsilon D_2\psi) - u_\varepsilon D_t\psi \right\} dx dt$$

$$+ \int_{\Pi_T} \psi(\vartheta_\varepsilon - f)\, d\sigma dt = 0, \tag{1.20}$$

for any smooth function ψ vanishing at $t = 0$ and $t = T$.

As before, we conclude that the functions $\vartheta_\varepsilon(x,t)$ converge to $\Theta(x_3,t)$ as $\varepsilon \to 0$, and that $u_\varepsilon = \Phi(\vartheta_\varepsilon)$ converge then to $U = \Phi(\Theta)$. Let us substitute the limit functions Θ and U into identity (1.20), then insert $\psi = \Psi(x_3,t)$ and transform the integral over the surface Π in this identity to the following integral over Ω:

$$\int_\Pi \Psi(\Theta - f)\, d\sigma = \int_\ell \left\{ \oint_\gamma \Psi(\Theta - f)\, d\sigma \right\} dx_3$$

$$= S_0 \int_\ell \Psi(\Theta - f)\, dx_3 = b_0 \int_\Omega \Psi(\Theta - f)\, dx,$$

where S_0 is the length of γ, and $b_0 = \frac{S_0}{a}$ is the ratio of this length to the area of the section of Ω by the plane $\{x_3 = \text{const.}\}$. We eventually arrive at the identity

$$\int_{\ell_T} \left\{ D_3\Theta D_3\Psi - U D_t\Psi + b_0\Psi(\Theta - f) \right\} dx_3 dt = 0. \tag{1.21}$$

The integral identity (1.21) together with the initial condition

$$U(x_3, 0) = U_0(x_3) \tag{1.22}$$

completely determine the solution of Problem (B_0).

It can be easily seen that (B_0) is a standard two-phase Stefan problem for the equation

$$D_t U - D_3^2\Theta + b_0\Theta = b_0 f, \qquad \Theta = \chi(U).$$

This problem was studied earlier, thus we confine ourselves here to the quasi-steady situation, with an infinite cylinder Ω, and $f(x_3, t) = F(x_3 - v_0 t)$.

The solution Θ of the quasi-steady problem is sought in the form

$$\Theta(x_3, t) = \omega(x_3 - v_0 t) \equiv \omega(z).$$

Substitution of this into the equation for Θ yields the ordinary differential equation

$$\frac{dA}{dz} - \frac{d^2\omega}{dz^2} + b_0\omega = b_0 F, \tag{1.23}$$

where $A = \Phi(\omega)$ is a discontinuous function of ω which has the meaning of the specific internal energy.

To complete the problem formulation, it is necessary to complement equation (1.23), understood in the sense of distributions, by asymptotic conditions as $|z| \to \infty$. A natural choice is to assume that the ingot temperature at infinity coincides with temperature of the surrounding medium:

$$\lim_{|z| \to \infty} |w(z) - F(z)| = 0. \tag{1.24}$$

As $t \to \infty$, the problem (1.23), (1.24) gives an asymptotic characterization of the temperature of an infinite ingot that moves uniformly, with a constant velocity v_0 with respect to the fixed slab of the furnace which has temperature $F(z)$. The coordinate z is connected with the fixed slab and x_3 with the ingot.

Problem (C). Determine a function $w(z)$ which satisfies equation (1.23) everywhere on the line $\mathbb{R} = \{z : |z| < \infty\}$ and the condition (1.24) at infinity.

Definition 3. A bounded continuous function $w(z)$, with the first derivatives locally summable on the line $\mathbb{R}$, and satisfying condition (1.24) as $|z| \to \infty$, is called the *generalized solution of Problem* (C), if for all smooth finite functions $\varphi(z)$

$$\int_{\mathbb{R}} \left\{ \left(v_0 A - \frac{dw}{dz} \right) \frac{d\varphi}{dz} + b_0 (F - w) \varphi \right\} dz = 0. \tag{1.25}$$

As before, by the *classical solution of Problem* (C), we shall mean the generalized solution $w(z)$ which vanishes at a finite number of points. Without loss of generality, we can assume $b_0 = 1$.

Theorem 3. *Let* $F(z) = F_1 = \text{const.} > 0$ *for* $z < 0$, $F(z) = F_2 = \text{const.} < 0$ *for* $z > 0$ *and let* $F_1 + F_2 > 0$.

Then for all $v_0 > 0$ *there exists a unique solution* $w(z)$ *of Problem* (C). *This solution is classical for* $v_* \in (0, v_0)$, *where* $v_* = |F_2| (1 + |F_2|)^{-1/2}$, *such that* $w(z) > 0$ *for* $z < z_0$ *and* $w(z) < 0$ *for* $z > z_0$,

$$z_0 = \frac{1}{\lambda_2} \log \frac{(\lambda_1 - \lambda_2)|F_2| - v_0}{\lambda_1 (F_1 - F_2)},$$

$$2\lambda_1 = v_0 + (v_0^2 + 4)^{1/2}, \quad 2\lambda_2 = v_0 - (v_0^2 + 4)^{1/2}.$$

For $v_0 > v_*$, $w(z)$ *is only the generalized solution of Problem* (C), *such that* $w(z) > 0$ *at* $z < z_1$, $w(z) = 0$ *at* $z_1 < z < z_2$ *and* $w(z) < 0$ *at* $z > z_2$, *where*

$$z_1 = \frac{1}{\lambda_2} \log \frac{F_1}{F_1 - F_2} > 0,$$

$$z_2 = z_1 + \frac{2 + |F_2|}{2|F_2|} - \frac{1}{2}(v_0^2 + 4)^{1/2}.$$

2. Existence and uniqueness of the generalized solution to Problem (A_0)

In proving the existence and uniqueness of the generalized solution to Problem (A_0) we shall proceed as in Section I.3 where we considered the multidimensional Stefan problem. Under the hypotheses of Theorem 1, let $\vartheta_\varepsilon(x,t)$ be a solution of Problem (A_ε), corresponding to the initial distribution $u_0(x)$ of the specific internal energy.

It is easy to show that the solution $\vartheta_\varepsilon(x,t)$ of Problem (A_ε) exists, is unique and satisfies the a priori bounds

$$\max\left\{\|D_t\vartheta_\varepsilon\|_{2,G_T},\ \max_{t\in(0,T)}\|D\vartheta_\varepsilon(t)\|_{2,G},\ \frac{1}{\varepsilon}\max_{t\in(0,T)}\|D_1\vartheta_\varepsilon(t),D_2\vartheta_\varepsilon(t)\|_{2,\Omega}\right\}\le M,$$

$$(2.1)$$

with the constant M dependent only on T and the quantity $m_0 = \|\vartheta_0\|_{2,G}^{(1)}$, $\vartheta_0 = \chi(u_0)$. This can be done in the usual way. The functions Φ and η in the representations (1.1) and (1.3) are approximated by smooth functions $\Phi^{(k)}$ and $\eta^{(k)}$ such that $\eta^{(k)}(x) = 1$ for $x\in\Omega$, and the functions $u_0(x)$ by smooth functions $u_0^{(k)}(x)$ such that $u_0^{(k)}(x) = U_0^{(k)}(x_3)$ for $x\in\Omega$.

Estimates analogous to (2.1) hold for the approximate solutions of the corresponding initial-boundary value problem. To prove this, we multiply the equation for $\vartheta_\varepsilon^{(k)}$ by the derivative $D_t\vartheta_\varepsilon^{(k)}$ and then integrate over the domain G at a fixed $t > 0$. The constant M, which bounds the above norms, is independent of k and ε. Indeed, by construction M is independent of k (the sequence $\{u_0^{(k)}(x)\}$ can be chosen so as provide uniform boundedness of the norms $\|\vartheta_0^{(k)}\|_{2,G}^{(1)}$ by $2m_0$), and the parameter $1/\varepsilon$ can enter M only as a multiplier in the term

$$\max\left\{\|D_1\vartheta_0^{(k)},D_2\vartheta_0^{(k)}\|_{2,\Omega}\right\}.$$

The functions $\vartheta_0^{(k)}$ are adjusted so as ensure that the above norm is equal to zero for all k.

The inequalities (2.2) for the approximate solutions $\vartheta_\varepsilon^{(k)}$ provide, for a fixed $\varepsilon > 0$, the possibility of selecting a subsequence $\{\vartheta_\varepsilon^{(k_n)}\}$ weakly convergent in $W_2^{1,1}(G_T)$ to a bounded function $\vartheta_\varepsilon(x,t) \in W_2^{1,1}(G_T)$ such that the sequence of functions $\{u_\varepsilon^{(k_n)}\}$, where $u_\varepsilon^{(k)} = \Phi_0^{(k)}(x,\vartheta_\varepsilon^{(k)})$, converges strongly in $L_2(G_T)$ to the bounded function $u_\varepsilon = \Phi_0(x,\vartheta_\varepsilon)$.

Obviously, the pair $\{u_\varepsilon,\vartheta_\varepsilon\}$ is a solution of Problem (A_ε) such that the estimate (2.1) holds. Thus, the above procedure can be repeated and subsequences of $\{\vartheta_\varepsilon\}$ and $\{u_\varepsilon\}$ can be selected which are convergent to bounded functions $\vartheta \in W_2^{1,1}(G_T)$, $u \in L_\infty(G_T)$, $\vartheta = \chi_0(x,u)$, weakly in $W_2^{1,1}(G_T)$ and strongly in $L_2(G_T)$, respectively. Because

$$\|D_1\vartheta_\varepsilon,D_2\vartheta_\varepsilon\|_{2,\Omega_T}\le M\varepsilon,$$

it follows that $\|D_1\vartheta,D_2\vartheta\|_{2,\Omega_T} = 0$ and, hence, $\vartheta(x,t) = \Theta(x_3,t)$ for $(x,t)\in\Omega_T$. The pair of functions $u(x,t), \vartheta(x,t)$ is the generalized solution of Problem (A_0) which is sought. To prove this, it suffices to take a test function $\psi(x,t)$, dependent only on t on the sections $\{x_3 = \text{const.}\}$ of Ω_T, in the integral identity for the solution $\{u_\varepsilon, \vartheta_\varepsilon\}$ of Problem (A_ε) and pass there to the limit as $\varepsilon \to 0$.

Lemma 1. *The bounded generalized solution of Problem (A_ε) is unique.*

Proof. The differences $\vartheta = \vartheta_1 - \vartheta_2$, $u = u_1 - u_2$ of two solutions of Problem (A_ε) satisfy the integral identity

$$\int_{Q_T} \vartheta\{D_t\psi + \Delta\psi\}\,dx dt + \int_{\Omega_T} U\{D_t\Psi - \mu\mathcal{J}\}\,dx dt = 0, \tag{2.2}$$

if $\psi \in W_2^{2,1}(G_T)$, and satisfies boundary condition (1.5). In (2.2) we have set

$$\mu = \Theta U^{-1}, \quad U = U_1 - U_2 = \Phi(\Theta_1) - \Phi(\Theta_2),$$

$$\mathcal{J} = a \oint_\gamma \frac{\partial \psi}{\partial n}\,ds - D_3^2\Psi.$$

By definition of Φ, it follows that

$$0 \le \mu(x_3, t) \le 1.$$

Let $\mu_k(x_3, t)$ be positive, sufficiently smooth functions such that

$$\mu_k(x_3, t) \ge \frac{1}{k}, \quad \|\mu - \mu_k\|_{2,l_T} \le \frac{1}{k}.$$

Let us consider the problem of determining a function $\psi_k(x,t)$, vanishing at $t = T$, satisfying boundary condition (1.5) and the equations

$$\begin{aligned} D_t\psi_k + \Delta\psi_k &= f & \text{for} \quad (x,t) &\in Q_T, \tag{2.3} \\ \psi_k(x,t) &= \Psi_k(x_3, t) & \text{for} \quad (x,t) &\in \Omega_T, \end{aligned}$$

$$-D_t\Psi_k + F = \mu_k\left\{D_3^2\Psi_k - a\oint_\gamma \frac{\partial\psi_k}{\partial n}\,ds\right\} \quad \text{for} \quad (x,t) \in \Pi_T, \tag{2.4}$$

with an arbitrary function $f(x,t)$ such that $f(x,t) = F(x_3, t)$ for $(x,t) \in \Omega_T$.

We shall assume that the functions $\Psi_k(x,t) \in H^{2+\alpha,(2+\alpha)/2}(\bar{Q}_T)$ have already been constructed. To obtain the basic energy estimate, we multiply equation (2.3) by $\Delta\psi$ (for simplicity we omit index k) and then integrate over Q at a fixed t, $0 < t < T$. We integrate the term $(D_t\psi - f)\Delta\psi$ by parts, and replace the multiplier $(D_t\Psi - F)$ of equation (2.4) in the integral over Π by $\mu\mathcal{J}$.

In the resulting integral over Π, the multiplier $\mathcal{J}$ depends only on x_3 and t; thus it can be transformed to the double integral

$$\int_\Pi \mu\mathcal{J}\frac{\partial\psi}{\partial n}\,d\sigma = \int_\ell \mu\mathcal{J}\left\{\oint_\gamma \frac{\partial\psi}{\partial n}\,ds\right\}dx_3$$

and, further, to an integral over the domain Ω. We have

$$\int_Q |\Delta\psi|^2 \, dx - \frac{1}{2}\frac{d}{dt}\int_Q |D\psi|^2 \, dx + \int_Q DfD\psi \, dx + a\int_\Omega \mu\mathcal{J}\left\{\oint_\gamma \frac{\partial\psi}{\partial n} \, ds\right\} dx = 0. \quad (2.5)$$

Let us now multiply the boundary condition (2.4) by $D_3^2\Psi$, integrate over Ω and apply integration by parts to the term $D_3^2\Psi \, (D_t\Psi - F)$. After taking the sum of the resulting equality and (2.5), by applying Hölder's and Young's inequalities, as well as Gronwall's lemma, we obtain the bound

$$\int_{\Omega_T} \mu_k \mathcal{J}_k^2 \, dxdt \leq (1 + e^T)\|Df\|_{2,G_T}^2 = M_1^2. \quad (2.6)$$

Let us substitute ψ_k into identity (2.2), to get

$$\int_{Q_T} \vartheta f \, dxdt + \int_{\Omega_T} U \, F \, dxdt = \Lambda, \qquad \text{with} \quad \Lambda = \int_{\Omega_T} U\mu_k^{1/2}\mathcal{J}_k\left\{\frac{\mu_k - \mu}{\mu_k}\right\} dxdt. \quad (2.7)$$

Estimating the absolute value of Λ from above with the use of Hölder's inequality, by the estimate (2.6), the boundedness of U and the properties of the functions μ_k, we infer that

$$|\Lambda| \leq \max_{(x,t)\in\Omega_T} |U|M_1 k^{-1/2}.$$

Hence, the right-hand side of (2.7) is equal to zero. By assuming $F(x_3, t) = 0$ on the boundary Π_T, we get the equality

$$\int_{Q_T} \vartheta f \, dxdt = 0,$$

equivalent to $\vartheta = 0$ almost everywhere in Q_T. Returning to the relation (2.7) and assuming $F(x_3, t) \neq 0$, we thus conclude that it is only possible that $U = 0$ almost everywhere on Π_T.

We now proceed to the construction of the functions $\psi_k(x, t)$.

Lemma 2. *Let* $f(x, t) = F(x_3, t)$ *for* $(x, t) \in \Pi_T$ *and* $f \in H^{\alpha,\alpha/2}(\bar{Q}_T)$. *Then there exists a unique solution* $\psi \in H^{2+\alpha,(2+\alpha)/2}(\bar{Q}_T)$ *of the initial-boundary value problem (1.5), (2.3), (2.4), which vanishes at* $t = T$.

Proof. Let us consider the one-parameter family of initial-boundary value problems (E_λ), $\lambda \in [0, 1]$, differing from the above problem in an additional additive term

$$(1 - \lambda)a\mu \oint_\gamma \frac{\partial\psi}{\partial n} \, ds$$

on the right-hand side of the boundary condition (2.4) imposed on Π_T. At $\lambda = 1$, (E_λ) coincides with the original problem, at $\lambda = 0$, it splits into two problems: the first, that of determining the function $\Psi(x_3, t)$ on the boundary Π_T as a solution of the equation

$$D_t\Psi + \mu D_3^2\Psi = F,$$

vanishing at $t = T$ and satisfying condition (1.5), and the second, that of determining a solution $\psi(x, t)$ of the inhomogeneous heat equation (2.3) in the domain Q_T, coinciding with $\Psi(x_3, t)$ on Π_T, satisfying condition (1.5) and vanishing at $t = T$.

Thus, the set of λ for which Problem (E_λ) has a solution is non-empty. We shall prove that the norm in $H^{2+\alpha,(2+\alpha)/2}(\bar{Q}_T)$ of any solution of Problem (E_λ) is bounded by a constant independent of λ.

To estimate the absolute value of such a solution, let us rewrite the boundary condition on Π_T in the form

$$D_t \Psi - \mu a \lambda \oint_\gamma \frac{\partial \psi}{\partial n} \, ds + \mu D_3^2 \Psi = F. \tag{2.8}$$

Suppose that the positive maximum of the function is achieved for a point (x^0, t^0) on the boundary Π_T. Then it is achieved at all points of Π which belong to the section $\gamma = \{x \,:\, x_3 = x_3^0 = \text{const.}\}$ at $t = t_0$. At these points,

$$D_t \Psi \leq 0, \quad D_3^2 \Psi \leq 0, \quad \oint_\gamma \frac{\partial \psi}{\partial n} \, ds > 0.$$

This follows from the Hopf-Zaremba-Giraud principle [92], since n is the normal vector to the surface Π, outward with respect to Q.

Therefore, the left-hand side of equation (2.8) is non-positive at the point of the positive maximum. Analogously, at any point corresponding to the negative minimum of the function ψ on the boundary Π_T, the left-hand side of equation (2.8) is non-negative. Hence, the functions

$$\{\psi - (T - t)(|f|_{Q_T}^{(0)} + 1)\} \quad \text{and} \quad \{\psi + (T - t)(|f|_{Q_T}^{(0)} + 1)\}$$

cannot achieve their positive maximum and negative minimum in the domain Q_T and on the boundary Π_T, respectively. This implies the bound

$$|\psi|_{Q_T}^{(0)} \leq T\left(|f|_{Q_T}^{(0)} + 1\right). \tag{2.9}$$

Let us now consider the boundary condition (2.8), treating it as a parabolic equation with the right-hand side

$$\tilde{F} = F + a\mu\lambda \oint_\gamma \frac{\partial \psi}{\partial n} \, ds.$$

According to Theorem 3 of Section I.2, we have

$$|\Psi|_{\Pi_T}^{(2+\alpha)} \leq c|\tilde{F}|_{\Pi_T}^{(\alpha)} \leq c\left(|f|_{Q_T}^{(\alpha)} + |\psi|_{Q_T}^{(1+\alpha)}\right).$$

On the other hand, by the same theorem we can conclude that the following estimate holds for the solution $\psi(x, t)$ of the inhomogeneous equation (2.3):

$$|\psi|_{Q_T}^{(2+\alpha)} \leq c\left(|f|_{Q_T}^{(\alpha)} + |\Psi|_{\Pi_T}^{(2+\alpha)}\right) \leq c\left(|f|_{Q_T}^{(\alpha)} + |\psi|_{Q_T}^{(1+\alpha)}\right). \tag{2.10}$$

Due to the interpolation inequalities (see Lemma 4, Section I.2),

$$|\psi|_{Q_T}^{(1+\alpha)} \leq \delta|\psi|_{Q_T}^{(2+\alpha)} + c\,\delta^{-1}|\psi|_{Q_T}^{(0)},$$

for sufficiently small $\delta > 0$, from (2.9) and (2.10) we can deduce the desired estimate

$$|\psi|_{Q_T}^{(2+\alpha)} \leq c|f|_{Q_T}^{(\alpha)},$$

which justifies the use of a parametric extension and so proves the existence of the solution to Problem (E_λ) at $\lambda = 1$. □

3. Existence of the classical solution to Problem (A_0)

The regularity of the generalized solutions to Problem (A_0) will be proved by a direct construction of the classical solution, as has already been done in Chapter V for the one-dimensional Stefan problem, rather than by considering the differential properties of the generalized solution obtained. By the uniqueness theorem, the classical solution we have constructed coincides with the generalized solution of Problem (A_0) on the interval $(0, T_*)$ of its existence.

The construction of the classical solution is itself based on using the Banach contraction theorem. To this end, a function space that should contain the free boundary Γ_T is fixed and a bounded subset $\mathcal{M}$ of the latter space is considered. Each point of the set $\mathcal{M}$ defines a surface $\tilde{\Gamma}_T$ contained in the hypersurface Π_T. The section $\{t = \text{const.}\}$ of the latter is a planar closed curve in Π which is contained in the section $\{x_3 = \text{const.}\}$, at the same time. For a given surface $\tilde{\Gamma}_T$ in $\mathcal{M}$, a solution of the heat equation is to be determined in the domain Q_T that satisfies initial condition (1.15), boundary condition (1.5), condition (1.12) everywhere on $\Pi_T \setminus \tilde{\Gamma}_T$ and vanishes on the surface $\tilde{\Gamma}_T$ (condition (1.13)).

The Stefan condition (1.14) is not assumed to be satisfied, it serves instead as a basis for construction of an operator $\mathcal{N}$ whose fixed points define the classical solution of Problem (A_0).

The main difficulty arises due to the discontinuity across $\tilde{\Gamma}_T$ exhibited by the tangent derivatives of the solution constructed in this way for the auxiliary problem. Hence, the appropriate normal derivatives of solution on the boundary Π_T, entering condition (1.12), in general exhibit unbounded growth close to $\tilde{\Gamma}_T$. We have proved in Chapter V that the analogous operator $\mathcal{N}$ in the one-dimensional Stefan problem "lifts" the regularity by $1/2$, i.e., it maps a bounded subset of $H^{(2+\alpha)/2}[0, T]$ into a bounded subset of $H^{(3+\alpha)/2}[0, T]$. In the present chapter we face a slightly different situation, nevertheless here also the regularity reserve of the operator $\mathcal{N}$ is sufficient to overcome the indicated difficulty and show the existence of the classical solution to Problem (A_0). The existence interval of the classical solution is small because there are no factors excluding the appearance of the mushy phase at $t = T_*$.

It is more convenient to consider the original problem in the new variables

$$\tau = t, \quad y_1 = x_1, \quad y_2 = x_2, \quad y_3 = q(x_3, R(t)), \tag{3.1}$$

where $R(t)$ is a function which defines the free boundary Γ_T of phase separation on surface Π_T. The function $q(s_1, s_2)$ of the class C^5 has been chosen so that $D_1 q \geq a_0 = \text{const.} > 0$ and it maps ℓ into itself for any fixed s_2 such that $|s_2| \leq 1/8$. The end-points $s_1 = \pm 1$ remain unchanged and q is linear with respect to both arguments for $|s_1| < 1/2$:

$$q(\pm 1, s_2) = \pm 1; \quad q(s_1, s_2) = s_1 - s_2, \quad |s_1| < 1/2.$$

The transformation (3.1) maps the domain Q_T and the surface Π_T into themselves, and the free boundary Γ_T on the surface Π_T into a fixed surface

$$\mathring{\gamma}_T = \mathring{\gamma} \times (0, T), \quad \mathring{\gamma} = \{x \in \Pi : x_3 = 0\}.$$

Let $\mathcal{M}$ be the set of functions $r(\tau) \in W_2^2(0, T)$ such that

$$r(0) = R_0, \quad \frac{dr}{d\tau}(0) = 0, \quad |r(\tau)| < 1/8, \quad \|r\|_{2,(0,T)}^{(2)} \leq 1.$$

Under the transformation (3.1) with $r \in \mathcal{M}$, the surface $\tilde{\Gamma}_T$ defined by $\{x \in \Pi, x_3 = r(t), t \in (0, T)\}$ is mapped onto the given surface $\mathring{\gamma}_T$. In the new variables, the heat equation is transformed to

$$D_\tau v - D_1^2 v - D_2^2 v - D_3(b D_3 v) + c D_3 v = 0, \tag{3.2}$$

where $b = (D_1 q)^2$, $c = \frac{dr}{d\tau} D_2 q + D_1^2 q$, and the boundary condition (1.12) is transformed to

$$-D_3(b D_3 V) + D_\tau V + a \oint_\gamma \frac{\partial v}{\partial n} ds + c D_3 V = 0. \tag{3.3}$$

Recalling the method applied in the construction of the solution, it seems more reasonable to pass from differential equations (3.2) and (3.3) to the corresponding integral identity.

Define

$$(v, w) = \int_G vw \, dy. \tag{3.4}$$

For a given function $r \in \mathcal{M}$, we shall consider the Problem (C_r) of determining a function $v(y, \tau)$ which fulfils conditions (1.5) and (1.15), vanishes on the curve $\mathring{\gamma}$ contained in the surface Π at all $\tau \in (0, T)$, and such that

$$(D_\tau v, \psi) + \sum_{i=1}^{2} \int_Q D_i v D_i \psi \, dy + (b D_3 \psi + c\psi, D_3 v) = 0 \tag{3.5}$$

for all functions $\psi \in W_2^{1,0}(G_T)$, such that $\psi(y, \tau) = \Psi(y_3, \tau)$ at $(y, \tau) \in \Omega_T$ and $\Psi(0, \tau) = 0$.

In (3.5) and everywhere following in this chapter, the values of all the functions in Ω_T, if defined in G_T, or traces of functions on Π_T, if defined in $\bar{Q}_T$, depend only on t and on y_3 (or x_3) and are denoted by the appropriate capital letters.

According to the Stefan condition (1.14), we define the operator $\mathcal{N}$ by the equality

$$\mathfrak{R}(r|\tau) = \int_0^\tau \left\{ D_3V(-0,t) - D_3V(+0,t) \right\} dt, \qquad (3.6)$$

where $v(y,\tau)$ is the solution of Problem (C_r).

We proceed to show that $\mathcal{N}$ maps the set $\mathcal{M}$ into the space $W_2^2(0,T)$. Let $R(\tau)$ be a fixed point of this mapping. Then the functions $R(t)$ and $\vartheta(x,t) \equiv v(y,\tau)$, where $v(y,\tau)$ is the solution of Problem (C_R), define the classical solution of Problem (A_0), provided that $\Theta(x_3,t)(x_3 - R(t)) > 0$ for $|x_3| < 1$.

3.1. Auxiliary Problem (C_r)

In constructing the solution of Problem (C_r), we shall use Galerkin's method.

Lemma 3. *For all $r \in \mathcal{M}$, there exists a unique solution $v(y,\tau)$ of Problem (C_r) such that*

$$\left\{ \|D_\tau v\|_{2,G_T}, \max_{\tau \in (0,T)} \|Dv(\tau)\|_{2,G} \right\} \leq M_1, \qquad (3.7)$$

$$\left\{ \|D_\tau^2 v\|_{2,G_T}, \max_{\tau \in (0,T)} \|DD_\tau v(\tau)\|_{2,G} \right\} \leq M_2. \qquad (3.8)$$

From now on we shall denote by M constants dependent only on T and $M_0 = |\vartheta_0|_Q^{(5+\alpha)}$, with $M(M_0,T) \leq M(M_0,T_0)$, whenever $T \leq T_0$. Because Theorem 2 has local character, we can always assume $T \leq 1$ and all constants M to depend only on M_0.

Proof. We shall take identity (3.5) as a basis for the construction of the Galerkin approximations

$$v^N = v_0(y) + \sum_{k=1}^N c_k(\tau)\psi_k(y),$$

where $v_0(y) = \vartheta_0(x)$ and $\{\psi_k\}_{k=1}^\infty$ represents a basis in the space $W_2^1(G)$ of functions $\psi(y)$ such that $\psi(y) = \Psi(y_3)$ for $y \in \Omega$ and $\Psi(0) = 0$. In the sequel, H will denote the space of all such functions.

Furthermore, let us take the eigenfunctions of the spectral problem

$$\begin{cases} -\Delta\psi = \lambda\psi & \text{for } y \in Q, \\ \partial\psi/\partial n = 0 & \text{for } y \in \partial G \setminus \partial\Omega, \\ \psi(y) = \Psi(y_3) & \text{for } y \in \Pi, \quad \Psi(0) = 0, \\ -D_3^2\Psi + a \oint_\gamma \dfrac{\partial\psi}{\partial n}\, ds = \lambda\Psi & \text{for } y \in \Pi, \end{cases} \qquad (3.9)$$

orthonormal with respect to (3.4), as a basis in the space H.

The operator L corresponding to problem (3.9) defines a scalar product $[\varphi, \psi]$, and the corresponding energy norm $\|D\psi\|_{2,G} = [\psi, \psi]^{1/2}$, in the space H by

$$(L\psi, \varphi) = \int_G D\psi D\varphi \, dy = [\varphi, \psi]. \tag{3.10}$$

The closure of H in the norm of $L_2(G)$ is a subspace of $L_2(G)$, including functions that are independent of y_1 and y_2 in the domain Ω and vanish on the intersection of Ω and the plane $\{y_3 = 0\}$. Each set in this space, bounded in the norm of $W_2^1(G)$, will be compact in $L^2(G)$ (more precisely, relatively compact). Therefore, to prove the existence of an orthonormal basis of the eigenfunctions of operator L, it suffices to show its strict positive definiteness [165]:

$$(L\psi, \psi) \geq c^{-1}(\psi, \psi), \qquad c = \text{const.} > 0.$$

This follows from the Poincaré inequality

$$\|\psi\|_{2,G} \leq c\|D\psi\|_{2,G},$$

because the functions ψ vanish on the intersection of $\{y_3 = 0\}$ and the domain Ω. The functions $c_k(\tau)$ can be deduced from the solution of the following linear Cauchy problem

$$\frac{dc_k}{d\tau} = \sum_{j=1}^{N} a_{jk}c_j + b_k, \quad c_k(0) = 0, \qquad k = 1, \dots, N, \tag{3.11}$$

with

$$-a_{jk} = \sum_{i=1}^{2}(D_i\psi_k, D_i\psi_j) + (bD_3\psi_k + c\psi_k, D_3\psi_j),$$

$$b_k = \sum_{i=1}^{2}(D_i^2 v_0, \psi_k) + (D_3(bD_3v_0) - cD_3v_0, \psi_k) - \int_\Omega a\left\{\oint_\gamma \frac{\partial v_0}{\partial n} ds\right\} \Psi_k \, dy.$$

In the expression for b_k, the derivatives of ψ_k in the appropriate integrals have been transformed into derivatives of the function v_0 via integration by parts.

Due to the choice of the function space which includes the functions $r(\tau)$, defining coefficients b and c in the identity (3.5), the coefficients $a_{jk}(\tau)$ and $b_k(\tau)$ of the linear ordinary differential system belong to the space $W_2^1(0,T)$. Hence, for any fixed N, there exists a unique solution $c_k \in W_2^2(0,T)$, $k = 1, \dots, N$, of the Cauchy problem (3.11).

We now derive estimates of the approximate solutions uniform with respect to N which justify the limit passage as $N \to \infty$. The first of these estimates (estimate (3.7) for the approximate solutions v^N) can be established in a standard way by multiplying the k-th equation of (3.11) by the derivative $D_\tau c_k$, taking the sum in k, and applying Hölder's and Young's inequalities, as well as Gronwall's lemma. We thus describe in detail only the derivation of the estimate (3.8) for the approximate solutions, analogous to (3.7).

To this end, let us differentiate each of the equations in system (3.11) with respect to t, multiply the result of differentiating the k-th equation by the second derivative $D_\tau^2 c_k$, take the sum in k, finally integrate over $\tau \in (0,t), t \le T$, to obtain after simple tranformations (we everywhere omit index N):

$$\left\|D_\tau^2 v\right\|_{2,G_t}^2 + \frac{1}{2}\sum_{i=1}^2\left\|D_\tau D_i v(t)\right\|_{2,G}^2 + \frac{1}{2}\left\|b^{1/2}D_\tau D_3 v(t)\right\|_{2,G}^2$$

$$= \frac{1}{2}\sum_{i=1}^2\left\|D_i D_\tau v(0)\right\|_{2,G}^2 + \frac{1}{2}\left\|b^{1/2}D_\tau D_3 v(0)\right\|_{2,G}^2 - \left(D_\tau b D_3 v(t), D_3 D_\tau v(t)\right)$$

$$+ \int_0^t \left\{ \frac{3}{2}\left(D_\tau b, |D_3 D_\tau v|^2\right) - \left(cD_\tau^2 v, D_3 D_\tau v\right) - \left(D_\tau c D_3 v, D_\tau^2 v\right)\right.$$

$$\left. + \left(D_\tau^2 b D_3 v, D_3 D_\tau v\right)\right\} d\tau. \tag{3.12}$$

The right-hand side of (3.12) can be estimated in a standard way by the positive terms of the left-hand side. In particular,

$$\int_1^t \left(D_\tau c D_3 v, D_\tau^2 v\right) d\tau \le \int_0^t \max_{y\in G}|D_\tau c|\,\|D_3 v(\tau)\|_{2,G}\,\|D_\tau^2 v(\tau)\|_{2,G}\,d\tau$$

$$\le M_1\|D_\tau^2 v\|_{2,G_t}\left(\int_0^t \max_{y\in G}|D_\tau c|^2\,d\tau\right)^{1/2} \le \lambda\|D_\tau^2 v\|_{2,G_t}^2 + \frac{M_1^2}{\lambda}\kappa\left(1 + r_0^2(T)\right).$$

where $r_0(T) = \|r\|_{2,[0,T]}^{(2)}$, $\kappa = $ const. > 0 depends only on the function q, defining the mapping (3.1), and $\lambda > 0$ is an arbitrary small number.

The other terms including derivatives of the functions b and c can also be estimated in exactly the same way.

It remains to find an estimate for the norms of the functions $D_i D_\tau v^N(y,0)$ in $L_2(G)$. Without any restriction, we can assume the constant R_0 that defines the initial position of the unknown boundary $x_3 = R(t)$ on surface Π is equal to 0. Then the mapping q at the initial time coincides with the identity (more precisely, it admits such a choice) and $b(0) = 1$. In view of equations (3.11),

$$\left\|D\left(D_\tau v^N(0)\right)\right\|_{2,G}^2 = \left\|Db^N\right\|_{2,G}^2 \equiv \mathcal{J},$$

where $b^N(y) = \sum_{k=1}^N b_k(0)\psi_k(y)$ and $b_k(0)$ represent the Fourier coefficients in the expansion of $\tilde{b}(y) = \Delta v_0(y)$ with respect to the basis $\{\psi_k\}_{k=1}^\infty$, $b_k(0) = (\tilde{b}, \psi_k)$. The latter equality follows from the compatibility conditions (1.17) at the initial time and from the vanishing of the derivative $D_\tau r(0)$, therefore $b|_{\tau=0} = 1$, $c|_{\tau=0} = 0$.

Thus

$$\mathcal{J} = \left[\sum_{k=1}^{N} b_k(0)\psi_k, \sum_{k=1}^{N} b_k(0)\psi_k\right] = \sum_{k=1}^{N} |b_k(0)|^2 \lambda_k,$$

where the scalar product $[\cdot, \cdot]$ is defined by (3.10). Since the function $\tilde{b} = \Delta v_0$ belongs to the space H (this is a consequence of the regularity of v_0 and the compatibility condition (1.18)),

$$\|D(\Delta v_0)\|_{2,G}^2 = [\tilde{b}, \tilde{b}] = \sum_{k=1}^{\infty} |b_k(0)|^2 [\psi_k, \psi_k] = \sum_{k=1}^{\infty} \lambda_k |b_k(0)|^2,$$

$$\mathcal{J} = \sum_{k=1}^{N} \lambda_k |b_k(0)|^2 \leq \|D(\Delta v_0)\|_{2,G}^2 \leq M_0.$$

By the above estimate and Gronwall's lemma, (3.12) implies the desired inequality (3.8) for the approximate solutions $v^N(y, \tau)$.

Due to estimates (3.7) and (3.8), there exists a subsequence of $\{v^N\}$, weakly convergent in the space $L_2(G_T)$, together with all derivatives entering (3.7) and (3.8), to a certain function $v(y, \tau)$. Certainly, $v(y, \tau)$ belongs to the space $L_2(0, T; H)$, satisfies the integral identity (3.5) and both estimates, (3.7) and (3.8).

The uniqueness of the solution of Problem (C_r) results from the integral identity for the difference v of two solutions after inserting there $\psi = v$, and using Hölder's inequality and Gronwall's lemma.   □

3.2. Differential properties of the solutions to Problem (C_r)

In the interior of Q_T, the regularity of the function $v(y, \tau)$ is determined by the differential properties of the coefficients b and c in the integral identity (3.5). In the standard way, we can prove that $v \in W_2^{4,2}(Q_T^{(\lambda)})$, where $Q_T^{(\lambda)} = Q^{(\lambda)} \times (0, T)$ and $Q^{(\lambda)}$ is an arbitrary domain in Q, with distance from the boundary Π larger than λ, $\lambda > 0$. It turns out that a similar statement remains true for domains whose closure contains surface Π_T, provided that the distance from such a domain to surface $\mathring{\gamma}_T$ is positive.

The above regularity of the solution to Problem (C_r) is required in the study of the operator $\mathcal{N}$ defined by (3.6). Wherever the differential properties of the solutions are rather poor (close to $\mathring{\gamma}_T$), the coefficients b and c in identity (3.5) assume an extremely simple form. Far from $\mathring{\gamma}_T$, the solution becomes "good" in this sense.

Let $Q^{(\lambda)}$ be the set of points in Q, with y_3-coordinate larger in absolute value than a positive constant λ. By Π^λ we shall denote the common part of the boundary $\partial Q^{(\lambda)}$ and the surface Π.

Lemma 4. *The solution $v(y, \tau)$ of Problem (C_r) belongs to the space $W_2^{4,2}(Q_T^{(\lambda)}) \cap W_2^{4,2}(\Pi_T^\lambda)$, where $\lambda > 3/8$, $Q_T^{(\lambda)} = Q^{(\lambda)} \times (0, T)$, $\Pi_T^{(\lambda)} = \Pi^{(\lambda)} \times (0, T)$ and the norm of the solution in the above space is bounded by a constant dependent only on M_0.*

Proof. To achieve more clarity, let us return to the original variables (x, t). By construction, the curve $\{x_3 = r(t)\}$ on surface Π does not exceed the set $\Pi \setminus \Pi^{(1/8)}$. Hence, the distance from the domains $Q^{(\lambda)}$ to the above curve is positive if $\lambda > \lambda_1 > 1/8$.

Expressed in the variables (x, t), the function $w(x, t) \equiv v(y, \tau)$ satisfies the integral identity

$$(D_t w, \psi) + (Dw, D\psi) = 0 \tag{3.13}$$

for all smooth functions $\psi(x, t)$ vanishing on the curve $\{x_3 = r(t)\} \subset \Pi$ at $t \in (0, T)$.

From the inequality (3.7), formulated for $w(x, t)$, we infer that

$$\left\{ \|D_t w\|_{2,G_T}, \max_{t \in (0,T)} \|Dw(t)\|_{2,G} \right\} \le M_3. \tag{3.14}$$

Thus, on the sections $\{t = \text{const.}\}$ in the interior of Q, $w(x, t)$ satisfies the Poisson equation

$$\Delta w = f, \tag{3.15}$$

with the right-hand side $f = D_t w$ in the space $L_2(Q)$. On the boundary Π, $w(x, t)$ coincides with a function $W(x_3, t)$ such that $D_3 W \in L_2(\Pi)$. On the remaining part of the boundary ∂Q, condition (1.5) holds.

We now represent w as the sum of two functions w_1 and w_2, where w_1 is a harmonic function in Q, coinciding with W on the boundary Π while w_2 satisfies the Poisson equation (3.15) with homogeneous Dirichlet condition on the surface Π. On the rest of the boundary ∂Q, functions w_1 and w_2 satisfy condition (1.5).

$$\left\| \frac{\partial w_1}{\partial n} \right\|_{2,\Pi}^2 \le c \left(\|D_3 W\|_{2,\Pi}^2 + \|Dw\|_{2,Q}^2 \right), \tag{3.16}$$

$$\left\| \frac{\partial w_2}{\partial n} \right\|_{2,\Pi}^2 \le c \|f\|_{2,Q}^2 = c \|D_t w\|_{2,Q}^2. \tag{3.17}$$

Estimate (3.17) follows from the elliptic regularity theory [138] and (3.16) is a consequence of Lemma 5 proved at the end of this section.

Lemma 5. *Let u be a harmonic function in Q, satisfying condition (1.5) on the subset $\partial Q \setminus \Pi$ of the boundary of Q and coinciding with function $U(x_3)$ on Π. Then*

$$\left\| \frac{\partial u}{\partial n} \right\|_{2,\Pi}^2 \le c \left(\|D_3 U\|_{2,\Pi}^2 + \|Du\|_{2,Q}^2 \right)$$

with the constant c dependent only on the shape of Q.

Taking the sum of estimates (3.16) and (3.17), then integrating in time, we get the bound

$$\left\| \frac{\partial w}{\partial n} \right\|_{2,\Pi_T}^2 \le c \left(\|D_t w\|_{2,Q_T}^2 + \|Dw\|_{2,G_T}^2 \right). \tag{3.18}$$

Returning to the integral identity (3.13), we see that everywhere on surface Π_T, except for the set $\bar{\Gamma}_T = \{x_3 = r(t), \, t \in (0, T)\}$, the function $w(x, t)$ satisfies

$$D_t W - D_3^2 W = -a \oint_\gamma \frac{\partial w}{\partial n} \, ds. \tag{3.19}$$

The higher regularity of the solution $w(x, t)$ follows by the "lifting" property of boundary condition (3.19). By (3.18), the right-hand side

$$F = -a \oint_\gamma \frac{\partial w}{\partial n} \, ds \tag{3.20}$$

of (3.19) belongs to the space $L_2(\ell_T)$. Due to local estimates for solutions of linear parabolic equations, $w(x, t)$ belongs to the space $W_2^{2,1}(\Pi_T^{(\lambda)})$, $\lambda > \lambda_1$ and $\|W\|_{2,\Pi_T^{(\lambda)}}^{(2)} \leq M_4$ (see Theorem 7, Section I.2).

$W_2^{2,1}(\Pi_T^{(\lambda)})$ can be embedded (cf., [176]) into the space $W_q^{2-1/q, 1-1/2q}(\Pi_T^{(\lambda)})$, $q = 8/3$, of functions $W(x_3, t)$ having fractional derivatives. In turn, functions in this space represent traces on Π_T of functions $w(x, t) \in W_q^{2,1}(Q_T)$. According to the local estimates of solutions to linear parabolic equations in the space $W_q^{2,1}(Q_T)$ with non-homogeneous boundary conditions on Π_T (cf., [137]), we can conclude that in a slightly smaller domain $Q_T^{(\lambda)}$, $\lambda > \lambda_2 > \lambda_1$, the function $w \in W_q^{2,1}(Q_T^{(\lambda)})$ and $\|w\|_{q, Q_T^{(\lambda)}}^{(2)} \leq M_5$, $q = 8/3$, $\lambda > \lambda_2$.

In view of the same embedding theorem [176], $DW \in L_4(\Pi_T^{(\lambda)})$. Hence, the right-hand side (3.20) of equation (3.19) on Π_T belongs to the space $L_4(\Pi_T^{(\lambda)})$. By Theorem 7 of Section I.2, in the domain $\Pi_T^{(\lambda)}$, $\lambda > \lambda_3 > \lambda_2$,

$$\|W\|_{4, \Pi_T^{(\lambda)}}^{(2)} \leq M_6, \quad \lambda > \lambda_3.$$

Since the space $W_4^{2,1}(\Pi_T^{(\lambda)})$ is contained in $W_p^{2-1/p, 1-1/2p}(\Pi_T^{(\lambda)})$ with $p = 16/3$ [176], by repeating the same considerations as above, we have

$$\|W\|_{P, Q_T^{(\lambda)}}^{(2)} \leq M_7, \quad p = 16/3, \quad \lambda > \lambda_4 > \lambda_3.$$

The space $W_p^{2,1}(Q_T^{(\lambda)})$ is embedded into $H^{1+\beta, (1+\beta)/2}(\overline{Q_T^{(\lambda)}})$ for some $\beta > 0$ (see Lemma 5, Section I.2). Thus,

$$|w|_{\Pi_T^{(\lambda)}}^{(1+\beta)} \leq M_8, \quad \lambda > \lambda_4.$$

Thus, the right-hand side F of (3.19) belongs to the space $H^{\beta, \beta/2}(\overline{\Pi_T^{(\lambda)}})$. By the local parabolic estimates, now already in Hölder spaces (see Theorem 6, Section I.2), $w(x, t)$ belongs to the space $H^{2+\beta, (2+\beta)/2}(\overline{\Pi_T^{(\lambda)}})$ on a smaller domain $\Pi_T^{(\lambda)}$, $\lambda > \lambda_5 > \lambda_4$, as well as to the space $H^{2+\beta, (2+\beta)/2}(\overline{Q_T^{(\lambda)}})$ with $\lambda > \lambda_6 > \lambda_5$. By repeating the above procedure twice, we obtain that $w \in H^{4+\beta,(4+\beta)/2}(\overline{Q_T^{(\lambda)}})$, $\lambda > \lambda_7 > \lambda_6$, and

$$|w|_{Q_T^{(\lambda)}}^{(4+\beta)} \le M_9, \quad \lambda > \lambda_7.$$

The assertion of the lemma follows by substitution of the variables, provided that $1/8 < \lambda_1 < \lambda_7 < 1/4$. $\qquad\square$

3.3. Proof of Theorem 2

We shall prove that the operator $\mathcal{N}$, defined by (3.6), is a contraction on the set $\mathcal{M}$, if the time interval $(0, T)$ is small enough.

Lemma 6. *For all $r \in \mathcal{M}$, the following estimates hold:*

$$\left\|\mathcal{N}(r_1|\tau) - \mathcal{N}(r_2|\tau)\right\|_{2,[0,T]}^{(2)} \le T^{1/4} M_{10} \|r_1 - r_2\|_{2,[0,T]}^{(2)}, \tag{3.21}$$

$$\left\|\mathcal{N}(r|\tau)\right\|_{2,[0,T]}^{(2)} \le T^{1/4} M_{11}. \tag{3.22}$$

Proof. Let $v_i(y, \tau)$ be a solution of Problem (C_{r_i}), corresponding to $r_i \in \mathcal{M}, i = 1, 2$. The appropriate integral identity for the difference $v = v_1 - v_2$ (correspondingly for $r = r_1 - r_2$) can be written in the form

$$(D_\tau v, \psi) + \sum_{i=1}^{2}(D_i v, D_i \psi) + \left(b^{(1)} D_3 \psi + c^{(1)} \psi, D_3 v\right) = (f_0, \psi), \tag{3.23}$$

where

$$b^{(1)} = b(y_3, r_1), \qquad c^{(1)} = c(y_3, r_1, D_\tau r_1),$$

$$f_0 = A_1 D_\tau r D_3 v_2 + r\left\{(A_2 + A_3 D_\tau r_1) D_3 v_2 + A_4 D_3^2 v_2\right\},$$

and the functions $A_i = A_i(y_3, r_1, r_2)$ are bounded. All the terms in the expression for f_0 are well-defined, because in the domain $Q_T \setminus Q_T^{(\lambda)}, \lambda > 3/8$, where in general the derivative $D_3^2 v_3$ does not appear, $A_1 = 1$, whereas the remaining coefficients A_i, $i = 2, 3, 4$, vanish. Outside the domain $Q_T \setminus Q_T^{(\lambda)}$, the function v_2 is sufficiently regular (see the preceding subsection).

In (3.23), let us insert the test function $\psi = D_\tau v$. By analogy to the inequalities we have obtained earlier,

$$\max_{\tau \in (0,T)} \left\|D_\tau v(\tau)\right\|_{2,G} \le M_{12} \|r\|_{2,[0,T]}^{(1)}. \tag{3.24}$$

To estimate the highest derivatives of the function $v(y, \tau)$, let us consider the identity (3.23) with basis functions $\psi_k(y)$ and differentiate this identity in time. By passing to arbitrary linear combinations of the above identities with coefficients dependent on time, we can derive an integral identity for the derivative $D_\tau v$. By substituting $\psi = D_\tau^2 v$, we obtain an equality almost coinciding with condition (3.12). The main difference consists of the presence of the additional term $(D_\tau f_0, D_\tau v)$, where $D_\tau f \in L_2(G_T)$ and

$$\|D_\tau f_0\|_{2, G_T} \leq M_{13}\|r\|_{2, [0, T]}^{(2)}.$$

Indeed,

$$D_\tau f_0 = A_5 D_\tau^2 r D_3 v_2 + A_6 r D_\tau D_3^2 v_2 + A_7 r D_3 v_2 D_\tau^2 r_1$$

$$+ r\{A_8 D_\tau D_3 v_2 + A_9 D_3^2 v_2 + A_{10} D_3 v_2\} + D_\tau r\{A_{11} D_\tau D_3 v_2 + A_{12} D_3^2 v_2 + A_{13} D_3^2 v_2\},$$

where $A_i = A_i(y_3, r_1, r_2, D_\tau r_1, D_\tau r_2)$ are bounded, and only the coefficients A_5 and A_8 are non-vanishing in the domain $Q_T \setminus Q_T^{(\lambda)}$, $\lambda > 3/8$.

The main difficulty in the expression $(D_\tau f_0, D_\tau^2 v)$ is related to estimating the first three terms. In particular, for the first term,

$$(A_5 D_\tau^2 r D_3 v_2, D_\tau^2 v) \leq \kappa \|D_\tau^2 v(\tau)\|_{2, G}^2 + \frac{1}{4\kappa} \max |A_5| \max_{t \in (0, T)} \|D_3 v_2(t)\|_{2, G}^2 |D_\tau^2 r(\tau)|^2,$$

where $\kappa > 0$ is an arbitrary small quantity, and the second term on the right-hand side of the inequality is summable on the interval $(0, T)$.

$$\left(\int_0^T |D_\tau^2 r(\tau)|^2 \, d\tau \right)^{1/2} \leq \|r\|_{2, [0, T]}^{(2)}.$$

The remaining terms in $(D_\tau f_0, D_\tau^2 v)$ can be estimated analogously. Thus, as in case of inequality (3.8), we have

$$\left\{ \|D_\tau^2 v\|_{2, G_T}, \max_{\tau \in (0, T)} \|D_\tau D v(\tau)\|_{2, G} \right\} \leq M_{14}\|r\|_{2, [0, T]}^{(2)}. \tag{3.25}$$

In terms of the original variables x, related to y by (3.1), where $R = r_1(t)$ and time t is fixed, we conclude that the derivative $z(x, t) = D_t v(y, t)$ in Q satisfies the Poisson equation with right-hand side

$$f = -D_t f_0 + D_t^2 v + D_t\{(c^{(1)} - D_3 b^{(1)}) D_3 v\} + D_3 D_t v D_t^2 q(x_3, r_1(t)),$$

and on the boundary Π coincides with function $Z = Z(x_3, t)$. By inequality (3.5),

$$\|f\|_{2, Q_T} \leq M_{15}\|r\|_{2, (0, T)}^{(2)},$$

$$\{\|Dz\|_{2, Q_T}, \|D_3 Z\|_{2, \Pi_T}\} \leq M_{16}\|r\|_{2, (0, T)}^{(2)}.$$

By using the same arguments as in deriving (3.18), we obtain the estimate

$$\left\| \frac{\partial}{\partial n}(D_\tau v) \right\|_{2, \Pi_T} \leq M_{17} \left\| \frac{\partial z}{\partial n} \right\|_{2, \Pi_T} \leq M_{18}(\|f\|_{2, Q_T} + \|Dz\|_{2, G_T}) \leq M_{19}\|r\|_{2, (0, T)}^{(2)}. \tag{3.26}$$

The regularity of the function $v(y, \tau)$ we have obtained ensures that the boundary condition

$$v = V(y_3, \tau), \quad D_\tau V - D_3(b^{(1)} D_3 V) + c^{(1)} D_3 V + a \oint_\gamma \frac{\partial v}{\partial n} \, ds = 0 \tag{3.27}$$

is satisfied on the surface Π_T outside $\mathring{\gamma}_T$. Furthermore, as in Lemma 4, it can be proved that $V \in W_2^{4,2}(\Pi_T^{(\lambda)})$, where $\lambda > 1/8$. In particular, the derivative $D_\tau D_3 V$ is defined on the boundaries $\{y_3 = \pm 1\}$ and, due to (1.5),

$$D_\tau D_3 V(\pm 1, \tau) = 0 \qquad \text{for} \quad \tau \in (0, T). \tag{3.28}$$

By condition (3.27) and estimates (3.25), (3.26), we can conclude that $D_\tau D_3^2 V \in L_2(\Pi_T^\pm)$, where Π_T^+ denotes the part of surface Π_T contained in the half-space $\{y_3 > 0\}$, whereas Π_T^- is contained in the half-space $\{y_3 < 0\}$ and

$$\left\| D_\tau D_3^2 V \right\|_{2, \Pi_T^\pm} \leq M_{20} \| r \|_{2, [0, T]}^{(2)}. \tag{3.29}$$

Let us now make use of the representation

$$\left| D_\tau D_3 V(\pm 0, \tau) \right|^2 = 2 \left| \int_0^{\pm 1} D_\tau D_3^2 V(y, \tau) D_\tau D_3 V(y, \tau) \, dy \right|,$$

where (3.28) has been taken into account, to deduce the bound

$$\int_0^T \left| D_\tau D_3 V(\pm 0, \tau) \right|^4 d\tau \leq 4 \max_{\tau \in (0,T)} \left\| D_\tau D_3 V(\tau) \right\|_{2, \Pi}^2 \left\| D_\tau D_3^2 V \right\|_{2, \Pi_T^\pm}^2$$

$$\leq M_{21} \left(\| r \|_{2, [0, T]}^{(2)} \right)^4.$$

Due to the following obvious relation

$$\int_0^T |w(\tau)|^2 \, d\tau \leq T^{1/2} \left(\int_0^T |w(\tau)|^4 \, d\tau \right)^{1/2},$$

the latter inequality implies the desired estimate (3.21) for the operator $\mathcal{N}$.

Estimate (3.22) admits a completely analogous proof. $\qquad\square$

Due to inequalities (3.21) and (3.22), the operator $\mathcal{N}$ is a contraction for sufficiently small T and maps the set $\mathcal{M}$ into itself, whenever $D_\tau \mathcal{N}(r|0) = 0$. The latter equality is a consequence of the continuity of the function $D_3 V_0(y_3)$. Hence, there exists a unique fixed point $R \in W_2^2(0, T)$ of the operator $\mathcal{N}$ which defines the classical solution of Problem (A_0) on a sufficiently small time interval $(0, T_*)$. The proof of Theorem 2 will be complete, if

$$y_3 V(y_3, \tau) > 0, \qquad 0 < |y_3| < 1$$

on the time interval $(0, T_*)$. This follows by the condition

$$y_3 V_0(y_3) > 0 \quad \text{if} \quad 0 < |y_3| < 1,$$

combined with (1.16), after taking a shorter interval $(0, T_*)$, if necessary. $\qquad\square$

3.4. Proof of Lemma 5

Let s be the length of an arc of the curve γ from a certain fixed point, let $\mathring{x}(s) = \left(\mathring{x}_1(s), \mathring{x}_2(s) \right)$ be the equation of curve γ on surface Π, contained in the plane $\{x_3 = $

const. $= z$} and let $\bar{e}(s)$ be the unit normal vector to surface Π in the point $\mathring{x}(s)$ of the curve γ. The equation

$$(x_1, x_2, z) = \mathring{x}(s) + n\bar{e}(s) \tag{3.30}$$

defines the orthogonal curvilinear coordinate system (s, n, z) in the domain Q close to the surface Π. In these coordinates, the function $w(s, n, z) \equiv u(x)$, periodic with respect to s, satisfies the elliptic equation

$$\frac{\partial^2 w}{\partial z^2} + \frac{\partial^2 w}{\partial n^2} + g_s \frac{\partial^2 w}{\partial s^2} + g_n \frac{\partial w}{\partial n} = 0 \tag{3.31}$$

in the domain $\mathcal{T} = \{(s, n, z) \ : \ s \in (0, S_0), \ n \in (0, n_0), \ |z| < 1\}$. In (3.31), we have used the notation

$$g_s^{-1} = \left(1 + n \left|\frac{d\bar{e}}{ds}(s)\right|^2\right), \quad g_n = g_s \frac{\partial g_s}{\partial n},$$

S_0 denotes the length of the curve γ and n_0 a sufficiently small number.

Under the mapping $(s, n, z) \to (x_1, x_2, x_3)$, the part of Q adjacent to Π corresponds to the domain $\mathcal{T}$, while, under the mapping $(x_1, x_2, x_3) \to (s, n, z)$, a certain subset $\mathcal{C}$ of the plane $\{n = 0\}$ corresponds to the surface Π.

Recall that a harmonic function is infinitely differentiable at the interior points of the domain where it satisfies the Laplace equation, and the norm of its derivatives of each order in an interior subdomain is bounded by any weak norm in the whole domain. In particular, all second derivatives of $u(x)$ on the surface $\{n = n_0\}$ in Q are uniformly bounded by a constant which depends only on n_0 and the norm $\|u\|_{2,Q}^{(1)}$. Since, further,

$$\|w\|_{2,\mathcal{T}}^{(1)} \le \text{const. } \|u\|_{2,Q}^{(1)}, \tag{3.32}$$

an analogous assertion is true for the second derivatives of the function $w(s, n, z)$ at $n = n_0$.

Let us differentiate equation (3.31) with respect to s, multiply by $\partial w / \partial s$ and integrate the result by parts in $\mathcal{T}$. By the equalities

$$\frac{\partial w}{\partial s} = \frac{\partial W}{\partial s} = 0 \quad \text{at} \quad n = 0$$

and due to the regularity of w at $n = n_0$, we deduce in a standard way that

$$\left\|\nabla \frac{\partial w}{\partial s}\right\|_{2,\mathcal{T}} \le \text{const. } \|u\|_{2,Q}^{(1)},$$

hence we eventually arrive at the estimate

$$\left\|\frac{\partial^2 w}{\partial s^2}\right\|_{2,\mathcal{T}} \le \text{const.} \|u\|_{2,Q}^{(1)}. \tag{3.33}$$

Let us now treat equation (3.31) in each section $\{s = \text{const.}\}$ as the two-dimensional Poisson's equation

$$\frac{\partial^2 w}{\partial z^2} + \frac{\partial^2 w}{\partial n^2} = f, \tag{3.34}$$

with $\quad -f = g_s \dfrac{\partial^2 w}{\partial s^2} + g_n \dfrac{\partial w}{\partial n}, \quad f \in L_2(T).$

As in the proof of Lemma 4, we shall decompose w into two components, w_1 and w_2, one of them, let it be w_1, satisfying Poisson's equation (3.34) in the intersection $T^{(s)}$ of the domain T with the plane $\{s = \text{const.}\}$ and vanishing on the boundary $C^{(s)}$, where $C^{(s)}$ represents the intersection of the boundary C with the plane $\{s = \text{const.}\}$. The second component is harmonic in $T^{(s)}$ and coincides with the function $W(z) = w(s, 0, z)$ on $C^{(s)}$.

The normal derivative of the function w_1 on the boundary $C^{(s)}$ is square-summable on $C^{(s)}$ and from [138] it follows that

$$\left\| \frac{\partial w_1}{\partial n} \right\|_{2, C^{(s)}} \le \text{const. } \|f\|_{2, T^{(s)}}. \tag{3.35}$$

By Hilbert's inversion formulae [169], an analogous norm of the harmonic function w_2 can be bounded by the norm in $L_2(C^{(s)})$ of the tangent derivative of the function W:

$$\left\| \frac{\partial w_2}{\partial n} \right\|_{2, C^{(s)}} \le \text{const. } \left\| \frac{\partial w}{\partial z} \right\|_{2, C^{(s)}}. \tag{3.36}$$

By integrating the inequalities (3.35) and (3.36) with respect to s and taking their sum, we eventually get the bound

$$\left\| \frac{\partial w}{\partial n} \right\|_{2, C} \le \text{const. } \left(\left\| \frac{\partial w}{\partial z} \right\|_{2, C} + \|f\|_{2, T} \right) \le \text{const. } \left(\|DU\|_{2, \Pi} + \|u\|_{2, Q}^{(1)} \right).$$

The assertion of the lemma follows from the last estimate after an appropriate transformation of the derivatives to the original variables. $\qquad\square$

4. The quasi-steady one-dimensional Stefan Problem (C)

We are going to study the structure of the generalized solution to Problem (C). We shall show that, for any bounded function $F(z)$, continuous except at $z = 0$, the zero level set of the generalized solution will consist of a finite number of closed intervals, possibly degenerating to single points. Due to the structure of the solution and the integral identity (1.25), we can impose conditions on the generalized solution at the ends of those intervals. The original problem separates into a sequence of autonomous problems to be solved successively. We shall prove the solvability of the latter problems in the simplest case of piecewise constant functions F.

Under the hypothesis that the generalized solution of Problem (C) exists, we proceed to study its structure. Let $F(z)$ be bounded, continuous everywhere except for $z = 0$ and let $zF(z) < 0$ for $z \ne 0$. In addition, we shall assume that

$$\lim_{z \to \infty} F(z) = F_2 < 0, \quad \lim_{z \to -\infty} F(z) = F_1 > 0. \tag{4.1}$$

By the continuity of the function $w(z)$, the set of z such that $w(z) \neq 0$ contains at most a countable number of open non-intersecting maximal intervals I_k. On each I_k, the function $A(w)$ is equal either to w (wherever $w > 0$) or to $w - 1$ (for $w < 0$).

Let supp $\varphi(z)$ be contained in the interval $I_k = (\alpha, \beta)$ and, for definiteness, let $w(z) > 0$ for $z \in I_k$. Then the identity (1.25) reduces to

$$\int_\alpha^\beta \left\{ \left(v_0 w - \frac{dw}{dz} \right) \frac{d\varphi}{dz} - (w - F)\varphi \right\} dz = 0.$$

By the definition of the generalized derivative, the latter equality means that the function $v_0 w - \frac{dw}{dz}$ is differentiable and its derivative is equal to $F - w$:

$$-\frac{d^2 w}{dz^2} + v_0 \frac{dw}{dz} + w = F \qquad \text{for} \quad z \in I_k. \tag{4.2}$$

If I_k does not contain the point $z = 0$, where F has a discontinuity of the first kind, then the solution $w(z)$ is twice continuously differentiable up to the points $z = \alpha$ and $z = \beta$.

In view of the maximum principle, neither the half-line $\mathbb{R}^- = \{z : z < 0\}$ contains the whole interval I_k on which $w(z) < 0$ nor $\mathbb{R}^+ = \{z : z > 0\}$ those I_n where $w(z) > 0$.

Indeed, let for instance $I_k = (\alpha, \beta)$ be a bounded interval in $\mathbb{R}^-$ on which $w(z) < 0$. The function $w(z)$, twice continuously differentiable on I_k, achieves there its minimum:

$$w(\alpha_*) = \min w(z), \quad \alpha_* \in [\alpha, \beta].$$

By definition of I_k, $w(\alpha) = w(\beta) = 0$, hence $\alpha_* \neq \alpha, \beta$. Therefore,

$$\frac{d^2 w}{dz^2}(\alpha_*) \geq 0, \quad \frac{dw}{dz}(\alpha_*) = 0, \quad w(\alpha_*) < 0,$$

in contradiction to equation (4.2), where $F(\alpha_*) > 0$.

If I_k is unbounded from below, i.e., $\alpha = -\infty$, then the asymptotic conditions (1.24) and (4.1) are to be imposed; they imply the existence of α_+ such that $w(\alpha_+) > 0$. The remaining considerations are unchanged, now referring to the interval (α_+, β).

We shall show that the following situations cannot occur on the half-line $\mathbb{R}^-$:
1) the intervals I_k and I_n, where $w(z) > 0$, have a common boundary point;
2) the interval $I = (\alpha, \beta)$ on which $w(z) > 0$ is adjacent to $J = (\beta, \gamma)$, where $w(z) \equiv 0$;
3) the boundary points, α or β, of the interval I on which $w(z) > 0$ represent the limits of the end-points α_n or β_n of the intervals $I_n = (\alpha_n, \beta_n)$, where $w(z) > 0$.

To prove the first statement, note that the function $A(w(z))$ has a removable discontinuity at the point $z = \alpha$ common to both intervals:

$$\lim_{z \to \pm \alpha} A(w(z)) = 0.$$

We can thus assume that $A(w) = w$ everywhere on $\bar{I}_k \cup \bar{I}_n$. The value $w = 0$ is minimal for $z \in \bar{I}_k \cup \bar{I}_n$ and cannot be achieved inside the above interval at $z = \alpha$, leading to a contradiction with the definition of I_k and I_n.

To prove the second statement, let us consider the identity (1.25) with test functions $\varphi(z)$ having support concentrated inside the interval $\mathcal{J} = (\beta, \gamma)$. Because $w(z) = 0$ on $\mathcal{J}$, the function $A(w)$ assumes at each point $z \in \mathcal{J}$ a certain value $\eta(z) \in [-1, 0]$. Then the identity (1.25) reduces to

$$\int_\beta^\gamma \left\{ v_0 \eta \frac{d\varphi}{dz} + F\varphi \right\} dz = 0,$$

which, as in the construction of (4.2), implies that the function $v_0 \eta$ has a generalized derivative equal to F:

$$v_0 \frac{d\eta}{dz} = F, \qquad z \in (\beta, \gamma). \tag{4.3}$$

Further, let us consider the identity (1.25) in the case of test functions with support close to a point $z = \beta$ and non-vanishing at that point, to deduce in a standard way that

$$\frac{dw}{dz}(\beta - 0) = -v_0 \eta(\beta + 0). \tag{4.4}$$

Indeed, let us split the integral in (1.25) into the sum of two integrals, one over the interval (α, β), the other over (β, γ) (we shall assume that φ is finite on the interval (α, γ)):

$$\int_\alpha^\beta \left\{ \left(v_0 w - \frac{dw}{dz} \right) \frac{d\varphi}{dz} + (F - w)\varphi \right\} dz + \int_\beta^\gamma \left\{ v_0 \eta \frac{d\varphi}{dz} + F\varphi \right\} dz = 0.$$

By transposition, using (4.2) and (4.3), we obtain the desired condition (4.4). Because $w(z) > 0$ at $z < \beta$, we have

$$\frac{dw}{dz}(\beta - 0) \le 0.$$

On the other hand, the function $\eta(z)$ is non-positive. Thus, (4.4) can be satisfied only in the case of vanishing left- and right-hand sides:

$$\frac{dw}{dz}(\beta - 0) = 0, \qquad \eta(\beta) = 0. \tag{4.5}$$

From (4.2) at the point $z = \beta$ we see that

$$\frac{d^2 w}{dz^2}(\beta - 0) = -f(\beta) < 0.$$

This together with the representation

$$w(z) = \frac{1}{2} \frac{d^2 w}{dz^2}(\beta - 0)(z - \beta)^2 + o(|z - \beta|^2)$$

implies the strict negativeness of $w(z)$ close to the point $z = \beta$ for $z < \beta$, contradicting the definition of the interval $\mathcal{J}$.

Suppose now that the last situation occurs: $w(z) > 0$ on the interval (α, β) and β is the limit point for the end-points (α_n, β_n) of the intervals I_n where $w(z) > 0$. From the regularity of $w(z)$ on I_n and (4.2), the integral identity (1.25), for finite functions

$\varphi(z)$ equal to 1 on the intervals (β, α_n) and vanishing at the points $z = \alpha, \beta_n$, can be written in the form

$$
0 = \int_\alpha^\beta \left\{ \left(v_0 \omega - \frac{d\omega}{dz} \right) \frac{d\varphi}{dz} + (F - \omega)\,\varphi \right\} dz + \int_\beta^{\alpha_n} (F - \omega)\,dz
$$
$$
+ \int_{\alpha_n}^{\beta_n} \left\{ \left(v_0 \omega - \frac{d\omega}{dz} \right) \frac{d\varphi}{dz} + (F - \omega)\,\varphi \right\} dz
$$
$$
= \int_\beta^{\alpha_n} (F - \omega)\,dz - \frac{d\omega}{dz}(\beta - 0) + \frac{d\omega}{dz}(\alpha_n + 0). \tag{4.6}
$$

By integrating equation (4.2) on the interval (α_n, β_n), we obtain

$$
-\frac{d\omega}{dz}(\beta_n - 0) + \frac{d\omega}{dz}(\alpha_n + 0) = \int_{\alpha_n}^{\beta_n} (F - \omega)\,dz. \tag{4.7}
$$

The function $\omega(z)$ is positive on the interior of I_n. Thus, at $z = \alpha_n$, its derivative is non-negative, and at $z = \beta_n$ the same derivative is non-positive:

$$
\frac{d\omega}{dz}(\alpha_n + 0) \geq 0, \quad \frac{d\omega}{dz}(\beta_n - 0) \leq 0.
$$

Hence, (4.7) implies the bound

$$
\left| \frac{d\omega}{dz}(\alpha_n + 0) \right| + \left| \frac{d\omega}{dz}(\beta_n - 0) \right| \leq \int_{\alpha_n}^{\beta_n} |F - \omega|\,dz.
$$

This, together with (4.6), yields the inequality

$$
\left| \frac{d\omega}{dz}(\beta - 0) \right| \leq \int_\beta^{\beta_n} |F - \omega|\,dz.
$$

From the arbitrariness of β_n, the fact that $\beta_n \to \beta$, and the boundedness of F and ω, the latter estimate implies that

$$
\frac{d\omega}{dz}(\beta - 0) = 0.
$$

We have returned to the situation considered in proving the second assertion. The equality obtained contradicts equation (4.2), the positiveness of the function $\omega(z)$ for $z < \beta$ and its vanishing at $z = \beta$.

In an analogous way we can consider the situations on the positive half-line $\mathbb{R}^+$. Therefore, the only possible cases are:

I.   There is a unique point z_0 such that $\omega(z) > 0$ for $z < z_0$, $\omega(z_0) = 0$, $\omega(z) < 0$ for $z > z_0$ (the classical solution).

II. There are points $z_2 > z_1 \geq 0$ such that $\omega(z) > 0$ for $z < z_1$, $\omega(z) = 0$ for $z \in (z_1, z_2)$ and $\omega(z) < 0$ for $z_1 > z_2$.

Another possible situation refers to $w(z) = 0$ on intervals $(-\infty, y_1)$ or (y_2, ∞), provided that

$$\int_{-\infty}^{0} |F(z)|\, dz < \infty \quad \text{or} \quad \int_{0}^{\infty} |F(z)|\, dz < \infty.$$

In our case, these integrals are infinite, hence the latter situation cannot occur.

In the first situation, the function $w(z)$ satisfies equation (4.2) everywhere except for the point $z = z_0$. At $z = z_0$ this function vanishes and satisfies the Stefan condition

$$\frac{dw}{dz}(z_0 - 0) - \frac{dw}{dz}(z_0 + 0) = v_0. \tag{4.8}$$

In the second situation, $w(z)$ satisfies equation (4.2) everywhere outside the segment (z_1, z_2), while on (z_1, z_2) where $w = 0$, the function $\eta(z)$ is unknown, to be determined from equation (4.3). At the boundary points z_1 and z_2, as in deriving (4.4), we have

$$w(z_1) = 0, \quad -\frac{dw}{dz}(z_1 - 0) = v_0 \eta(z_1 + 0), \tag{4.9}$$

$$w(z_2) = 0, \quad -\frac{dw}{dz}(z_2 + 0) = v_0(\eta(z_2 - 0) + 1). \tag{4.10}$$

In each of the above cases, conditions (1.24) are to be imposed at infinity.

As a matter of fact, the second of conditions (4.9) splits into two independent conditions. Indeed, $w(z)$ is positive for $z < z_1$ and $w(z_1) = 0$. Thus,

$$\frac{dw}{dz}(z_1 - 0) \leq 0.$$

On the other hand, $\eta(z)$ is non-positive for $z \geq z_1$ and the second equation in (4.9) is possible only if

$$\frac{dw}{dz}(z_1 - 0) = 0, \qquad \eta(z_1 + 0) = 0. \tag{4.11}$$

Therefore, the point z_1 and the function $w(z)$ for $z < z_1$ are determined in the second case by the autonomous problem (4.2), (1.24) subject to the first of conditions (4.9) and (4.11). After the point z_1 has been found, the function $\eta(z)$ is defined for $z > z_1$ by equation (4.3) and the second of conditions (4.11) as

$$\eta(z) = \frac{1}{v_0} \int_{z_1}^{z} F(\xi)\, d\xi.$$

The point z_2 and function $w(z)$ for $z > z_2$ can be determined from equation (4.2), condition (1.24) and conditions (4.10), with

$$\eta(z_2 - 0) = \frac{1}{v_0} \int_{z_1}^{z_2} F(\xi)\, d\xi.$$

We now proceed to the proof of Theorem 4. First of all, we show that for small v_0 the first case always occurs.

Let $z_0 > 0$ and solve equation (4.2) for $z < z_0$ and $z > z_0$, to obtain

$$w(z) = F_1 + c_1 e^{\lambda_1 z} \qquad \text{for} \quad z \in (-\infty, 0),$$

$$w(z) = \frac{F_1 - F_2}{\lambda_1 - \lambda_2}(\lambda_1 e^{\lambda_2 z} - \lambda_2 e^{\lambda_1 z}) + F_2 + c_1 e^{\lambda_1 z} \qquad \text{for} \quad z \in (0, z_0),$$

$$w(z) = F_2 + c_2 e^{\lambda_2 z} \qquad \text{for} \quad z \in (z_0, \infty).$$

Here,

$$2\lambda_1 = v_0 + (v_0^2 + 4)^{1/2} > 0, \ 2\lambda_2 = v_0 - (v_0^2 + 4)^{1/2} < 0.$$

The constants c_1 and c_2 are given by the condition $w(z_0 - 0) = w(z_0 + 0) = 0$:

$$c_1 e^{\lambda_1 z_0} = \frac{F_1 - F_2}{\lambda_1 - \lambda_2}(\lambda_2 e^{\lambda_1 z_0} - \lambda_1 e^{\lambda_2 z_0}) - F_2, \qquad c_2 e^{\lambda_2 z_0} = -F_2.$$

The Stefan condition (4.8) is then an equation for determining the point z_0 :

$$e^{\lambda_2 z_0} = -\frac{(\lambda_1 - \lambda_2)F_2 + v_0}{\lambda_1(F_1 - F_2)}. \tag{4.12}$$

Because $\lambda_2 < 0$, the positive root z_0 of equation (4.12) always exists, if

$$0 < \frac{|F_2|(\lambda_1 - \lambda_2) - v_0}{\lambda_1(F_1 - F_2)} \le 1. \tag{4.13}$$

At $v_0 = 0$, $\lambda_1 = \lambda_2$ and (4.13) follows from the hypothesis $F_1 + F_2 > 0$. Since all the expressions considered are continuous, (4.13) is also true for small positive values of v_0.

By a simple calculation we can see that the left inequality (4.13) is satisfied for all $v_0 > 0$, and the right inequality, for

$$0 \le v_0 \le 2|F_2|(1 - |F_2|^2)^{-1/2} = v^*, \tag{4.14}$$

whenever $0 < |F_2| < 1$, and for all $v_0 \ge 0$ if $|F_2| \ge 1$.

The derivative $\frac{dw}{dz}(z_0 - 0)$ is an increasing function of the parameter v_0 :

$$\frac{dw}{dz}(z_0 - 0) = \frac{2 + |F_2|}{2}v_0 + \frac{|F_2|}{2}(v_0^2 + 4)^{1/2} \equiv g(v_0). \tag{4.15}$$

For $v_0 = v_*$, with $v_* = |F_2|(1 + |F_2|)^{-1/2}$, the equality $g(v_*) = 0$ holds. Thus, the first case is realized in the interval $(0, \min(v_*, v^*))$, since for $v > v_*$ (4.15) specifies positive values of the derivative $\frac{dw}{dz}(z_0 - 0)$, but the latter is impossible due to positiveness of the function $w(z)$ for $z < z_0$ and the equality $w(z_0) = 0$. Because $v_* < v^*$ for $|F_2| \le 1$, the above interval coincides with $(0, v_*)$.

For $v_0 > v^*$, the second case is realized. The function $w(z)$ for $z < z_1$ and the point z_1 itself can be determined from the autonomous problem (4.2), (1.24) and the first conditions on (4.9) and (4.11) :

$$w(z) = F_1 + c_1 e^{\lambda_1 z} \qquad \text{for} \quad z \in (-\infty, 0).$$

$$w(z) = \frac{F_1 - F_2}{\lambda_1 - \lambda_2}(\lambda_1 e^{\lambda_2 z} - \lambda_2 e^{\lambda_1 z}) + F_2 + c_1 e^{\lambda_1 z} \qquad \text{for} \quad z \in (0, z_1).$$

As above, the constant c_1 is given by the condition $w(z_1) = 0$:

$$c_1 e^{\lambda_1 z_1} = |F_2| + \frac{F_1 - F_2}{\lambda_1 - \lambda_2} \left(\lambda_2 e^{\lambda_1 z_1} - \lambda_1 e^{\lambda_2 z_1} \right)$$

and the point z_1 follows from the condition (4.11) :

$$z_1 = \frac{1}{\lambda_2} \log \frac{|F_2|}{F_1 - F_2} > 0.$$

For $z > z_1$, the function $\eta(z)$ can be determined from equation (4.3) and the second of conditions (4.11) :

$$\eta(z) = \frac{F_2}{v_0}(z - z_1).$$

By solving the problem (4.2), (1.24) and (4.10), we find the point z_2 and the function $w(z)$ for $z > z_2$:

$$w(z) = F_2 \left(1 - e^{\lambda_2 (z - z_2)} \right) \qquad \text{for} \quad z \in (z_2, \infty),$$

where

$$z_2 = z_1 - (1/F_2) g(v_0).$$

Modelling of binary alloy crystallization

by I.G. Götz and A.M. Meirmanov

Within the phenomenological theory, the binary alloy solidification (crystallization) problem is as usual formulated in the following way: in a given domain Ω determine a smooth surface $\Gamma(t)$ (phase transition boundary) which splits Ω into two subdomains $\Omega^+(t)$ and $\Omega^-(t)$, occupied by the liquid and solid phases, respectively. Whether liquid or solid, the state of a medium at a given point is specified by the phase equilibrium diagram (see Figure A.1) according to the values of temperature ϑ and admixture concentration c : if $\vartheta > \varphi^+(c)$, the medium is in liquid state, if $\vartheta < \varphi^-(c)$, the state is solid.

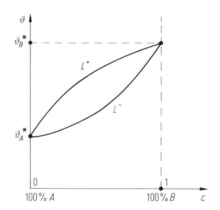

Figure A.1

$L^\pm = \{(\vartheta, c) \ : \ \vartheta = \varphi^\pm(c), 0 < c < 1\} = \{(\vartheta, c) \ : \ c = f^\pm(\vartheta), \vartheta_A^* < \vartheta < \vartheta_B^*\}$
are referred to as the liquidus and solidus curves, respectively. In each of the domains $\Omega^\pm(t)$, the temperature ϑ satisfies the heat equation

$$a\frac{\partial \vartheta}{\partial t} = \operatorname{div}(\kappa \nabla \vartheta) \tag{A.1}$$

and the admixture concentration c satisfies the diffusion equation

$$\frac{\partial c}{\partial t} = \operatorname{div}(D \nabla c), \tag{A.2}$$

where a, κ, D are the heat capacity, thermal conductivity and diffusion coefficient, respectively. As a rule these are assumed to be constant within fixed phases, in general different for different phases. In particular, $a = a^+ = $ const. in the liquid phase, $a = a^- = $ const. in the solid and $a^+ \neq a^-$.

On the free boundary $\Gamma(t)$, the conditions expressing the energy balance (Stefan condition),

$$[U]\, V_n = -\left[\kappa\, \frac{\partial \vartheta}{\partial n}\right], \tag{A.3}$$

and those for the mass balance,

$$[c]\, V_n = -\left[D\, \frac{\partial c}{\partial n}\right], \tag{A.4}$$

are imposed, where $[U] = U^+ - U^-$; $U^\pm$ is the limit of the enthalpy U on $\Gamma(t)$, taken from the side of $\Omega^\pm(t)$, respectively; V_n is the normal velocity of $\Gamma(t)$.

The temperature is assumed to be continuous across $\Gamma(t)$,

$$[\vartheta] = 0 \tag{A.5}$$

and the limit values of the concentration are given by the phase equilibrium diagram,

$$\varphi^+(c^+) = \varphi^-(c^-) = \vartheta^+ = \vartheta^-. \tag{A.6}$$

The model is completed by two conditions on the boundary of Ω prescribed for ϑ and c, and by initial conditions for ϑ, c and the phase transition boundary. In the above model, $\Gamma(t)$ is a strong discontinuity surface for the concentration c and the enthalpy U, thus it seems natural to refer to it as to strongly discontinuous motion (SDM) and to treat its solutions as classical ones (by analogy to the Stefan problem).

The initial-boundary value problem (A.1)–(A.6) (problem (SDM)) is rather complex, the existence of its classical solution has so far been proved only locally in time [230], [231]. Besides, in [5], [232] strange solutions were obtained, with the temperature on both sides of the crystallization boundary below the value given by solidus curve L^- on the phase equilibrium diagram. Those solutions were interpreted as reflecting a supercooling phenomenon, although, despite satisfying the system (A.1)–(A.6), they did not represent a solution of problem (SDM), since $\Gamma(t)$ could not be characterized there as the set separating $\Omega^-(t)$ where $\vartheta < \varphi^-(c)$ and $\Omega^+(t)$ where $\vartheta > \varphi^+(c)$.

So, there is no proof of the global in time existence of the classical solution and in some situations such a solution simply does not exist (in our opinion, this is the essence of the results of [9], [10]). By analogy to the Stefan problem [5], one should expect that in the case of a binary alloy solidification problem also the process is described by a larger class of generalized solutions which includes the classical solutions as a subclass. Various approaches to this problem are developed in [233]–[238].

Our main purpose here is to construct an axiomatic model within the class of generalized motions (further referred to as (GM)) under a minimal number of generally accepted thermodynamic postulates. As for the Stefan problem [5], such a model admits the existence of a mushy phase, where the temperature of the medium is equal to the melting temperature and the enthalpy may assume any value from the segment connecting the appropriate limit values which correspond to the liquid and solid phases.

For the crystallization of a binary alloy, the thermodynamic state of the system is described by seven parameters: the enthalpy U, the temperature ϑ, the entropy S, the concentrations c_A, c_B and the chemical potentials μ_A, μ_B for the components A and B, respectively. We shall assume that these parameters satisfy the Gibbs identity

$$dU = \vartheta \, dS + \mu_A \, dc_A + \mu_B \, dc_B \qquad \text{with} \quad c_A + c_B = 1,$$

and that the enthalpy is dependent only on the temperature,

$$U = \alpha\vartheta + \beta \qquad \text{with} \quad \alpha, \beta = \text{const.} \tag{A.7}$$

By the Gibbs-Duhem theorem [6], which states that

$$\mu_A \equiv \Psi = U - \vartheta S - c\mu, \qquad \text{where} \quad c = c_B, \ \mu = \mu_B - \mu_A, \tag{A.8}$$

it is thus possible to select two independent parameters, for instance the temperature ϑ and the potential μ which define all the remaining parameters and, hence, completely characterize the thermodynamic state of the system.

In terms of ϑ and μ treated as the independent variables, the Gibbs identity assumes the form

$$d\psi = -S \, d\vartheta - c \, d\mu, \tag{A.9}$$

which implies that

$$S = -\frac{\partial\Psi}{\partial\vartheta}, \qquad c = -\frac{\partial\Psi}{\partial\mu}.$$

In particular, treating (A.8) as a first-order equation with respect to Ψ, with the function U given by (A.7), we infer that

$$\Psi(\vartheta, \mu) = \Psi_0\left(\frac{\mu}{\vartheta}\right)\vartheta - \alpha\vartheta \log\vartheta + \beta \tag{A.10}$$

in each phase, with an arbitrary function $\Psi_0(\tau)$.

The first axiom of the problem (GM) states that the identity (A.9) and representation (A.7) hold both in the liquid and solid phases, and that the phase transition from the liquid to solid state (or the reverse) proceeds at local thermodynamic phase equilibrium, i.e., the temperature ϑ and the chemical potentials μ_A, μ_B of the liquid phase are equal to the corresponding parameters of the solid.

The latter statement reflects the existence of a curve $\mu = \Phi(\vartheta)$ in the (ϑ, μ)–plane, separating the domains Π^+ and Π^- such that the values of ϑ and μ in Π^+ define the liquid phase, whereas those in Π^- define the solid; on the above curve, $\Psi^+(\vartheta, \mu) = \Psi^-(\vartheta, \mu)$.

As far as the form of Ψ in each phase is concerned, we have

$$\mu = \Phi(\vartheta) \equiv \vartheta \, \varphi(\vartheta). \tag{A.11}$$

The dependence of ϑ on μ, inverse to (A.11), defines the melting temperature $\vartheta = \vartheta^*(\mu)$. We shall assume that in each phase the enthalpy is an increasing function of temperature and that it increases at the transition from the solid phase to the liquid, i.e.,

either $\quad \alpha^+ \ge \alpha^- > 0, \quad \vartheta \ge \vartheta_0 = -\dfrac{[\beta]}{[\alpha]}$ $\quad$ or $\quad \alpha^- \ge \alpha^+ > 0, \quad \vartheta \le \vartheta_0.$

According to the construction of the problem (GM), we assume that there is a fold located along the phase equilibrium curve (A.11) which conceals the mushy phase. One of the approaches to smoothing the fold consists of passing to variables ϑ and c. To this purpose, from the equations

$$c = -\frac{\partial \Psi^{\pm}}{\partial \mu}$$

we first find the value $\mu = \vartheta g^{\pm}(c)$ and then substitute it into (A.11), to obtain

$$\vartheta = (\varphi^{-1} \circ g^{\pm})(c) \equiv \varphi^{\pm}(c).$$

In the phase equilibrium diagram (see Figure A.1), $\{\vartheta > \varphi^+(c), \ 0 < c < 1\}$ (we shall use also the notation Π^+) corresponds to the domain Π^+ in the variables ϑ, μ, and the solidus curve L^- corresponds to the boundary of the fold (A.11) from the side of the solid phase.

It is easy to recover the phase equilibrium curve (A.11) form the appropriate diagram in (ϑ, c)–plane. To this end, let us differentiate the identity

$$\vartheta \Psi_0^+(\varphi(\vartheta)) - \alpha^+ \vartheta \log \vartheta + \beta^+ = \vartheta \Psi_0^-(\varphi(\vartheta)) - \alpha^- \vartheta \log \vartheta + \beta^-$$

with respect to ϑ, to obtain by virtue of (A.7)

$$\frac{d\varphi}{d\vartheta} = \frac{1}{\vartheta^2} \frac{U^+ - U^-}{c^+ - c^-} = \frac{1}{\vartheta^2} \frac{[U]}{[c]}.$$

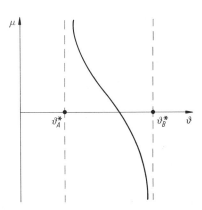

Figure A.2

We shall assume that the solidus and liquidus curves in the phase equilibrium diagram approach the lines $c = 0$ and $c = 1$ at different angles, other than $k\pi/2$, k an integer. Then the function φ in (A.11) has a logarithmic singularity at the points ϑ_A^* and ϑ_B^*, and the potential μ is unbounded along (A.11) when ϑ approaches these values. An example of the phase equilibrium curve in the (ϑ, μ)–plane is shown in Figure A.2.

The domain Π^* in the (ϑ, c)–plane, bounded by the solidus and liquidus curves, corresponds to the mushy phase. How can the thermodynamic parameters Ψ, U, S and μ be extended to the mushy phase? First of all, it is natural to assume the identity (A.9) and representation (A.8) for the potential Ψ. Another natural hypothesis concerns the existence of a process in which a binary mixture passes from the liquid state to the solid (or the reverse transition occurs) via an intermediate mixed state (in the domain Π^* of variables ϑ, c, the continuous curve connecting the liquidus and solidus curves corresponds to this process) so that the chemical potentials μ_A and μ_B (or potentials μ and Ψ) vary continuously. Thus, the potentials μ and Ψ are well-defined and continuous for parameters ϑ, c in the domain Π^*.

The continuity hypothesis is essential for ensuring that the potentials μ and Ψ can be uniquely recovered in the domain Π^*. To this end it is enough to consider the segments $I_\vartheta = \{(\vartheta, c) : \vartheta = \text{const.}\} \subset \Pi^*$ connecting the liquidus and solidus curves. In the (ϑ, μ)–plane, a certain point of the phase equilibrium curve (A.11) corresponds to the segment I_ϑ, i.e., the value of μ is constant along I_ϑ in Π^*. By (A.9) and (A.11) it follows also that Ψ is constant on I_ϑ.

By the equality $F = \Psi + c\mu$, together with the continuity of the potentials Ψ and μ, the free energy F is continuous in the domain Π. We can rewrite (A.9) in the form

$$dF = -S \, d\vartheta + \mu \, dc,$$

to conclude that the entropy S also is continuous on the liquidus and solidus curves. Indeed, the latter identity is equivalent to the equalities

$$\mu = \frac{\partial F}{\partial c}, \qquad S = -\frac{\partial F}{\partial \vartheta}.$$

Let

$$F = F^*, \ \mu = \mu^* \ \text{in} \ \Pi^* \quad \text{and} \quad F = F^+, \ \mu = \mu^+ \ \text{in} \ \Pi^+.$$

We have

$$F^+(\vartheta, f^+(\vartheta)) = F^*(\vartheta, f^+(\vartheta))$$

and

$$\mu^*(\vartheta, f^+(\vartheta)) = \frac{\partial F}{\partial c}(\vartheta, f^+(\vartheta)) = \frac{\partial F^+}{\partial c}(\vartheta, f^+(\vartheta)) = \mu^+(\vartheta, f^+(\vartheta)).$$

Differentiating the relation $F^+ = F^*$ along the liquidus curve, we obtain the equality

$$\frac{\partial F^+}{\partial \vartheta} + \mu^+ \frac{\partial f^+}{\partial \vartheta} = \frac{\partial F^*}{\partial \vartheta} + \mu^* \frac{\partial f^+}{\partial \vartheta},$$

which implies the continuity of S across the liquidus curve. The behaviour on the solidus curve can be considered in a similar way.

The continuity of S and the relation $U = F + \vartheta S$ imply that the entropy is continuous everywhere in Π. By the obvious corollary from the Gibbs identity,

$$\frac{\partial}{\partial \vartheta}\left(\frac{\mu}{\vartheta}\right) + \left(\frac{1}{\vartheta^2}\right)\frac{\partial U}{\partial c} = 0$$

and the definition of μ in the mushy phase, we infer that the enthalpy is linearly dependent on the concentration in the mushy phase,

$$U = \alpha^- \vartheta + \beta^- + \frac{[\alpha]\vartheta + [\beta]}{f^+(\vartheta) - f^-(\vartheta)}(c - f^-(\vartheta)), \qquad (A.12)$$

where $f^+(\vartheta) < c < f^-(\vartheta)$, $\vartheta_A^* < \vartheta < \vartheta_B^*$.

The second axiom (A.2) of the model (GM) postulates that all states which correspond to values of the parameters ϑ and c from Π^*, are physically feasible, (A.9) and (A.7) hold in Π^*, and the chemical potentials μ_A, μ_B are continuous on Π.

The potential μ is unbounded at the extreme points of the phase equilibrium curve, thus it turns out more convenient to take ν given by $\vartheta\varphi(\nu) = \mu$ as a test function. Referring to the special form of the potential Ψ and the concentration c, we can see that the corresponding potential ν depends only on c. By construction, $\nu = \vartheta$ in the mushy phase, on the boundary Π^*, as well as on the liquidus and solidus curves L^+ and L^-,

$$\nu = \vartheta = \varphi^+(c) \text{ on } L^+ \quad \text{and} \quad \nu = \vartheta = \varphi^-(c) \text{ on } L^-.$$

Hence,

$$\nu = \varphi^+(c) \text{ in } \Pi^+ \quad \text{and} \quad \nu = \varphi^-(c) \text{ in } \Pi^-.$$

From a more detailed analysis we can see that the fold, located along the curve (A.11) in (ϑ, μ)–plane, has not been completely smoothed in the (ϑ, c)–plane: at the points $(\vartheta_A^*, 0)$ and $(\vartheta_B^*, 1)$, the mushy states that corresponded to transitions of pure materials A and B remained mixed. In the new variables U and ν, the fold has been completely opened out. In the (U, ν)–plane, the liquid phase is transformed onto the domain

$$\Pi^+ = \{U \geq \alpha^+\nu + \beta^+, \; \vartheta_A^* \leq \nu \leq \vartheta_B^*\},$$

the solid phase onto the domain

$$\Pi^- = \{U \leq \alpha^-\nu + \beta^-, \; \vartheta_A^* \leq \nu \leq \vartheta_B^*\},$$

and the mushy phase onto

$$\Pi^* = \{\alpha^-\nu + \beta^- \leq U \leq \alpha^+\nu + \beta^+, \; \vartheta_A^* \leq \nu \leq \vartheta_B^*\}.$$

The mushy state of pure materials A and B is transformed there onto the segments $\{\vartheta_A^* = \nu\}$ and $\{\vartheta_B^* = \nu\}$ in Π^*.

Figures A.3 and A.4 show typical dependencies of ϑ and c upon the enthalpy U (at $\nu = $ const.) and upon the potential ν (at $U = $ const.). By construction, the functions ϑ and c are continuous, while their first derivatives have discontinuities of the first kind. The discontinuity points of the derivatives of c shown in Figure A.4 move along the liquidus and solidus curves so that the function $c(U, \nu)$ increases with ν for all t.

The thermodynamic state of a binary mixture is characterized by two independent parameters (ϑ and c, if $0 < c < 1$, otherwise U and ν). Therefore, for a complete description of the system evolution we need two additional equations, expressing energy and mass balances. Those equations should remain valid for all values of the independent thermodynamic parameters in Ω for all t.

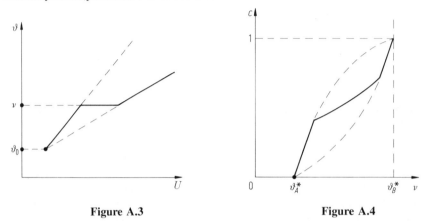

Figure A.3   Figure A.4

Since the model (GM) represents an extension of (SDM), within domains occupied by pure liquid and solid phases the energy balance law should coincide with the heat equation (A.1), and the mass balance law with the diffusion equation (A.2).

For the reference Stefan problem [5], the above construction leads to the equation

$$\frac{\partial U}{\partial t} = \text{div}\,(\kappa \nabla \vartheta), \tag{A.13}$$

with the dependence of the temperature ϑ on the enthalpy U at a given value of ν depicted in Figure A.3. If $\nu = \text{const.}$, (the melting temperature of pure material is constant), then the coefficient κ in (A.13) is defined in the liquid and solid phases; in the mushy phase this coefficient can only be assumed to be bounded, with $\nabla \vartheta \equiv 0$. If $\nu = \nu(x,t)$, the thermal conductivity has to be extended to the mushy phase. Since no additional indications on how to construct such an extrapolation are available, we shall apply the simplest method, assuming that κ is a linear function of U, continuously connecting the values κ^+ and κ^-.

Equation (A.13) has now been completely defined. According to its physical sense, the function U in the evolution term of (A.13) can exhibit discontinuities of the first kind. Then, as in [15], this equation can be interpreted in the weak sense as the integral identity

$$\int_{\Omega_T} \left\{ -U \frac{\partial \zeta}{\partial t} + \kappa\,\nabla \vartheta \nabla \zeta \right\} dx dt = 0 \tag{A.14}$$

to be satisfied for all smooth finite functions ζ in $\Omega_T \times (0,T)$. Clearly, if there is a smooth surface $\Gamma(t)$ in Ω which splits Ω into subdomains $\Omega^\pm(t)$ occupied by the liquid ($U \geq \alpha^+ \nu + \beta^+$) and solid phases ($U \leq \alpha^- \nu + \beta^-$), appropriately, then the temperature ϑ satisfies equation (A.1) in $\Omega^\pm(t)$ and the Stefan condition (A.3) on $\Gamma(t)$.

As for the Stefan problem, we shall assume that in the problem (GM) equation (A.13) is satisfied, with the coefficient κ equal to κ^+ and to κ^- in Π^+ and Π^-, respectively, linearly dependent on U (if U and ν are taken as independent variables) or on c (with ϑ and c treated as the independent variables) in the domain Π^* and continuous on the whole of Π. With ϑ and c treated as the independent variables, the enthalpy depends on the temperature according to (A.7) in Π^+ and Π^-, and according to (A.12) in Π^*. For U and ν being the independent variables, $\vartheta = \vartheta(U, \nu)$ is given by the diagram in Figure A.3.

Let us now consider the mass balance law in the form of equation (A.2). By an extension of the coefficient D onto the mushy phase (analogous to the construction in [18]), we at once eliminate the classical solutions of the problem (SDM). It is known [15] that all solutions of equation (A.2) are continuous for $D \geq D_0 = \text{const.} > 0$, while the concentration c in the classical solution has a finite jump on the phase transition surface $\Gamma(t)$.

Actually, the diffusion equation for an admixture has a form different from (A.2). Let us assume the existence of a mass flux vector $\mathbf{j}$ such that

$$\frac{\partial c}{\partial t} = \text{div } \mathbf{j}.$$

Then, by virtue of the Gibbs formula $\vartheta dS = dU + \mu \, dc$ and by the entropy production principle for any thermally isolated bulk volume ω,

$$\frac{d}{dt} \int_\omega S \, dx \geq 0, \quad \mathbf{j} \cdot \mathbf{n} \, |_{\partial\omega} = 0, \quad \nabla\vartheta \cdot \mathbf{n} \, |_{\partial\omega} = 0,$$

combined with Onsager's symmetry principle [6], one can derive the representation

$$\mathbf{j} = \Lambda\nabla\left(\frac{\mu}{\vartheta}\right) = \lambda\nabla\nu$$

which underlies the admixture diffusion equation

$$\frac{\partial c}{\partial t} = \text{div } (\lambda\nabla\nu). \tag{A.15}$$

The latter equation conforms with (A.2) in the liquid and solid phases, provided that

$$\lambda^\pm \frac{d\varphi^\pm}{dc}(c) = D^\pm.$$

In the mushy phase, the coefficient λ admits an extension as a linear function of the enthalpy, continuous for all values of the independent parameters. Equation (A.15), analogously to (A.13), is to be understood in the generalized sense of an appropriate integral identity. If ϑ and ν are continuous in Ω_T in the solutions of the latter identity, and there exists a smooth surface $\Gamma(t)$ in Ω which separates the domains $\Omega^\pm(t)$ occupied by the liquid and solid phases, then (A.15) induces the diffusion equation (A.2) for the concentration in each of the domains $\Omega^\pm(t)$ and the mass balance condition (A.4) on the boundary $\Gamma(t)$.

Conditions (A.5) and (A.6) follow as a consequence of the continuity of ϑ and ν. Hence, the model (GM) comprehends (SDM) as a special case. The axiom which completes the construction of the problem (GM) by giving a characterization of the coefficients κ and λ is based on the assumption that equations (A.13) and (A.15) are fulfilled everywhere in Ω_T.

The problem (GM) is more general than (SDM). This follows already in the simplest one-dimensional case when $\Omega = (0,1)$. We now recall the main results proved in [239] in this respect.

In the one-dimensional stationary problem, the functions U, ν and ϑ are to be determined which satisfy the system of equations

$$\frac{d}{dx}\left\{\kappa(U,\nu)\frac{d\vartheta}{dx}\right\} = 0, \quad \frac{d}{dx}\left\{\lambda(U,\nu)\frac{d\nu}{dx}\right\} = 0 \tag{A.16}$$

in sense of distributions in Ω, subject to the conditions

$$\vartheta(i) = \vartheta^i, \quad \nu(i) = \nu^i, \quad i = 0, 1. \tag{A.17}$$

For a given solution U, ν, ϑ, the admixture concentration c is defined by the above state equations.

Theorem 1. *Assume that the system can be only either in liquid or solid state on the boundary of Ω, i.e., $(\vartheta^0 - \nu^0)(\vartheta^1 - \nu^1) \neq 0$ and that the strictly monotone functions $f^\pm(\tau)$ are twice continuously differentiable on the segment $[X, Y]$ which includes the values ν^0 and ν^1, with*

$$(\pm 1)(f^\pm)''(\tau) \geq 0 \quad \text{for} \quad \tau \in [X, Y].$$

Then there exists at least one generalized solution of problem (A.16), (A.17) such that $\vartheta, \nu \in W_2^1(\Omega)$. The entire domain Ω admits a partition into at most three segments occupied by different phases, with the interior segment, if there is one, filled by the mushy phase. If

$$\frac{d^2}{d\tau^2}f^\pm(\tau) \equiv 0,$$

then there is no mushy phase at all.

Theorem 2. *Let the hypotheses of Theorem 1 hold, with $|\vartheta^1 - \vartheta^0| = \alpha \neq 0$ and let*

$$\frac{d^2}{d\tau^2}f^+(\vartheta^0) \neq 0 \quad \text{for } \nu^0 < \vartheta^0, \qquad \frac{d^2}{d\tau^2}f^-(\vartheta^0) \neq 0 \quad \text{for } \nu^0 > \vartheta^0,$$

$$\frac{d^2}{d\tau^2}f^+(\vartheta^1) \neq 0 \quad \text{for } \nu^1 < \vartheta^1, \qquad \frac{d^2}{d\tau^2}f^-(\vartheta^1) \neq 0 \quad \text{for } \nu^1 > \vartheta^1.$$

Then there exists $\varepsilon = \varepsilon(\alpha)$ such that, for $|\vartheta^0 - \nu^0| < \varepsilon$ and $|\vartheta^1 - \nu^1| < \varepsilon$, the solution of problem (A.16), (A.17) describes the crystallization process with a mushy phase arising.

References

[1] R. Alexander, P. Manselly and K. Miller: Moving finite elements for the Stefan problem in two dimensions. Atti Accad. Naz. Lincei Rend. Cl. Sci. Fis. Mat. Natur. *(8), 67* (1979), no. 1-2, 57–61 (1980).

[2] S.N. Antontsev, A.V. Kazhikhov and V.N. Monakhov: *Boundary value problems of non-homogeneous fluid mechanics.* Nauka, Novosibirsk (1983). *(Russian)*

[3] D.R. Atthey: A finite difference scheme for melting problems. J. Inst. Math. Appl. *13* (1974), 353–366.

[4] D.R. Atthey: A finite difference scheme for melting problems based on the method of weak solutions. In: [183], 182–191.

[5] N.A. Avdonin: *A mathematical description of crystallization processes.* Zinatne, Riga (1980). *(Russian)*

[6] R.D. Bachelis and V.G. Melamed: Solution by the straight-line method of a quasi-linear two-phase problem of the Stefan type with weak constraints on the input data of the problem. Zh. Vychisl. Mat. i Mat. Fiz. *12* (1972), 828–829 = U.S.S.R. Comput. Math. and Math. Phys. *12* (1972), 342–343.

[7] R.D. Bachelis, V.G. Melamed and D.B. Shlaifer: Solution of Stefan's problem by the method of lines. Zh. Vychisl. Mat. i Mat. Fiz. *9* (1969), 113–126 = U.S.S.R. Comput. Math. and Math. Phys. *9* (1969), 113–126.

[8] J. Baumeister, K.-H. Hoffmann and P. Jochum: Numerical solution of a parabolic free boundary problem via Newton's method. J. Inst. Math. Appl. *25* (1980), 99–109.

[9] B.V. Bazalii: On an existence proof for the two-phase Stefan problem. In: Mathematical analysis and probability theory. Naukova Dumka, Kiev (1978), 7–11. *(Russian)*

[10] B.V. Bazalii: Stability of smooth solutions of the two-phase Stefan problem. Dokl. Akad. Nauk SSSR *262* (1982), 265–269 = Soviet Math. Dokl. *25* (1982), 46–50.

[11] B.V. Bazalii and V.Yu. Shelepov: On a quasi-stationary thermophysical free boundary problem. In: Boundary value problems of thermophysics. Naukova Dumka, Kiev (1975), 45–56. *(Russian)*

[12] B.V. Bazalii and V.Yu. Shelepov: The asymptotic behaviour of the solution of a Stefan problem. Dokl. Akad. Nauk Ukrain. SSR Ser. A, *no. 12* (1978), 1059–1061. *(Russian)*

[13] B.V. Bazalii and V.Yu. Shelepov: Stabilization rate estimates for the solution of Stefan problem. Dokl. Akad. Nauk Ukrain. SSR Ser. A, *no. 4* (1981), 4–6. *(Russian)*

[14] A.E. Berger, H. Brezis and J.C.W. Rogers: A numerical method for solving the problem $u_t - \Delta f(u) = 0$. RAIRO Modél. Math. Anal. Numér. *13* (1979), 297–312.

[15] L. Bers, F. John and M. Schechter: *Partial differential equations.* Interscience, New York (1964).

[16] M. Bertsch, P. de Mottoni and L.A. Peletier: Degenerate diffusion and the Stefan problem. Nonlinear Anal. *8* (1984), 1311–1336.

[17] G.I. Bizhanova: Application of a small parameter method to solving a problem of Stefan type. In: Abstracts of the All-Union Conference on Asymptotic Methods in the Theory of Singularly Perturbed Equations. Part 2. Nauka. Alma-Ata (1979), 191–192. *(Russian)*

[18] G.I. Bizhanova: Stabilization of the solution of Stefan's second boundary value problem. Izv. Akad. Nauk Kazakh. SSR Ser. Fiz.-Mat. *5* (1980), 12–17. *(Russian)*

[19] D. Blanchard and M. Frémond: The Stefan problem: computing without the free boundary. Internat. J. Numer. Methods Engrg. *20* (1984), 757–771.

[20] C. Bonacina, G. Comini, A. Fasano and M. Primicerio: Numerical solution of phase change problems. Int. J. Heat Mass Transfer *16* (1973), 1825–1832.

[21] R. Bonnerot and P. Jamet: A second order finite element method for the one-dimensional Stefan problem. Internat. J. Numer. Methods Engrg. *8* (1974), 811–820.

[22] R. Bonnerot and P. Jamet: Numerical computation of the free boundary for the two-dimensional Stefan problem by space-time finite elements. J. Comput. Phys. *25* (1977), 163–181.

[23] R. Bonnerot and P. Jamet: A conservative finite element method for one-dimensional Stefan problems with appearing and disappearing phases. J. Comput. Phys. *41* (1981), 357–388.

[24] G. Borgioli, E. DiBenedetto and M. Ughi: Stefan problems with nonlinear boundary conditions: the polygonal method. ZAMM *58* (1978), 539–546.

[25] M.A. Borodin: An existence theorem for the solution of a single-phase quasi-stationary Stefan problem. Dokl. Akad. Nauk Ukrain. SSR Ser. A, *no. 7* (1976), 582–585. *(Russian)*

[26] M.A. Borodin: A one-phase quasi-stationary Stefan problem. Dokl. Akad. Nauk Ukrain. SSR Ser. A, *no. 9* (1977), 775–777. *(Russian)*

[27] M.A. Borodin: On the solvability of a two-phase quasistationary Stefan problem. Dokl. Akad. Nauk Ukrain. SSR Ser. A, *no. 2* (1982), 3–5. *(Russian)*

[28] M.A. Borodin: Solvability of a two-phase nonstationary Stefan problem. Dokl. Akad. Nauk SSSR *263* (1982), 1040–1042 = Soviet Math. Dokl. *25* (1982), 469–471.

[29] M.A. Borodin and U. Felgenhauer: A one-phase quasilinear Stefan problem. Dokl. Akad. Nauk Ukrain. SSR Ser. A, *no. 2* (1978), 99–101. *(Russian)*

[30] M.A. Borodin and U. Felgenhauer: An axially symmetric single-phase Stefan problem. In: Mat. Fiz. *24* (1978), 74–76. *(Russian)*

[31] A. Bossavit and A. Damlamian: Homogenization of the Stefan problem and application to magnetic composite media. IMA J. Appl. Math. *27* (1981), 319–334.

[32] B.M. Budak and M.Z. Moskal: On the classical solution of the multidimensional Stefan problem. Dokl. Akad. Nauk SSSR *184* (1969), 1263–1266 = Soviet Math. Dokl. *10* (1969), 219–223.

[33] B.M. Budak and M.Z. Moskal: Classical solution of the multidimensional multi-front Stefan problem. Dokl. Akad. Nauk SSSR *188* (1969), 9–12 = Soviet Math. Dokl. *10* (1969), 1043–1046.

[34] B.M. Budak and M.Z. Moskal: On the classical solution of the Stefan problem with boundary conditions of the first kind for the multidimensional heat equation in a coordinate parallelepiped. In: Solutions of Stefan problems. Izd. MGU, Moscow (1970/71), 87–133. *(Russian)*

[35] L.A. Caffarelli: The smoothness of the free surface in a filtration problem. Arch. Rational Mech. Anal. *63* (1976), 77–86.

[36] L.A. Caffarelli: The regularity of elliptic and parabolic free boundaries. Bull. AMS *82* (1976), 616–618.

[37] L.A. Caffarelli: The regularity of free boundaries in higher dimensions. Acta Math. *139* (1977), 155–184.

[38] L.A. Caffarelli: Some aspects of the one-phase Stefan problem. Indiana Univ. Math. J. *27* (1978), 73–77.

[39] L.A. Caffarelli and L.C. Evans: Continuity of the temperature in two-phase Stefan problems. In: [84], 380–382.

[40] J.R.Cannon and E. DiBenedetto: On the existence of weak-solutions to an n-dimensional Stefan problem with nonlinear boundary conditions. SIAM J. Math. Anal. *11* (1980), 632–645.

[41] J.R.Cannon and J. Douglas, Jr: The stability of the boundary in a Stefan problem. Ann. Scuola Norm. Sup. Pisa Cl. Sci. *21* (1967), 83–91.

[42] J.R.Cannon, J. Douglas, Jr and C.D. Hill: A multi-boundary Stefan problem and the disappearance of phases. J. Math. Mech. *17* (1967), 21–33.

[43] J.R.Cannon and A. Fasano: A nonlinear parabolic free boundary problem. Ann. Mat. Pura Appl. *92* (1977), 119–149.

[44] J.R.Cannon, D.B. Henry and D.B. Kotlow: Continuous differentiability of the free boundary for weak solutions of the Stefan problem. Bull. AMS *80* (1974), 45–48.

[45] J.R.Cannon, D.B. Henry and D.B. Kotlow: Classical solutions of the one-dimensional two-phase Stefan problem. Ann. Mat. Pura Appl. *107* (1975), 311–341.

[46] J.R.Cannon and C.D. Hill: Existence, uniqueness, stability, and monotone dependence in a Stefan problem for the heat equation. J. Math. Mech. *17* (1967), 1–19.

[47] J.R.Cannon and C.D. Hill: Remarks on a Stefan problem. J. Math. Mech. *17* (1967), 433–441.

[48] J.R.Cannon and C.D. Hill: On the infinite differentiability of the free boundary in a Stefan problem. J. Math. Anal. Appl. *22* (1967), 385–397.

[49] J.R.Cannon, C.D. Hill and M. Primicerio: The one-phase Stefan problem for the heat equation with boundary temperature specification. Arch. Rational Mech. Anal. *39* (1970), 270–274.

[50] J.R.Cannon and M. Primicerio: Remarks on the one-phase Stefan problem for the heat equation with the flux prescribed on the fixed boundary. J. Math. Anal. Appl. *35* (1971), 361–373.

[51] J.R.Cannon and M. Primicerio: A two-phase Stefan problem with temperature boundary conditions. Ann. Mat. Pura Appl. *88* (1971), 177–191.

[52] J.R.Cannon and M. Primicerio: A two-phase Stefan problem with flux boundary conditions. Ann. Mat. Pura Appl. *88* (1971), 193–205.

[53] J.R.Cannon and M. Primicerio: A two-phase Stefan problem: regularity of the free boundary. Ann. Mat. Pura Appl. *88* (1971), 217–228.

[54] J.R.Cannon and M. Primicerio: A Stefan problem involving the appearance of a phase. SIAM J. Math. Anal. *4* (1973), 141–148.

[55] C.Y. Chan: Continuous dependence on the data for a Stefan problem. SIAM J. Math. Anal. *1* (1970), 282–287.

[56] O.M. Chekmareva: On the application of the Cauchy integral to the study of Stefan-type problems. In: Problems of mathematical physics. Nauka, Leningrad (1976), 193–197. *(Russian)*

[57] J.R. Ciavaldini: Analyse numérique d'un problème de Stefan à deux phases par une méthode d'éléments finis. SIAM J. Numer. Anal. *12* (1975), 464–487.

[58] M. Ciment and R.B. Guenther: Numerical solutions of a free boundary value problem for parabolic equations. Appl. Anal. *4* (1974), 39–62.

[59] A.B. Crowley: Numerical solution of Stefan problems. Int. J. Heat Mass Transfer *21* (1978), 215–219.

[60] A.B. Crowley and J.R. Ockendon: A Stefan problem with a non-monotone boundary. J. Inst. Math. Appl. *20* (1977), 269–281.

[61] A. Damlamian: Some results on the multi-phase Stefan problem. Comm. Partial Differential Equations *2* (1977), 1017–1044.

[62] A. Damlamian: Homogénéisation du problème de Stefan. C.R. Acad. Sci. Paris, *Sér. A-289* (1979), 9–11.

[63] A. Damlamian: The homogenization of the Stefan problem and related topics. In: [143], Vol. I, 267–275.

[64] A. Damlamian: How to homogenize a nonlinear diffusion equation: Stefan's problem. SIAM J. Math. Anal. *12* (1981), 306–313.

[65] A. Damlamian and N. Kenmochi: Le problème de Stefan avec conditions latérales variables. Hiroshima Math. J. *10* (1980), 271–293.

[66] I.I. Danilyuk: On a variant of the two-phase Stefan problem. Dokl. Akad. Nauk Ukrain. SSR Ser. A, *no. 9* (1973), 783–787. *(Russian)*

[67] I.I. Danilyuk: On the crystallization process in pattern formation. In: Mat. Fiz. *17* (1975), 99–111. *(Russian)*

[68] I.I. Danilyuk: On the two-phase quasistationary Stefan problem. Dokl. Akad. Nauk Ukrain. SSR Ser. A, *no. 1* (1982), 6–10. *(Russian)*

[69] I.I. Danilyuk: On the multidimensional single-phase quasistationary Stefan problem. Dokl. Akad. Nauk Ukrain. SSR Ser. A, *no. 1* (1984), 13–17. *(Russian)*

[70] I.I. Danilyuk and S.V. Salei: On a variant of the two-phase Stefan problem with heat sources. Dokl. Akad. Nauk Ukrain. SSR Ser. A, *no. 11* (1975), 972–976. *(Russian)*

[71] A. Datzeff: Sur le problème linéaire de Stefan. In: Mém. Sci. Phys. Fasc. *69*. Gauthiers-Villars, Paris (1970).

[72] E. DiBenedetto and R.E. Showalter: A pseudo-parabolic variational inequality and the Stefan problem. MRC, Madison (Wisconsin), Technical Report *no. 2100* (1980).

[73] S.S. Domalevskii: Influence of thermoelastic stresses on an erosion of electric contacts. Izv. Vuzov. Elektromekhanika *1* (1978), 14–17. *(Russian)*

[74] S.S. Domalevskii, E.I. Kim and S.N. Kharin: A model of thermoelastic destruction of electrodes at an impulse discharge. Izv. Akad. Nauk Kazakh. SSR Ser. Fiz.-Mat. *5* (1975), 9–14. *(Russian)*

[75] G. Duvaut: Résolution d'un problème de Stefan. C. R. Acad. Sci. Paris, *Sér. A-276* (1973), 1461–1463.

[76] G. Duvaut: The solution of a two-phase Stefan problem by a variational inequality. In: [183], 173–181.

[77] G. Duvaut: Two phases Stefan problem with varying specific heat coefficients. An. Acad. Brasil. Ciênc. *47* (1975), 377–380.

[78] G. Duvaut: Stefan problem for two-phases varying. Memórias de Math. Univ. Fed. Rio de Janeiro *51* (1975).

[79] C.M. Elliott and J.R. Ockendon: *Weak and variational methods for moving boundary problems.* Pitman Res. Notes Math. Ser. *59*, London (1982).

[80] A. Fasano and M. Primicerio: General free-boundary problems for the heat equation. Part I. J. Math. Anal. Appl. *57* (1977), 694–723.

[81] A. Fasano and M. Primicerio: General free-boundary problems for the heat equation. Part II. J. Math. Anal. Appl. *58* (1977), 202–231.

[82] A. Fasano and M. Primicerio: General free-boundary problems for the heat equation. Part III. J. Math. Anal. Appl. *59* (1977), 1–14.

[83] A. Fasano and M. Primicerio: Free boundary problems for nonlinear parabolic equations with nonlinear free boundary conditions. J. Math. Anal. Appl. *72* (1979), 247–273.

[84] A. Fasano and M. Primicerio (Eds.): *Free boundary problems: theory and applications*, Vols. I, II. Pitman Res. Notes Math. Ser. *78, 79*, Boston (1983).

[85] A. Fasano and M. Primicerio: A parabolic-hyperbolic free boundary problem. SIAM J. Math. Anal. *17* (1986), 67–73. — Mushy regions with variable temperature in melting processes. Boll. Un. Mat. Ital. B *4* (1985), 601–626.

[86] A. Fasano, M. Primicerio and S. Kamin: Regularity of weak solutions of one-dimensional two-phase Stefan problems. Ann. Mat. Pura Appl. *115* (1977), 341–348.

[87] A. Fasano, M. Primicerio and L.I. Rubinstein: A model problem for heat conduction with a free boundary in a concentrated capacity. J. Inst. Math. Appl. *26* (1980), 327–347.

[88] U. Felgenhauer: On a one-phase non-stationary Stefan problem. Dokl. Akad. Nauk Ukrain. SSR Ser. A, *no. 1* (1981), 30–32. *(Russian)*

[89] M. Frémond: Variational formulation of the Stefan problem. Coupled Stefan problem. Frost propagation in porous media. In: Proc. Int. Conf. on Computational Methods in Nonlinear Mechanics. Austin (Texas) (1974), 341–350.

[90] M. Frémond: Diffusion problems with free boundaries. Autumn course on applications of analysis to mechanics. ICTP, Trieste (1976).

[91] A. Friedman: *Partial differential equations of parabolic type*. Prentice-Hall, Englewood Cliffs, N.J. (1964).

[92] A. Friedman: The Stefan problem in several space variables. Trans. AMS *132* (1968), 51–87. Correction: Ibid. *142* (1969), 557.

[93] A. Friedman: One dimensional Stefan problems with non-monotone free boundary. Trans. AMS *133* (1968), 89–114.

[94] A. Friedman: Analyticity of the free boundary for the Stefan problem. Arch. Rat. Mech. Anal. *61* (1976), 97–125.

[95] A. Friedman: *Variational principles and free boundary problems*. John Wiley, New York (1982).

[96] A. Friedman and R. Jensen: A parabolic quasi-variational inequality arising in hydraulics. Ann. Scuola Norm. Sup. Pisa Cl. Sci. *2* (1975), 421–468.

[97] A. Friedman and D. Kinderlehrer: A one phase Stefan problem. Indiana Univ. Math. J. *24* (1975), 1005–1035.

[98] K.K. Golovkin: Two classes of inequalities for sufficiently smooth functions of n variables. Dokl. Akad. Nauk SSSR *138* (1961), 22–25 = Soviet Math. Dokl. *2* (1961), 510–513.

[99] S.P. Gorodnichev: Improving the accuracy refinement of an integral method for the solution of one-phase Stefan problems. Izv. Akad. Nauk Kazakh. SSR Ser. Fiz.-Mat. *5* (1977), 16 - 21. *(Russian)*

[100] D. Greenspan: A particle model of the Stefan problem. Comp. Math. Appl. Mech. Eng. *13* (1978), 95–104.

[101] S.G. Grigor'ev: A package of applied codes for approximate numerical and analytical solution of Stefan problems. In: Packages of applied computer programs. Methods and advances. Nauka, Novosibirsk (1981), 201–205. *(Russian)*

[102] S.G. Grigor'ev, V.N. Kosolapov, M.A. Pudovkin and V.A. Chugunov: A scheme of the generalized method of integral relations for one-phase Stefan problems and its applications. In: Applied problems of theoretical and mathematical physics. Izd. Latv. Univ. im. P. Stuchki, Riga (1980), 43–52. *(Russian)*

[103] R.S. Gupta and D. Kumar: Variable time step methods for a one-dimensional Stefan problem with mixed boundary condition. Int. J. Heat Mass Transfer *24* (1981), 251–259.

[104] E.I. Hanzawa: Classical solution of the Stefan problem. Tohoku Math. J. *33* (1981), 297–335.

[105] K.-H. Hoffmann: Monotonie bei Zweiphasen-Stefan-Problemen. Numer. Funct. Anal. Optim. *1* (1979), 79–112.

[106] Y. Ichikawa and N. Kikuchi: A one-phase multi-dimensional Stefan problem by the method of variational inequalities. Internat. J. Numer. Methods Engrg. *14* (1979), 1197–1220.

[107] R. Jensen: Smoothness of the free boundary in the Stefan problem with supercooled water. Illinois J. Math. *22* (1978), 623–629.

[108] J.W. Jerome: Existence and approximation of weak solutions of the Stefan problem with nonmonotone nonlinearities. In: Numerical analysis, Dundee 1975, Lecture Notes in Math. *506*, Springer-Verlag, Berlin (1976), 148–165.

[109] J.W. Jerome and M.E. Rose: Error estimates for the multi-dimensional two-phase Stefan problem. Math. Comp. *39 (160)* (1982), 377–414.

[110] I.A. Kaliev: Two-phase Stefan problem with homogeneous boundary conditions. In: Proceedings of 19th All-Union Research Stud. Conf. Math. Izd. Novosib. Univ., Novosibirsk (1981), 35–39. *(Russian)*

[111] I.A. Kaliev and A.M. Meirmanov: The structure of the generalized solution of a one-dimensional Stefan problem. Dinamika Sploshn. Sredy *64* (1984), 24–47. *(Russian)*

[112] S.L. Kamenomotskaya: On the Stefan problem. Mat. Sb. (N.S.) *53* (1961), 489–514. *(Russian)*

[113] E.M. Kartashov and B.Ya. Lyubov: Analytical methods of solving boundary value problems with moving boundaries for the heat equation. Izv. Akad. Nauk SSSR Energetika i Transport *6* (1974), 83–112. *(Russian)*

[114] H. Katz: A large time expansion for the Stefan problem. SIAM J. Appl. Math. *32* (1977), 1–20.

[115] H. Kawarada and M. Natori: On numerical solutions of the Stefan problem I. Mem. Numer. Math. *1* (1974), 43–54.

[116] H. Kawarada and M. Natori: On numerical solutions of the Stefan problem II. Mem. Numer. Math. *2* (1975), 1–20.

[117] A.A. Kavokin: On an estimate of the free boundary location in a Stefan-type problem. In: Mathematics and mechanics. MViSSO Kazakh. SSR, Alma-Ata, *vol. 7* (1971), 95–97. *(Russian)*

[118] A.A. Kavokin: On a Stefan problem with nonlinear boundary condition. In: Mathematics and mechanics. MViSSO Kazakh. SSR, Alma-Ata, *vol. 8* (1973), 56–61. *(Russian)*

[119] A.A. Kavokin: On asymptotic continuity of the solution to the Stefan problem with respect to boundary condition at small times. Izv. Akad. Nauk Kazakh. SSR Ser. Fiz.-Mat. *1* (1976), 63–66. *(in Russian)*

[120] J. Kern: A simple and apparently safe solution to the generalized Stefan problem. Int. J. Heat Mass Transfer *20* (1977), 467–474.

[121] S.N. Kharin: Mathematical models of heat and mass transfer in electric contacts. In: Electric contacts. Nauka, Moscow (1975), 5–14. *(Russian)*

[122] N.V. Khusnutdinova: On the behaviour of solutions to the Stefan problem at infinite time growth. Dinamika Sploshn. Sredy *2* (1969), 168–177. *(Russian)*

[123] N. Kikuchi and Y. Ichikawa: Numerical methods for a two-phase Stefan problem by variational inequalities. Internat. J. Numer. Methods Engrg. *14* (1979), 1221–1239.

[124] E.I. Kim and G.I. Bizhanova: A study of the second boundary value Stefan problem at small times. Vestnik Akad. Nauk Kazakh. SSR *6* (1981), 76–86. *(Russian)*

[125] E.I. Kim and G.I. Bizhanova: On a certain class of integro-differential equations. Vestnik Akad. Nauk Kazakh. SSR *5* (1982), 38–48. *(Russian)*

[126] E.I. Kim, L.A. Omel'chenko and S.N. Kharin: *Mathematical models of thermal processes at electric contacts.* Nauka, Alma-Ata (1977). *(Russian)*

[127] D. Kinderlehrer and L. Nirenberg: The smoothness of the free boundary in the one phase Stefan problem. Comm. Pure Appl. Math. *31* (1978), 257–282.

[128] D. Kinderlehrer and L. Nirenberg: Hodograph methods and the smoothness of the free boundary in the one phase Stefan problem. In: [222], 57–69.

[129] D. Kinderlehrer and G. Stampacchia: *An introduction to variational inequalities and their applications.* Academic Press, New York (1980).

[130] A.N. Kolmogorov and S.V. Fomin: *Elements of the theory of functions and functional analysis.* 3rd rev. ed. Nauka, Moscow (1972). (English ed. (2 vols.): Graylock Press, Rochester, New York (1957), (1961).)

[131] A.S. Kronrod: On functions of two variables. Uspekhi Mat. Nauk. *5* (1950), 24–134. *(Russian)*

[132] S.N. Kruzhkov: On some problems with unknown boundaries for the heat conduction equation. Prikl. Matem. Mekh. *31* (1967), 1009–1020 = J. Appl. Math. Mech. *31* (1967), 1014–1024.

[133] S.N. Kruzhkov: A class of problems with an unknown boundary for the heat equation. Dokl. Akad. Nauk SSSR *178* (1968), 1036–1038 = Soviet Phys. Dokl. *13* (1968), 101–103.

[134] S.N. Kruzhkov: On a fundamental a priori estimate for solutions of a quasilinear parabolic equation. Izv. Akad. Nauk UzSSR Ser. Fiz.-Mat. Nauk *3* (1972), 16–20. *(Russian)*

[135] A.A. Lacey and M. Shillor: The existence and stability of regions with superheating in the classical two-phase one-dimensional Stefan problem with heat sources. IMA J. Appl. Math. *30* (1983), 215–230.

[136] A.A. Lacey and A.B. Tayler: A mushy region in a Stefan problem. IMA J. Appl. Math. *30* (1983), 303–313.

[137] O.A. Ladyzhenskaya, V.A. Solonnikov and N.N. Uraltseva: *Linear and quasilinear equations of parabolic type.* Nauka, Moscow (1967). (English ed.: Transl. Math. Monographs, Vol. 23, AMS, Providence (1968).)

[138] O.A. Ladyzhenskaya and N.N. Uraltseva: *Linear and quasilinear elliptic equations.* Nauka, Moscow (1973). (English ed.: Mathematics in Science and Engineering (ed. by R. Bellman), Vol. 46, Academic Press, New York (1968).)

[139] E.M. Landis: *Second order equations of elliptic and parabolic types.* Nauka, Moscow (1971). *(Russian)*

[140] E. Langcham: The nature of the mushy region in Stefan problems with Joule heating. In: [184], 256–257.

[141] J. Li-Shang: Existence and differentiability of the solution of a two-phase Stefan problem for quasilinear parabolic equations. Acta Math. Sinica *15* (1965), 749–764 = Chinese Math. - Acta *7* (1965), 481–496.

[142] E. Magenes: Topics in parabolic equations: Some typical free boundary problems. Lab. Anal. Numer. CNR, Pavia. Report *no. 130* (1977).

[143] E. Magenes (Ed.): *Free boundary problems*, Vols. I, II. Ist. Naz. Alta Mat. Francesco Severi, Roma (1980).

[144] E. Magenes: Problemi di Stefan bifase in piu variabili spaziali. Lab. Anal. Numer. CNR, Pavia. Report *no. 309* (1983).

[145] E. Magenes, C. Verdi and A. Visintin: Semigroup approach to a Stefan problem with non-linear flux. Atti Accad. Naz. Lincei Rend. Cl. Sci. Fis. Mat. Natur. *(8)*, *75* (1983), no. 1-2, 24–33.

[146] O.G. Martynenko and I.A. Solov'ev: Some solutions of one-phase and one-dimensional Stefan problems. In: Analysis and Optimization Methods for Transport Processes. Izd. Lykov Inst. Heat Mass Transfer AN BSSR (1979), 198–201. *(Russian)*

[147] A.M. Meirmanov: Generalized solution of the Stefan problem in a medium with concentrated capacity. Dinamika Sploshn. Sredy *10* (1972), 85–101. *(Russian)*

[148] A.M. Meirmanov: The multi-phase Stefan problem for quasilinear parabolic equations. Dinamika Sploshn. Sredy *13* (1973), 74–86. *(Russian)*

[149] A.M. Meirmanov: Unique solvability and asymptotic behaviour as $t \to \infty$ in a Stefan-type problem arising from hydraulics. Dinamika Sploshn. Sredy *30* (1977), 98–111. *(Russian)*

[150] A.M. Meirmanov: On classical solvability of the Stefan problem. Dokl. Akad. Nauk SSSR *249* (1979), 1309–1312 = Soviet Math. Dokl. *20* (1979), 1426–1429.

[151] A.M. Meirmanov: On the classical solution of the multidimensional Stefan problem for quasilinear parabolic equations. Mat. Sb. (N.S.) *112* (1980), 170–192 = Math. USSR - Sb. *40* (1981), 157–178.

[152] A.M. Meirmanov: On close to one-dimensional solutions of the two-dimensional two-phase Stefan problem. Dinamika Sploshn. Sredy *50* (1981), 135–149. *(Russian)*

[153] A.M. Meirmanov: The Stefan problem: an approximate modelling. Uspekhi Mat. Nauk. *36*, no. 4, (1981), 201. *(Russian)*

[154] A.M. Meirmanov: An example of non-existence of a classical solution of the Stefan problem. Dokl. Akad. Nauk SSSR. *258* (1981), 547–549 = Soviet Math. Dokl. *23* (1981), 564–566.

[155] A.M. Meirmanov: On two-dimensional, self-similar solutions of filtration problems with free boundaries for compressible liquid. In: Numerical methods of continuum mechanics: mathematical modelling *12* (1980), 56–64. *(Russian)*

[156] A.M. Meirmanov: Approximate models of the two-phase Stefan problem ("concentrated capacity"). Dinamika Sploshn. Sredy *52* (1981), 56–77. *(Russian)*

[157] A.M. Meirmanov: Periodic solutions of the two-phase Stefan problem. Dinamika Sploshn. Sredy *57* (1982), 35–40. *(Russian)*

[158] A.M. Meirmanov: The structure of a generalized solution of the Stefan problem. Periodic solutions. Dokl. Akad. Nauk SSSR *272* (1983), 789–791 = Soviet Math. Dokl. *28* (1983), 440–443.

[159] A.M. Meirmanov: The structure of the generalized solution of the quasi-steady one-dimensional Stefan problem. Differentsial'nye Uravneniya *20* (1984), 882–885. *(Russian)*

[160] A.M. Meirmanov and V.V. Pukhnachev: Lagrangian coordinates in the Stefan problem. Dinamika Sploshn. Sredy *47* (1980), 90–111. *(Russian)*

[161] G.H. Meyer: The numerical solution of multidimensional Stefan problems–a survey. In: [222], 73–89.

[162] G.H. Meyer: A numerical method for two-phase Stefan problems. SIAM J. Numer. Anal. *8* (1971), 555–568.

[163] G.H. Meyer: Multidimensional Stefan problems. SIAM J. Numer. Anal. *10* (1973), 522–538.

[164] G.H. Meyer: One-dimensional parabolic free boundary problems. SIAM Rev. *19* (1977), 17–34.

[165] S.G. Mikhlin: *Variational methods in mathematical physics.* Nauka, Moscow (1977). *(Russian)*

[166] F. Milinazzo and G.W. Bluman: Numerical similarity solutions to Stefan problems. ZAMM *55* (1975), 423–429.

[167] C. Miranda: *Partial differential equations of elliptic type.* Springer-Verlag, Berlin (1970).

[168] B.D. Moiseenko and A.A. Samarskii: An economic continuous calculation scheme for the Stefan multidimensional problem. Zh. Vychisl. Mat. i Mat. Fiz. *5* (1965), 816–827 = U.S.S.R. Comput. Math. and Math. Phys. *5 (5)* (1965), 43–58.

[169] V.N. Monakhov: *Free boundary problems for systems of elliptic type.* Nauka, Novosibirsk (1977). *(Russian)*

[170] M. Mori: Numerical solution of the Stefan problem by the finite element method. Mem. Numer. Math. *2* (1975), 35–44.

[171] M. Mori: Stability and convergence of the finite element method for solving the Stefan problem. Publ. RIMS, Kyoto Univ. *12* (1976), 539–563.

[172] M. Mori: A finite element method for the two phase Stefan problem in one space dimension. Publ. RIMS, Kyoto Univ. *13* (1977), 723–753.

[173] B.I. Nalimov and V.V. Pukhnachev: *Non-stationary motions of an ideal liquid with a free boundary.* Izd. Novosib. Univ., Novosibirsk (1975). *(Russian)*

[174] M. Niezgódka and I. Pawlow: A generalized Stefan problem in several space variables. Appl. Math. Optim. *9* (1983), 193–224.

[175] M. Niezgódka, I. Pawlow and A. Visintin: On multi-phase Stefan type problems with nonlinear flux at the boundary in several space variables. Lab. Anal. Numer. CNR, Pavia. Report *no. 293* (1981).

[176] S.M. Nikol'skii: *Approximation of functions of several variables and embedding theorems.* Nauka, Moscow (1969). (English ed.: Grundlehren Math. Wiss. *205*, Springer, Berlin–New York (1975).)

[177] L. Nirenberg: *Topics in nonlinear functional analysis.* Courant Inst. Math. Sci., New York (1974).

[178] J.A. Nitsche: Finite element approximation to the one-dimensional Stefan problem. In: Recent advances in numerical analysis. Academic Press, New York (1978), 119–142.

[179] J.A. Nitsche: Approximation des eindimensionalen Stefan-Problems durch finite Elemente. In: Proc. Internat. Congress of Mathematicians, Helsinki (1978), 923 –928.

[180] J.A. Nitsche: Finite element approximations for free boundary problems. In: Proc. of TICOM, 2nd Int. Conf. on Computational Methods in Nonlinear Mechanics, Austin (Texas) (1979), 26–29.

[181] T. Nogi: A difference scheme for solving two-phase Stefan problem of heat equation. Publ. RIMS, Kyoto Univ. *16* (1980), 313–341.

[182] J.R. Ockendon: Numerical and analytic solutions of moving boundary problems. In: [222], 129–145.

[183] J.R. Ockendon and W.R. Hodgkins (Eds.): *Moving boundary problems in heat flow and diffusion.* Clarendon Press, Oxford (1975).

[184] O.A. Oleinik: A method of solution of the general Stefan problem. Dokl. Akad. Nauk SSSR *135* (1960), 1054–1057 = Soviet Math. Dokl. *1* (1960), 1350–1354.

[185] L.V. Ovsyannikov: *Lectures on the foundations of gas dynamics.* Nauka, Moscow (1981). *(Russian)*

[186] I. Pawlow: A variational inequality approach to a generalized two-phase Stefan problem in several space variables. Ann. Mat. Pura Appl. *133* (1982), 333–373.

[187] M. Primicerio: Stefan-like problems with space-dependent latent heat. Meccanica–J. Ital. Assoc. Theoret. Appl. Mech. *5* (1970), 187–190.

[188] M. Primicerio: Problemi di diffusione a frontiera libera. Boll. Un. Mat. Ital. A *18* (1981), 11–68.

[189] M. Primicerio: Mushy regions in phase-change problems. In: Applied nonlinear functional analysis (ed. by R. Gorenflo and K.-H. Hoffmann). Lang, Frankfurt (1983), 251–269.

[190] M.A. Pudovkin, S.G. Grigor'ev and V.N. Kosolapov: Application of a generalized method of integral relations in the study of a process of forming the cryo-hydratic shielding in rock cavities. In: Teplomassoobmen VI. Izd. Inst. Teplomassoobmena AN BSSR, *vol. IV*, 32–37, Minsk (1980). *(Russian)*

[191] M.A. Pudovkin, A.N. Salamatin and V.A. Chugunov: Solution of the two-phase Stefan problem for a cylindrical domain. In: Issledovaniya po prikladnoj matematike, *vol. 3*, 97–107. Izd. Kazan. Univ., Kazan (1975). *(Russian)*

[192] V.V. Pukhnachev: On a Stefan problem arising from modelling an electric blow-up in conductors. In: Trudy Seminara S.L. Soboleva, *no. 2*, 69–82. Izd. Inst. Mat. SO AN SSSR, Novosibirsk (1976). *(Russian)*

[193] V.V. Pukhnachev: Occurence of a singularity in the solution of a Stefan problem. Differentsial'nye Uravneniya *16* (1980), 492–500 = Differential Equations *16* (1980), 313–318.

[194] E.V. Radkevich and A.S. Melikulov: On the solvability of the two-phase quasistationary crystallization problem. Dokl. Akad. Nauk SSSR *265* (1982), 58–62 = Soviet Phys. Dokl. *27* (1982), 540–542.

[195] H. Rasmussen: An approximate method for solving two-dimensional Stefan problems. Letters Heat Mass Transfer *4* (1977), 273–277.

[196] J.C.W. Rogers, A.E. Berger and M. Ciment: The alternating phase truncation method for numerical solution of a Stefan problem. SIAM J. Numer. Anal. *16* (1979), 563–587.

[197] L.I. Rubinstein: *The Stefan Problem.* Zvaigzne, Riga (1967). (English ed.: Transl. Math. Monographs, Vol. *27*, AMS, Providence (1971).)

[198] L.I. Rubinstein: The Stefan problem: comments on its present state. J. Inst. Math. Appl. *27* (1979), 739–750.

[199] L.I. Rubinstein: Analyticity of the free boundary for the one-phase Stefan problem with strong nonlinearity. Boll. Un. Mat. Ital. Suppl. *1* (1981), 47–68.

[200] S.V. Salei: On the global solvability of a Stefan-type problem. Dokl. Akad. Nauk Ukrain. SSR Ser. A, *no. 6* (1979), 424–428. *(Russian)*

[201] S.V. Salei: On a free boundary problem with heat sources. In: Boundary value problems of mathematical physics. Naukova Dumka, Kiev (1979), 172–192. *(Russian)*

[202] S.V. Salei: On the convergence rate of the free boundary to its limit value in a Stefan problem. Dokl. Akad. Nauk Ukrain. SSR Ser. A, *no. 6* (1981), 19–23. *(Russian)*

[203] P.E. Sobolevskii: Local and non-local existence theorems for nonlinear second order parabolic equations. Dokl. Akad. Nauk SSSR *136* (1961), 292–295 = Soviet Math. Dokl. *2* (1961), 63–66.

[204] D.G. Schaeffer: A new proof of the infinite differentiability of the free boundary in the Stefan problem. J. Diff. Equations *20* (1976), 266–269.

[205] R.A. Seban: A comment on the periodic freezing and melting of water. Int. J. Heat Mass Transfer *14* (1971), 1862–1864.

[206] V.V. Shapovalenko: Computation of the temperature field and crystallization front in a sloped cylindrical ingot. In: Mathematical physics. Naukova Dumka, Kiev. *22* (1977), 98–102. *(Russian)*

[207] B. Sherman: A general one-phase Stefan problem. Quart. Appl. Math. *28* (1970), 377–382.

[208] B. Sherman: Limiting behavior in some Stefan problems as the latent heat goes to zero. SIAM J. Appl. Math. *20* (1971), 319–327.

[209] B. Sherman: General one-phase Stefan problems and free boundary problems for the heat equation with Cauchy data prescribed on the free boundary. SIAM J. Appl. Math. *20* (1971), 555–570.

[210] N.I. Shmulev: Periodic solutions of the first boundary value problem for parabolic equations. Mat. Sb. (N.S.) *66* (1965), 398–410. *(Russian)*

[211] A.D. Solomon: An easily computable solution to a two-phase Stefan problem. Solar Energy *25* (1979), 525–528.

[212] M. Štĕdrý and O. Vejvoda: Time periodic solutions of a one-dimensional two-phase Stefan problem. Ann. Mat. Pura Appl. *127* (1981), 67–78.

[213] L.N. Tao: The Stefan problem with arbitrary initial and boundary conditions. Quart. Appl. Math. *36* (1978), 223–233.

[214] L.N. Tao: The analyticity of solutions of the Stefan problem. Arch. Rat. Mech. Anal. *72* (1980), 285–301.

[215] D.A. Tarzia: Sur le problème de Stefan à deux phases. C.R. Acad. Sci. Paris, *Sér. A-288* (1979), 941–944.

[216] D.A. Tarzia: Una revision sobre problemas de frontera movil y libre para la ecucion del calor. El problema de Stefan. Separata de Math. Notal, Univ. Nac. Rosario. *29* (1981/82), 147–241.

[217] A.B. Tayler: The mathematical formulation of Stefan problems. In: [183], 120–137.

[218] A.V. Uspenskii: Method of rectifying the fronts for many-front one-dimensional Stefan-type problems. Dokl. Akad. Nauk SSSR *172* (1967), 61–64 = Soviet Phys. Dokl. *12, no. 1* (1967), 26–29.

[219] M.I. Vishik: On some inequality for boundary values of functions harmonic in a ball. Uspekhi Mat. Nauk *6* (1951), 165–166. *(Russian)*

[220] A. Visintin: Sur le problème de Stefan avec flux non linéaire. Boll. Un. Mat. Ital. C *18* (1981), 63–86.

[221] A. Visintin: The Stefan problem for a class of degenerate parabolic equations. In: [84], 419–430.

[222] D.G. Wilson, A.D. Solomon and P.T. Boggs (Eds.): *Moving boundary problems*. Academic Press, New York (1978).

[223] K. Yosida: *Functional analysis*. Springer-Verlag, Berlin (1974).

Supplementary references

[224] J.C.W. Rogers and A.E. Berger: Some properties of the nonlinear semigroup for the problem $u_t - \Delta f(u) = 0$. Nonlinear Anal. *8* (1984), 909–939.

[225] I.G. Götz and B.B. Zaltzman: Nonincrease of mushy region in a non-homogeneous Stefan problem. Quart. Appl. Math. to appear.

[226] S.N. Kruzhkov: First-order quasilinear equations in several independent variables. Mat. Sb. (N.S.) *81* (1970), 228–255 = Math. USSR - Sb. *10* (1970), 217–243.

[227] I.G. Götz: Existence of a generalized solution to the Stefan problem with non-constant melting temperature. In: Some applications of functional analysis to problems of mathematical physics. Izd. of the Inst. of Mathematics of the Siberian Branch of the USSR Academy of Sciences. (1986), 57–64.

[228] B. Zaltzman: The correctness of exact and approximate models in the two-phase Stefan problem, Ph.D. Thesis, Novosibirsk State University. *(Russian)*

[229] R.N. Nochetto: A class of non-degenerate two-phase Stefan problems in several space variables. Comm. Partial Differential Equations, *12* (1987), 21–45.

[230] A.G. Petrova: Local solvability of a thermo-diffusive Stefan-like problem. Dinamika Sploshn. Sredy *58* (1982). *(Russian)*

[231] E.V. Radkevich, and A.S. Melikulov : *Boundary value problems with free boundary*. FAN, Tashkent (1988). *(Russian)*

[232] G.P. Ivantsov: "Diffusive" supercooling in binary alloy solidification. Dokl. Akad. Nauk SSSR *81* (1951), 179–182. *(Russian)*

[233] A.B. Crowley and J.R. Ockendon: On the numerical solution of an alloy solidification problem. Int. J. Heat Mass Transfer *22* (1979), 941–947.

[234] A. Bermudez and C. Saguez: Mathematical formulation and numerical solution of an alloy solidification problem. In: [84], 237–247.

[235] V. Alexiades, D.G. Wilson, A. Solomon: Modeling binary alloy solidification processes. Preprint ORNL/CSD-117, Oak Ridge (1983).

[236] J.D.P. Donnelly: A model for nonequilibrium thermodynamic processes involving phase changes. J. Inst. Math. Appl. *24* (1979), 425–438.

[237] S. Luckhaus and A. Visintin: Phase transition in multicomponent systems. Manuscripta Math. *43* (1983), 261–288.

[238] A.B. Crowley: Numerical solution of alloy solidification problems revisited. In: Free boundary problems: applications and theory (ed. by A. Bossavit, A. Damlamian, M. Frémond). Pitman Res. Notes Math. Ser. *120* (1985), 122–131.

[239] I.G. Götz: Crystallization modelling for binary alloys. In: Problems of hydrodynamics and heat & mass transfer with free boundaries. Novosibirsk University (1987), 61–67.

Index

Walter de Gruyter
Berlin · New York

Computational Modelling
of Free and Moving Boundary Problems

Proceedings of the First International Conference,
held 2–4 July 1991, Southampton, U. K.

Edited by L. C. Wrobel and C. A. Brebbia

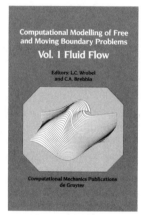

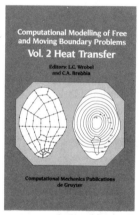

Vol. 1: Fluid Flow

1991. 464 pp. 15,5 x 23 cm.
Cloth DM 198,–
ISBN 3-11-013172-2

Vol. 2: Heat Transfer

1991. 332 pp. 15,5 x 23 cm.
Cloth DM 198,–
ISBN 3-11-013173-0

Set price: Cloth DM 336,– ISBN 3-11-013174-9

The contributions are classified in the following sections:
Vol. 1: Flow through porous media · Wave propagation · Cavitational flow · Free surface
flow · Mathematical problems and computational techniques.
Vol. 2: Solidification and melting · Metal casting and welding · Electrical / Electromagnetic
problems · Scientific applications.